JAN - 5 2008
P9-ELX-337

Oakland Public Library
Oakland, CA
www.oaklandlibrary.org

ALSO BY MARIO LIVIO

The Golden Ratio: The Story of Phi, the World's Most Astonishing Number

The Accelerating Universe: Infinite Expansion, the Cosmological Constant, and the Beauty of the Cosmos

The Equation That Couldn't Be Solved

How Mathematical Genius Discovered the Language of Symmetry

Mario Livio

Simon & Schuster
NEW YORK · LONDON · TORONTO · SYDNEY

SIMON & SCHUSTER
Rockefeller Center
1230 Avenue of the Americas
New York, NY 10020

Copyright © 2005 by Mario Livio
All rights reserved, including the right of reproduction
in whole or in part in any form.

SIMON & SCHUSTER and colophon are registered trademarks
of Simon & Schuster, Inc.

For information regarding special discounts for bulk purchases,
please contact Simon & Schuster Special Sales:
1-800-456-6798 or business@simonandschuster.com.

Designed by Paul Dippolito

Manufactured in the United States of America

1 3 5 7 9 10 8 6 4 2

Library of Congress Cataloging-in-Publication Data
Livio, Mario.
The equation that couldn't be solved :
how mathematical genius discovered the language of symmetry / Mario Livio.
p. cm.
Includes bibliographical references and index.
1. Group theory—History. 2. Galois theory—History.
3. Galois, Evariste, 1811–1832. 4. Symmetric functions—History.
5. Symmetry—Mathematics—History. 6. Diophantine analysis—History.
I. Title.

QA174.2.L58 2005
512'.2'09—dc22 2005044123

ISBN-13: 978-0-7432-5820-3
ISBN-10: 0-7432-5820-7

To Sofie

Contents

Preface

Ever since my high school days I have been fascinated by Évariste Galois. The fact that a twenty-year-old could invent an exciting new branch of mathematics has been a source of true inspiration. By the end of my undergraduate years, however, the young French romantic had also become a source of deep frustration. What else can you feel when you realize that even by the age of twenty-three you have not accomplished anything of comparable magnitude? The concept introduced by Galois—*group theory*—is recognized today as the "official" language of all symmetries. And, since symmetry permeates disciplines ranging from the visual arts and music to psychology and the natural sciences, the significance of this language cannot be overemphasized.

The list of people who have contributed directly and indirectly to this book could in itself fill more than a few pages. Here I will only mention those without whose help I would have had a hard time completing the manuscript. I am grateful to Freeman Dyson, Ronen Plesser, Nathan Seiberg, Steven Weinberg, and Ed Witten for conversations on the role of symmetry in physics. Sir Michael Atiyah, Peter Neumann, Joseph Rotman, Ron Solomon, and especially Hillel Gauchman, provided insights and critical comments on mathematics in general and on Galois theory in particular. John O'Connor and Edmund Robertson helped with the history of mathematics. Simon Conway Morris and David Perrett pointed me in the right direction in topics related to evolution and evolutionary psychology. I had fruitful discussions with Ellen Winner on the topic of creativity. Philippe Chaplain, Jean-Paul Auffray, and Norbert Verdier provided me with invaluable materials and information on Galois. Victor Liviot helped me to understand Galois's autopsy report. Stefano Corazza, Carla Cacciari, and Letizia Stanghellini provided useful information on the mathematicians from Bologna. Ermanno Bianconi was equally helpful concerning the mathematicians from San Sepolcro. Laura Garbolino, Livia Giacardi, and Franco Pastrone provided me with essential materials on the history of mathematics. Patrizia Moscatelli and Biancastella Antonio provided important documents from the library of the University of Bologna. Arild Stubhaug helped me to understand some aspects of Niels Abel's life and provided important documents, as did Yngvar Reichelt.

I am extremely grateful to Patrick Godon and Victor and Bernadette Liviot for their help with translations from French, to Tommy Wiklind and Theresa Wiegert for translations from Norwegian, and to Stefano Casertano, Nino Panagia, and Massimo Stiavelli for their assistance with translations from Italian and Latin. Elisabeth Fraser and Sarah Stevens-Rayburn provided me with invaluable bibliographic and linguistic help. The manuscript could not have been brought to print without the skillful preparation work by Sharon Toolan and the drawings by Krista Wildt.

The research and writing associated with a book of this scope put an inevitable burden on family life. Without the continuous support and infinite patience of my wife, Sofie, and my children, Sharon, Oren, and Maya, I could not have even dreamed of ever bringing the book to completion. I hope that my mother, Dorothy Livio, whose entire life has revolved and is still revolving around music, will enjoy this book on symmetry.

Finally, my sincere gratitude goes to my agent, Susan Rabiner, for her incredible work and encouragement, to my editor at Simon & Schuster, Bob Bender, for his professionalism and unrelenting support, and to Johanna Li, Loretta Denner, Victoria Meyer, and the entire team at Simon & Schuster for their help in producing and promoting this book.

The Equation That
Couldn't Be Solved

– ONE –

Symmetry

An inkblot on a piece of paper is not particularly attractive to the eye, but if you fold the paper before the ink dries, you may get something that looks like figure 1 that is much more intriguing. In fact, the interpretation of similar inkblots forms the basis for the famous Rorschach test developed in the 1920s by the Swiss psychiatrist Hermann Rorschach. The declared purpose of the test is to somehow elicit the hidden fears, wild fantasies, and deeper thoughts of the viewers interpreting the ambiguous shapes. The actual value of the test as an "x-ray of the mind" is vehemently debated in psychological circles. As Emory University psychologist Scott Lilienfeld once put it, "Whose mind, that of the client or the examiner?" Nevertheless, there is no denial of the fact that images such as that in figure 1 convey some sort of attractive and fascinating impression. Why?

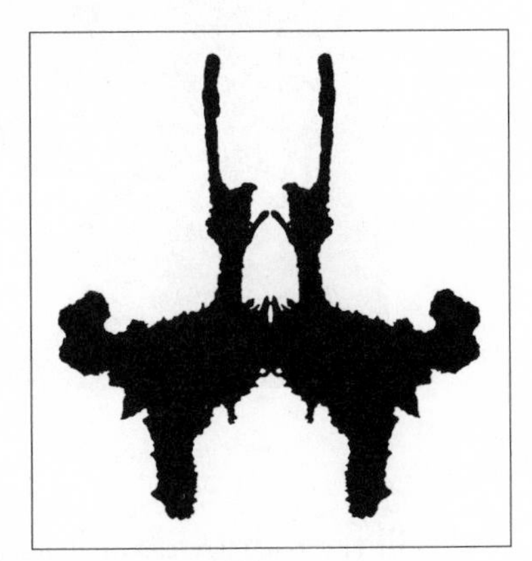

Figure 1

Is it because the human body, most animals, and so many human artifacts possess a similar bilateral symmetry? And why do all those zoological features and creations of the human imagination exhibit such a symmetry in the first place?

Most people perceive harmonious compositions such as Botticelli's *Birth of Venus* (figure 2) as symmetrical. Art historian Ernst H. Gombrich even notes that the "liberties which Botticelli took with nature in order to achieve a graceful outline add to the beauty and harmony of the design." Yet mathematicians will tell you that the arrangements of colors and forms in that painting are not symmetric at all in the mathematical sense. Conversely, most nonmathematical viewers do not perceive the pattern in figure 3 as symmetrical, even though it actually is symmetrical according to the formal mathematical definition. So what is symmetry really? What role, if any, does it play in perception? How is it related to our aesthetic sensibility? In the scientific realm, why has symmetry become such a pivotal concept in our ideas about the cosmos around us and in the fundamental theories attempting to explain it? Since symmetry spans such a wide range of disciplines, what "language" and what "grammar" do we use to describe and characterize symmetries and their attributes, and how was that universal language invented? On a lighter note, can symmetry provide an answer to the all-important question posed in the title of one of the songs of rock star Rod Stewart—"Do Ya Think I'm Sexy?"

Figure 2

I will try to provide at least partial answers to all of these questions and many more. Along the way, I hope that the story as a whole will depict both the humanistic side of mathematics and, even more importantly, the human side of mathematicians. As we shall see, symmetry is the paramount tool for bridging the gap between science and art, between psychology and mathematics. It permeates objects and concepts ranging from Persian carpets to the molecules of life, from the Sistine Chapel to the sought-after "Theory of Everything." Yet group theory, the mathematical language that describes the essence of symmetries and explores their properties, did not emerge from the study of symmetries at all. Rather, this astonishingly unifying idea of modern thought emanated from a most unlikely source—an equation that couldn't be solved. The dramatic and tortuous history of this equation is an essential part of this intellectual saga. At the same time, this tale will shed light on the loneliness of genius and on the tenacity of the human intellect in the face of seemingly insurmountable challenges. I have put a tremendous effort into trying to solve the two-centuries-old mystery of the death of the protagonist of this story—the brilliant mathematician Évariste Galois. I believe that I have come closer to the truth than was ever possible before.

Figure 3

The witty playwright George Bernard Shaw once said, "The reasonable man adapts himself to the world; the unreasonable one persists in trying to adapt the world to himself. Therefore all progress depends on the unreasonable man." In this book we shall encounter many unreasonable men and women. The creative process, by its very nature, seeks uncharted intellectual and emotional terrain. Brief forays into mathematical abstraction will offer a peek into the very nature of creativity. I begin with a concise exploration of the wonderland of symmetries.

IMMUNITY TO CHANGES

The word *symmetry* has ancient roots, coming from the Greek *sym* and *metria,* which translate into "the same measure." When the Greeks labeled a work of art or an architectural design symmetric, they meant that one could identify some small piece of the work, such that the dimensions of all the other parts contained that piece a precise number of times (the parts were "commensurable"). This early definition corresponds more to our modern notion of proportion than to symmetry. Nevertheless, the great philosophers Plato (428/427–348/347 BC) and Aristotle (384–322 BC) were quick to associate symmetry with beauty. In Aristotle's words, "The chief forms of beauty are orderly arrangement [in Greek *taxis*], proportion [*symmetria*], and definiteness [*horis-*

menon], which are revealed in particular by mathematics." Following in the Greeks' footsteps, the identification of symmetry with "due proportion" was subsequently propagated by the influential Roman architect Vitruvius (ca. 70–25 BC), and it persisted all the way through the Renaissance. In his *De Architectura Libri Decem (Ten Books on Architecture),* literally the architectural bible in Europe for centuries, Vitruvius writes:

> The design of a temple depends on symmetry, the principles of which must be carefully observed by the architect. They are due to proportion. Proportion is a correspondence among the measures of the members of an entire work, and of the whole to a certain part selected as standard. From this result the principles of symmetry.

The modern meaning of symmetry (first introduced in the late eighteenth century) in the precise mathematical sense is really "immunity to a possible change." Or, as mathematician Hermann Weyl (1885–1955) once put it, "A thing is symmetrical if there is something you can do to it so that after you have finished doing it it looks the same as before." Examine for example the verses

Is it odd how asymmetrical
Is "symmetry"?
"Symmetry" is *asymmetrical.*
How odd it is.

This stanza remains unchanged if read word by word from the end to the beginning—it is symmetrical with respect to backward reading. If you envision the words as being arranged like beads along a string, you could regard this reverse reading as a sort of (not literal) mirror reflection of the stanza. This stanza does not change when mirror-reflected in the above sense—it is symmetrical with respect to such mirror reflection. Alternatively, if you prefer to think in terms of reading the poem out loud, then the backward reading corresponds to a time reversal, somewhat like rewinding a videotape (again, not literally, because the individual sounds are not reversed). Phrases with this property are called *palindromes.*

The invention of palindromes is generally attributed to Sotades the Obscene of Maronea, who lived in the third century BC in Greek-dominated Egypt. Palindromes have been extremely popular with many word-play wizards such as the Englishman J. A. Lindon, and with the superb recreational-mathematics author Martin Gardner. One of Lindon's amusing word-unit palindromes reads: "Girl, bathing on Bikini, eyeing boy, finds boy eyeing bikini on bathing girl." Other palindromes are symmetric with respect to back-to-front reading letter by letter—"Able was I ere I saw Elba" (attributed jokingly to Napoleon), or the title of a famous *NOVA* program: "A Man, a Plan, a Canal, Panama."

Surprisingly, palindromes appear not just in witty word games but also in the structure of the male-defining Y chromosome. The Y's full genome sequencing was completed only in 2003. This was the crowning achievement of a heroic effort, and it revealed that the powers of preservation of this sex chromosome have been grossly underestimated. Other human chromosome pairs fight damaging mutations by swapping genes. Because the Y lacks a partner, genome biologists had previously estimated that its genetic cargo was about to dwindle away in perhaps as little as five million years. To their amazement, however, the researchers on the sequencing team discovered that the chromosome fights withering with palindromes. About six million of its fifty million DNA letters form palindromic sequences—sequences that read the same forward and backward on the two strands of the double helix. These copies not only provide backups in case of bad mutations, but also allow the chromosome, to some extent, to have sex with itself—arms can swap position and genes are shuffled. As team leader David Page of MIT has put it, "The Y chromosome is a hall of mirrors."

Of course, the most familiar example of mirror-reflection symmetry is that of the bilateral symmetry that characterizes the animal kingdom. From butterflies to whales, and from birds to humans, if you reflect the left half in a mirror you obtain something that is almost identical to the right half. I will, for the moment, ignore the small if tantalizing external differences that do exist, and also the fact that neither the internal anatomy nor the functions of the brain possess bilateral symmetry.

To many, the word *symmetry* is actually assumed to mean bilateral symmetry. Even in *Webster's Third New International Dictionary,* one of the definitions reads: "Correspondence in size, shape, and relative position of parts that are on opposite sides of a dividing line or median plane." The precise mathematical description of reflection symmetry uses the same concepts. Take a drawing of a bilaterally symmetric butterfly and mark a straight line down the middle of the figure. If you flip the drawing over keeping the central line in place, perfect overlapping will occur. The butterfly remains unchanged—invariant—under reflection about its central line.

Bilateral symmetry is so prevalent in animals that it can hardly be due to chance. In fact, if you think of animals as vast collections of trillions and trillions of molecules, there are infinitely more ways to construct asymmetrical configurations out of these building blocks than symmetrical ones. The pieces of a broken vase can lie in a pile in many different assortments, but there is only one arrangement in which they all fit together to reproduce the intact (and usually bilaterally symmetric) vase. Yet the fossil record from the Ediacara Hills of Australia shows that soft-bodied organisms (*Spriggina*) that date back to the Vendian period (650 to 543 million years ago) already exhibited bilateral symmetry.

Since life forms on Earth were shaped by eons of evolution and natural selection, these processes must have somehow preferred bilateral or mirror symmetry. Of all the different guises animals could have taken, bilaterally symmetrical ones had superiority. There is no escape from the conclusion that this symmetry was a likely outcome of biological growth. Can we understand the cause for this particular predilection? We can at least try to find some of its engineering roots in the laws of mechanics. One key point here is the fact that all directions on the surface of the Earth were not created equal. A clear distinction between up and down (*dorsal* and *ventral* in animals, in the biological jargon) is introduced by the Earth's gravity. In most cases what goes up must come down, but not the other way around. Another distinction, between front and back, is a result of animal locomotion.

Any animal moving relatively rapidly, be it in the sea, on land, or in the air, has a clear advantage if its front is different from its rear end. Having all the sensory organs, the major detectors of light, sound, smell, and taste, in the front clearly helps the animal in deciding where to go and how to best get there. A frontal "radar" also provides an early warning against potential dangers. Having the mouth in the front can make all the difference between reaching lunch first or not. At the same time, the actual mechanics of movement (especially on land and in the air) under the influence of the Earth's gravitational force have generated a clear difference between bottom and top. Once life emerged from the sea and onto the land, some sort of mechanical devices—legs—had to develop to carry the animal around. No such appendages were needed at the top, so the difference between top and bottom became even more pronounced. The aerodynamics of flying (still under Earth's gravity) coupled with the requirements for a landing gear plus some means of movement on the ground combined to introduce top-bottom differences in birds.

Here, however, comes an important realization: *There is nothing major in the sea, on the ground, or in the air, to distinguish between left and right.* The hawk looking to the right sees just about the same environment it sees to the left. The same is not true about up and down—up is where the hawk flies even higher into the sky, while down is where it lands and builds its nest. Political puns aside, there really is no big difference between left and right on Earth, because there are no strong horizontal forces. To be sure, the Earth's rotation around its axis and the Earth's magnetic field (the fact that Earth acts on its surroundings like a bar magnet) do introduce an asymmetry. However, these effects are not nearly as significant at the macroscopic level as those of gravity and rapid animal motion.

The description so far explains why bilateral symmetry of living organisms makes sense mechanically. Bilateral symmetry is also economical—you get two organs for the price of one. How this symmetry or lack thereof emerged from evolutionary biology (the genes) or even more fundamentally from the laws of physics is a more difficult question, to parts of which I shall return in chapters 7 and 8. Here let me note that many multicellular animals have an early embryonic body that lacks bilateral symmetry. The driving force behind the modification of the "original plan" as the embryo grows may indeed be mobility.

Not all animate nature lives in the fast lane. Life forms that are anchored in one place and are unable to move voluntarily, such as plants and sessile animals, do have very different tops and bottoms, but no distinguishable front and back or left and right. They have symmetry similar to that of a cone—they produce symmetrical reflections in any mirror passing through their central, vertical axis. Some animals that move very slowly, such as jellyfish, have a similar symmetry.

Obviously, once bilateral symmetry had developed in living creatures, there was every reason to keep it intact. Any loss of an ear or an eye would make an animal much more vulnerable to a predator sneaking up on it unnoticed.

One may always wonder whether the particular standard configuration nature has endowed humans with is the optimal one. The Roman god Janus, for example, was the god of the gates and of new beginnings, including the first month (January) of the year. Accordingly, he is always depicted in art with two faces, one in the front facing forward (symbolically toward the coming year) and one at the back of the head (toward the year that has passed). Such an arrangement in humans, while useful for some purposes, would have left no space for the parts of the brain that are responsible for the nonsensory systems. In his wonderful book *The New Ambidextrous Universe,* Martin Gardner tells the story of a Chicago entertainer who had a routine discussing the advantages of having various sensory organs at unusual spots on the body. Ears under the armpits, for instance, would be kept warm in the cold Chicago winters. Clearly, other shortcomings would be associated with such a configuration. The hearing of armpit ears would be seriously impaired unless you kept your arms raised all the time.

Science-fiction movies invariably feature aliens that are bilaterally symmetric. If extraterrestrial intelligent creatures that have evolved biologically exist, how likely are they to possess reflection symmetry? Quite likely. Given the universality of the laws of physics, and in particular the laws of gravity and motion, life forms on planets outside the solar system face some of the same environmental challenges that life on Earth does. The gravitational force still holds everything on the surface of the planet and creates a significant discrimination between up and down. Locomotion similarly separates the front end from the rear. E.T. is or was most likely ambidextrous. This does not mean, however, that any delegation of visiting aliens would look anything like us. Any civilization sufficiently evolved to engage in interstellar travel has likely long passed the merger of an intelligent species with its far superior computational-technology-based creatures. A computer-based superintelligence is most likely to be microscopic in size.

Some of the capital letters in the alphabet are among the numerous human-created objects that are symmetric with respect to mirror reflections. If you hold a sheet of paper with the letters A, H, I, M, O, T, U, V, W, X, Y up to a mirror, the letters look the same. Words (or even entire phrases) constructed from these letters and printed vertically, such as the not-too-deep instruction

Y
O
U

M
A
Y

W
A
X

I
T

T
I
M
O
T
H
Y

remain unchanged when mirror reflected. The Swedish pop-music group AᗺBA, whose music inspired the successful musical *Mamma Mia,* introduced a trick into the spelling of its name that makes it mirror symmetric (MAMMA MIA written vertically is also mirror symmetric). A few letters, such as B, C, D, E, H, I, K, O, X, are symmetric with respect to reflection in a mirror that bisects them horizontally. Words composed of these letters, such as COOKBOOK, BOX, CODEX, or the familiar symbols for hugs and kisses, XOXO, remain unchanged when held upside down to a mirror.

The importance of mirror-reflection symmetry to our perception and aesthetic appreciation, to the mathematical theory of symmetries, to the laws of physics, and to science in general, cannot be overemphasized, and I will return to it several times. Other symmetries do exist, however, and they are equally relevant.

THE FROLIC ARCHITECTURE OF SNOW

The title of this section is taken from "The Snowstorm" by the American poet and essayist Ralph Waldo Emerson (1803–82). It expresses the bewilderment one feels upon discerning the spectacular shapes of snowflakes (figure 4). While the common phrase "no two snowflakes are alike" is actually not true at the naked-eye level, snowflakes that have formed in different environments are indeed different. The famous astronomer Johannes Kepler (1571–1630), who discovered the laws of planetary motion, was so impressed with the marvels of snowflakes that he devoted an entire treatise, *The Six-Cornered Snowflake,* to the attempt to explain the symmetry of snowflakes. In addition to mirror-reflection symmetry, snowflakes possess *rotational symmetry*—you can rotate them by certain angles around an axis perpendicular to their plane (passing through the center) and they remain the same. Due to the properties and shape of water molecules, snowflakes have typically six (almost) identical corners. Consequently, the smallest rotation angle (other than no rotation at all) that leaves the shape unchanged is one in which each corner is displaced by one "step": 360 ÷ 6 = 60 degrees. The other angles that lead to an indistinguishable final figure are simple multiples of this angle: 120, 180, 240, 300, 360 degrees (the last one returns the snowflake to its original position and is equivalent to no rotation at all). Snowflakes therefore have sixfold rotational symmetry. By comparison, starfish have fivefold rotational symmetry; they can be rotated by 72, 144, 216, 288, and 360 degrees with no discernable difference. Many flowers, such as the chrysanthemum, the English daisy, and the tickseed (coreopsis), display an approximate rotational symmetry. They look essentially the same when rotated by any angle (figure 5). Symmetry, when combined with rich colors and intoxicating smells, is an underlying property that gives flowers their universal aesthetic appeal. Perhaps no one has expressed better the associative relationship between flowers and works of art than the painter James McNeill Whistler (1834–1903):

Figure 4

> The masterpiece should appear as the flower to the painter—perfect in its bud as in its bloom—with no reason to explain its presence—no mission to fulfill—a joy to the artist, a delusion to the philanthropist—a puzzle to the botanist—an accident of sentiment and alliteration to the literary man.

What is it in a symmetric pattern that provokes such an emotional response? And is this truly the same excitement that is stimulated by works of art? Note that even if the answer to the latter question is an unequivocal yes, this does not necessarily bring us any closer to answering the first question. The answer to the question, What is it in works of art that provokes an emotional response? is far from clear. Indeed, what quality is shared by such different masterpieces as Jan Vermeer's *Girl with a Pearl Earring*, Pablo Picasso's *Guernica*, and Andy Warhol's *Marilyn Diptych*? Clive Bell (1881–1964), an art critic and member of the Bloomsbury group (which, by the way, included novelist Virginia Woolf), suggested that the one quality common to all true works of art was what he called "significant form." By this he meant a particular combination of lines, colors, forms, and relations of forms that stirs our emotions. This is not to say that all works of art evoke the same emotion. Quite the contrary: every work of art may evoke an entirely different emotion. The commonality is in the fact that all works of art do evoke some emotion. If we were to accept this aesthetic hypothesis, then symmetry may simply represent one of the components of this (rather vaguely defined) significant form. In this case, our reaction to symmetric patterns may not be too different (even if less intense perhaps) from our broader aesthetic sensibility. Not all agree with such an assertion. Aesthetics theorist Harold Osborne had this to say about the human response to symmetry of individual elements or objects, such as snowflakes: "They can arouse interest, curiosity and admiration. But visual interest in them is short-lived and superficial: In contrast to the impact of an artistic masterpiece, perceptual attention soon wanders, never goes deep. There is no enhancement of perception." Actually, as I will show

Figure 5

in the next chapter and in chapter 8, symmetry has much to do with perception. For the moment, however, let me concentrate on the purely aesthetic "value" of symmetry.

Dartmouth College psychologists Peter G. Szilagyi and John C. Baird conducted a fascinating experiment in 1977 that was intended to explore the quantitative relationship between the amount of symmetry in designs and aesthetic preference. Twenty undergraduate students (the most common subjects of experimental psychology) were asked to perform three simple tasks. In the first, they were invited to arrange eight squares with a black dot at their centers inside a row of eighteen cells, each of a size equal to that of the squares (figure 6a). The instructions to the subjects were to arrange the pieces in a manner that they found "visually pleasing." Each piece had to cover one cell entirely, and all the squares had to be used. The second and third tasks were similar in nature. In the second, eleven pieces had to be arranged in a 5 × 5 grid (figure 6b). In the third, twelve cubes had to be fitted into holes in a three-dimensional transparent structure consisting of three horizontal planes, each containing nine square holes (figure 6c). The results showed an unambiguous aesthetic preference for symmetrical designs. For instance, 65 percent of the subjects created perfect mirror-reflection-symmetric patterns in the first task. In fact, symmetry was the primary component in the designs of most subjects (in one, two, and three dimensions), with perfect symmetry being the most favored condition.

The association between symmetry and artistic taste emerged not just in experiments, but also in a more speculative theory of aesthetics developed by the famous Harvard mathematician George David Birkhoff (1884–1944). Birkhoff is best known for proving in 1913 a famous

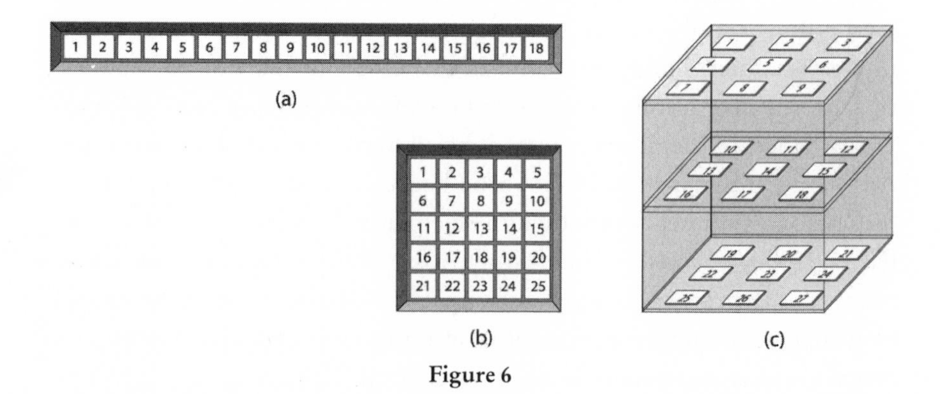

Figure 6

geometric conjecture formulated by the French mathematician Henri Poincaré, and for his ergodic theorem (published in 1931–32)—a contribution of paramount significance for the theory of gases and for probability theory. During his undergraduate days, Birkhoff started to be intrigued by the structure of music, and around 1924 he expanded his interests to aesthetics in general. In 1928, he spent half a year traveling extensively in Europe and the Far East in an attempt to absorb as much art, music, and poetry as he could. His efforts to develop a mathematical theory of aesthetic value culminated in the publication of *Aesthetic Measure* in 1933. Birkhoff specifically discusses the intuitive feeling of value evoked by works of art, which is "clearly separable from sensuous, emotional, moral, or intellectual feeling." He separates the aesthetic experience into three phases: (1) the effort of attention necessary for perception; (2) the realization that the object is distinguished by a certain order; (3) the appreciation of value that rewards the mental effort. Birkhoff further assigns quantitative measures to the three stages. The preliminary effort, he suggests, increases in proportion to the complexity of the work (denoted by C). Symmetries play a key role in the order (denoted by O) characterizing the object. Finally, the feeling of value is what Birkhoff calls the "aesthetic measure" (denoted by M) of the work of art.

The essence of Birkhoff's theory can be summarized as follows. Within each class of aesthetic objects, such as ornaments, vases, pieces of music, or poetry, one can define an order O and a complexity C. The aesthetic measure of any object in the class can then be calculated simply by dividing O by C. In other words, Birkhoff proposed a formula for the feeling of aesthetic value: $M = O \div C$. The meaning of this formula is: For a given degree of complexity, the aesthetic measure is higher the more order the object possesses. Alternatively, if the amount of order is specified, the aesthetic measure is higher the less complex the object. Since for most practical purposes, the order is determined primarily by the symmetries of the object, Birkhoff's theory heralds symmetry as a crucial aesthetic element.

Birkhoff was the first to admit that the precise definitions of his elements O, C, and M were tricky. Nevertheless, he made a valiant attempt to provide detailed prescriptions for the calculation of these measures for a variety of art forms. In particular, he started with simple geometrical shapes such as those in figure 7, continued with ornaments and Chinese vases, proceeded to harmony in the diatonic musical scale, and concluded with the poetry of Tennyson, Shakespeare, and Amy Lowell.

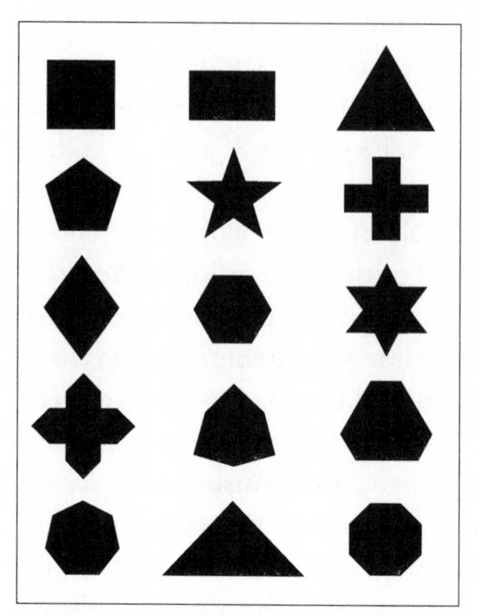

Figure 7

No one, especially not Birkhoff himself, would claim that the intricacies of aesthetic pleasure could be reduced entirely to a mere formula. However, in Birkhoff's words, "In the inevitable analytic accompaniment of the creative process, the theory of aesthetic measure is capable of performing a double service: it gives a simple, unified account of the aesthetic experience, and it provides means for the systematic analysis of typical aesthetic fields."

Returning now from this brief detour into the land of aesthetics to the specific case of rotational symmetry, we note that one of the simplest rotationally symmetric figures in the plane is a circle (figure 8a). If you rotate it around its center through, say, 37 degrees, it remains unchanged. In fact you can rotate it through *any* angle around a perpendicular axis through its center and you will not notice any difference. The circle therefore has an infinite number of rotational symmetries. These are not the only symmetries the circle possesses. Reflections in all the axes that cut along a diameter (figure 8b) also leave the circle unchanged.

The same system can, therefore, have multiple symmetries, or be symmetric under a variety of *symmetry transformations.* Rotating a perfect sphere about its center, using an axis running in any direction, leaves it looking precisely the same. Or examine, for instance, the equilateral (all sides the same) triangle in figure 9a. We are allowed neither to change the shape or size of this triangle, nor to move it about. What

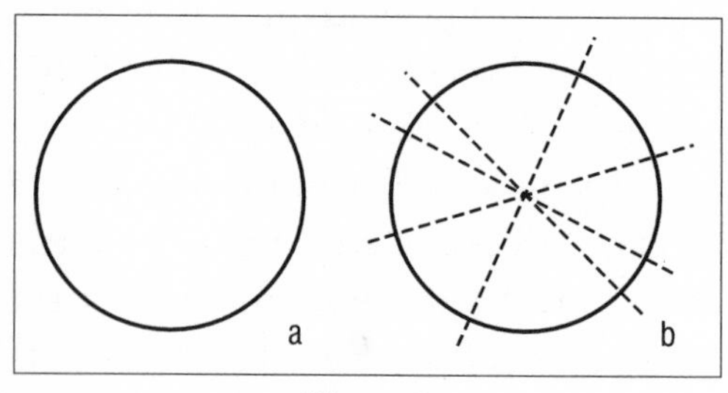

Figure 8

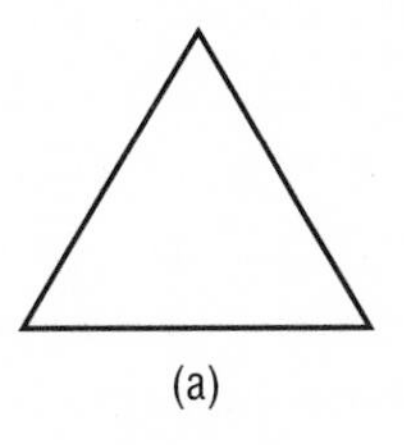

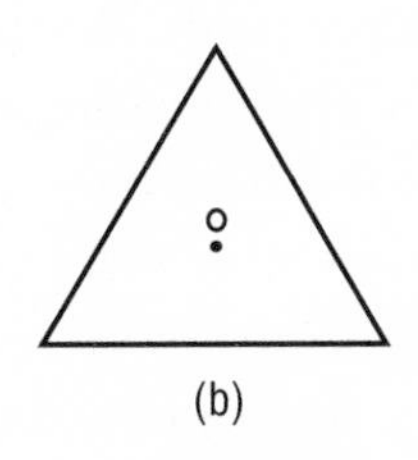

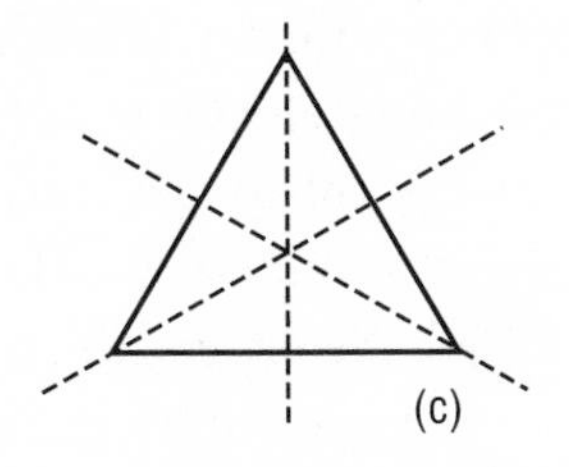

Figure 9

transformations could we apply to it to leave it unchanged? We could rotate it by 120, 240, and 360 degrees around an axis perpendicular to the plane of the figure and passing through point O (figure 9b). These transformations do interchange the locations of the vertices, but if you turn your back while somebody is performing these rotations you won't notice anything different. Note that a rotation by 360 degrees is equivalent to doing nothing at all, or rotating by zero degrees. This is known as the *identity transformation.* Why bother to define such a transformation at all? As we shall see later in the book, the identity transformation plays a similar role to that of the number zero in the arithmetic operation of addition or the number one in multiplication—when you add zero to a number or multiply a number by one, the number remains unchanged. We can also mirror-reflect the triangle about the three dashed lines in figure 9c. There are, therefore, precisely six symmetry transformations—three rotations and three reflections—associated with the equilateral triangle.

What about combinations of some of these transformations, such as a reflection followed by a rotation? Don't they add to the number of symmetries of the triangle? I shall return to this question in the context of the language of symmetries. For the moment, however, another important symmetry awaits exposition.

MORRIS, MOZART, AND COMPANY

One of the most familiar of all symmetric patterns is that of a repeating, recurring motif. From friezes of classical temples and pillars of palaces to carpets and even birdsong, the symmetry of repeating patterns has always produced a very comforting familiarity and a reassuring effect. An elementary example of this type of symmetry was presented in figure 3.

The symmetry transformation in this case is called *translation,* meaning a displacement or shift by a certain distance along a certain line. The pattern is called symmetric if it can be displaced in various directions without looking any different. In other words, regular designs in which the same theme repeats itself at fixed intervals possess translational symmetry. Ornaments that are symmetric under translation can be traced all the way back to 17,000 BC (the Paleolithic era). A mammoth-ivory bracelet found in the Ukraine is marked with a repeating zigzag pattern. Other translation-symmetric designs are found in a variety of art forms ranging from medieval Islamic tiling in the Alhambra palace in Granada, Spain (figure 10a), through Renaissance typography, to the drawings of the fantastic Dutch graphic artist M. C. Escher (1898–1972; figure 10b). Nature also provides examples of translation-symmetric creatures, such as the centipedes, in which identical body segments may repeat as many as 170 times.

The Victorian artist, poet, and printer William Morris (1834–96) was a prolific producer of decorative art. Much of his work is literally the embodiment of translational symmetry. Early in life, Morris became fascinated by medieval architecture, and at age twenty-seven he started a firm of decorators that later became famous as Morris and Company. In a strong reaction to the increasing industrialism in nineteenth-century

a

Figure 10 b

a Figure 11

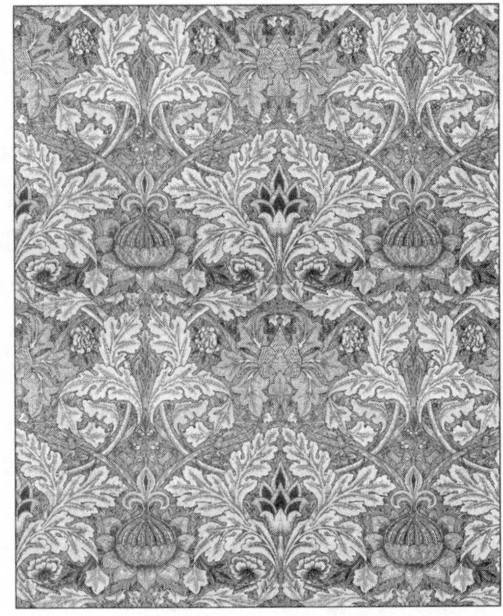

b

England, Morris looked for ways to revive artistic craftsmanship and to revitalize the splendor of the decorative arts of the Middle Ages. Morris and Company, and later the Kelmscott Press founded by Morris in 1890, designed spectacular tiles, tableware, textiles, and illustrated manuscripts in medieval design. But it was in wallpaper design where Morris first achieved his incredible mastery of translation-symmetric repeating patterns. A couple of his sumptuous themes are shown in figure 11. While Morris's designs may not have been any more innovative than those of some of his contemporaries, such as Christopher Dresser or A. W. N. Pugin, his influence and legacy have been enormous. Morris himself was interested in promoting arts and crafts and not in the mathematics of symmetry. In *The Beauty of Life* he summarized his socio-aesthetic philosophy this way:

> You may hang your walls with tapestry instead of whitewash or paper; or you may cover them with mosaic; or have them frescoed by a great painter: all this is not luxury, if it be done for beauty's sake, and not for show: it does not break our golden rule: Have nothing in your houses which you do not know to be useful or believe to be beautiful.

An interesting question is whether symmetry with respect to translation, and indeed reflection and rotation too, is limited to the visual arts, or may be exhibited by other artistic forms, such as pieces of music. Evidently, if we refer to the sounds, rather than to the layout of the written musical score, we would have to define symmetry operations in terms other than purely geometrical, just as we did in the case of the palindromes. Once we do that, however, the answer to the question, Can we find translation-symmetric music? is a resounding yes. As Russian crystal physicist G. V. Wulff wrote in 1908: "The spirit of music is rhythm. It consists of the regular, periodic repetition of parts of the musical composition . . . the regular repetition of identical parts in the whole constitutes the essence of symmetry." Indeed, the recurring themes that are so common in musical composition are the temporal equivalents of Morris's designs and symmetry under translation. Even more generally, compositions are often based on a fundamental motif introduced at the beginning and then undergoing various metamorphoses.

Simple examples of symmetry under translation in music include the opening measures in Mozart's famous Symphony no. 40 in G Minor (figure 12), as well as the entire structure of some common musical forms. In the former example you can see the translational symmetry not only within each line of the score (where the short declining gestures are marked), but also between the first line and the second (denoted by *a* and *b*). In terms of overall design, if we use the symbols A, B, and C to describe entire sections of a movement, then the pattern for a rondo as a whole, for instance, can be expressed as ABACA or ABACABA, where the translational symmetry is apparent. Mozart's association with objects of mathematics should come as no surprise. His sister, Nannerl, recalled that he once covered the walls of the staircase and of all the rooms in their house with numbers, and when no space remained, he moved on to the walls of a neighboring house. Even the margins of

Figure 12

Mozart's manuscript for the Fantasia and Fugue in C Major contain calculations of the probability to win the lottery. No wonder then that British musicologist and composer Donald Tovey identified the "beautiful and symmetrical proportions" of Mozart's compositions as one of the key reasons for their popularity.

Another great composer known for his obsession with numbers, mental games, and their use in complex musical form was Johann Sebastian Bach (1685–1750). Both reflection and translation feature frequently in Bach's music on many levels. An example encompassing reflection by a horizontal "mirror" is the opening of Bach's Two-Part Invention no. 6 in E Major (see figure 13). Imagine a mirror in the space between the two score lines. The ascending trend marked by line *a* is reflected (half a bit later) by the descending trend *b*, and the entire gesture is reflected and repeated again slightly later (starting at *d*). Another example is provided by the entire large-scale structure of one of Bach's most notable works, the famous *Musical Offering*. The composition consists of these musical forms:

Ricercar   5 Canons   Trio Sonata   5 Canons   Ricercar

It exhibits reflection symmetry (obviously not sound by sound).

Ricercar (from *ricercare*—"to research, or seek out") was an old term used loosely for any type of prelude, usually in fugal style. The great humanitarian, physician, and philosopher Albert Schweitzer (1875–1965) was also a great Bach enthusiast. In his book *J. S. Bach* he notes: "The word [ricercar] signifies a piece of music in which we have to seek something—namely a theme." The *Musical Offering* also contains ten canons, which, by construction, involve the operation of translation. In any canon (the word means "rule"), one melodic strand

Figure 13

determines the rule (in terms of melodic line or rhythm) for the second or more voices. The second voice follows at some fixed interval of time—a *temporal translation.* A simple, familiar example is

Row row row your boat
Gently down the stream
Merrily merrily merrily merrily
Life is but a dream,

where the second voice starts when the first reaches the word "gently."

The story surrounding the *Musical Offering* is in itself truly fascinating. Three years before his death, Bach was on his way to Berlin to visit his daughter-in-law Johanna Maria Dannemann (wife of the composer Carl Philipp Emanuel Bach), who was at the time expecting a child. Exhausted from the long journey, the aged composer made a stop at Potsdam, then the seat of King Frederick the Great of Prussia, who also employed Carl Philipp Emanuel. The news of Bach's arrival at the royal palace prompted the king to cancel a planned evening concert featuring himself playing on the flute in favor of an impromptu series of recitals by Bach on seven new fortepianos. Gottfried Silbermann, the master organ builder of the German baroque, constructed these instruments. Following a virtuoso performance in seven different rooms of the palace, Bach offered to his delighted audience to improvise a fugue on a theme His Royal Highness would suggest. Upon returning home, Bach developed the *Musical Offering* from that improvised fugue. He added to it a set of magnificently complex canons and a trio sonata and elaborated upon the other contrapuntal movements. The sonata featured a flute (King Frederick's instrument), a violin, and continuo (keyboard and cello). For the title of the *Offering,* the ever word-playful Bach chose *Regis iussu cantio et reliqua canonica arte resoluta* (*Upon the King's Demand the Theme and Additions Resolved in Canonic Style*), which forms the acronym RICERCAR.

There are even more symmetries in the *Musical Offering.* In Canon I (the *Crab Canon*), each violin plays the other's part backward, resulting in reflection symmetry (of the score) in a vertical mirror. Finally, canons in general were considered at the time to be some sort of symmetry puzzles. The composer provided the theme, but it was the musicians' task to figure out what type of symmetry operation he had in mind for the theme to be performed. In the case of the *Musical Offering,* Bach accompanied the last two canons before the trio sonata with the inscription "*Quaerendo inventis,*" meaning "Seek, and ye shall find." As we shall see in chapter 7, this is not very different conceptually from the puzzle posed to us by the universe—it lies in all its glory open to inspection—for us to find the underlying patterns and symmetries. Even the uncertainties and ambiguities involved in the attempts to uncover the "theory of everything" may have an analogy in Bach's intellectual challenge. You see, one of the canons in the *Musical Offering* has three possible solutions.

Translation and reflection can be combined into one symmetry operation known as *glide reflection.* The footprints generated by an alternating left-right-left-right walk exhibit glide-reflection symmetry (figure 14). The operation consists simply of a translation (the glide), followed by a reflection in a line parallel to the direction of the displacement (the dashed line in the figure). Equivalently, you could look at glide reflection as a mirror reflection followed by a translation parallel to the mirror. Glide-reflection symmetry is common in classical friezes, and also in the ceramics of native Americans in New Mexico. Whereas patterns that are translation symmetric tend to convey an impression of motion in one direction, glide-reflection-symmetric designs create a snakelike visual sensation. Real snakes achieve these patterns by alternately contracting and relaxing muscle groups on both sides of their body—when they contract a group on the right, the corresponding group on the left is relaxed, and vice versa.

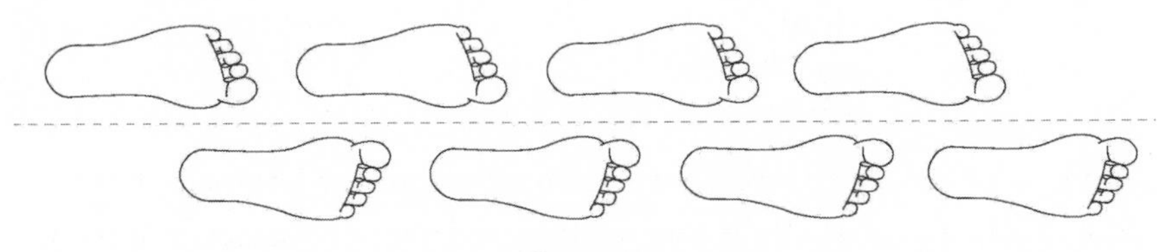

Figure 14

We have by now encountered all the rigid transformations that result in symmetries in two dimensions. The word *rigid* simply means that after the transformation every two points end up the same distance apart as they were to begin with—we cannot shrink figures, inflate them, or deform them.

In three-dimensional space, in addition to the symmetry under translation, rotation, reflection, and glide reflection, we can find yet another symmetry known as *screw symmetry.* This is the type of symmetry of a corkscrew, where rotation about some axis is combined with translation along that axis. Some stems of plants, where the leaves appear at regular intervals after completing the same fraction of a full circle around the stem, possess this symmetry. Are these all the symmetries that exist? Definitely not.

ALL ARE EQUAL, BUT . . .

The arts and sciences are chock-full of fascinating examples of symmetry under the operations of translation, rotation, reflection, and glide reflection, and we shall return to some of those in later chapters. An interesting transformation that is not geometrical in nature involves *permutations*—the different rearrangement of objects, numbers, or concepts. For example, to test the wear of four different brands of tires you may want to schematize a strategy that will ensure that you interchange the positions of all the tires every month for four months, with every tire occupying every position. If you label the brands A, B, C, D and the positions FL (front left), FR (front right), RL (rear left), and RR (rear right), then the four-month plan may look something like this:

MONTH	FL	FR	RL	RR
First	A	B	C	D
Second	B	A	D	C
Third	C	D	A	B
Fourth	D	C	B	A

Each row or column represents a permutation of the letters A, B, C, D. Note that to accomplish the desired test, no row or column should contain the same label twice. Squares of the 4 × 4 type presented here are known as *Latin squares,* and they were studied extensively by the famous Swiss mathematician Leonhard Euler (1707–83). Incidentally, you may get a kick out of solving the following popular eighteenth-century card puzzle: Arrange all the jacks, queens, kings, and aces from a deck of cards in a square, so that no suit or value would appear twice in any row, column, or the two main diagonals. In case you are having trouble with this baroque brainteaser, I show a solution in appendix 1.

Permutations feature in such diverse circumstances as the changing of partners in Scottish folk dancing and shuffled decks of cards. The main concern of the operation of permutations is not so much with which object lies where, as with which object takes the place of which. For example, in the permutation: 1 2 3 4 → 4 1 3 2, the number 1 was replaced by 4, 2 was replaced by 1, 3 stayed put, and 4 was replaced by 2. This is usually denoted by

$$\begin{pmatrix} 1234 \\ 4132 \end{pmatrix}$$

where each number in the upper row is replaced by the number directly underneath. The same permutation operation could have been written as

$$\begin{pmatrix} 3214 \\ 3142 \end{pmatrix}$$

because precisely the same replacements took place, and the order in which the numbers are written is not important. You may wonder how can a system be symmetric (i.e., not change) under permutations? Evidently, if you have ten books on a shelf and they are all different, any permutation that is not the identity (leaving the books untouched) will change the order. However, if you have three copies of the same book, for example, clearly some permutations will leave the order unchanged. The English essayist and critic Charles Lamb (1775–1834), known for his self-revealing observations of life, had a rather strong opinion on some such book "rearrangements." He writes, "The human species, according to the best theory I can form of it, is composed of two distinct races, the men who borrow, and the men who lend. . . . Your borrowers of books—those mutilators of collections, spoilers of the symmetry of shelves, and creators of odd volumes."

Symmetry under permutation can appear in more abstract circumstances. Examine the contents of the phrase "Rachel is David's cousin." The meaning will remain unchanged if we interchange *David* with *Rachel.* The same is not true for the phrase "Rachel is David's daughter." Similarly, the equality between two quantities, $a = b$, is symmetric under the transposition of a and b, since $b = a$ is the same relation. While this may seem trivial, the relation "greater than" (commonly denoted by the symbol >) does not have this property. The relation $a > b$ means "a is greater than b." Permuting the letters results in $b > a$, "b is greater than a," and the two relations are mutually exclusive.

Various mathematical formulae can also be symmetric under permutations. The value of the expression $ab + bc + ca$ (where ab means "a times b" and so on) remains unchanged under any permutation of the letters a, b, c. As I shall discuss in more detail later, there are precisely six possible permutations of three letters, including one (the first below) that is the identity, mapping each letter into itself:

$$\begin{pmatrix} abc \\ abc \end{pmatrix}\begin{pmatrix} abc \\ acb \end{pmatrix}\begin{pmatrix} abc \\ bca \end{pmatrix}\begin{pmatrix} abc \\ cab \end{pmatrix}\begin{pmatrix} abc \\ cba \end{pmatrix}\begin{pmatrix} abc \\ bac \end{pmatrix}$$

You can easily check that the above expression is unaltered by these permutations. For instance, the third permutation changes a into b, b into c, and c into a. The entire formula therefore changes to read: $bc + ca + ab$. However, since in whichever order we either multiply numbers or add them up the result is always the same, the new expression is equal to the original one.

People playing roulette in a casino provide an interesting case of symmetry under permutations. The roulette is composed of a rotating wheel in which eighteen numbered red slots, eighteen black slots, and two green slots commonly labeled 0 and 00 are marked. A white ball is dropped onto the spinning wheel, and after rolling rapidly around the rim a few times, it bounces around and eventually lands and comes to rest in one of the slots. When the wheel is mechanically perfect, the game of roulette is absolutely symmetric under any permutation of the players. Everybody has precisely the same chance to win or lose irrespective of whether they are casino rats or novices, experts in probability theory or village idiots. The expectation to win (rather to lose, about 5.3 cents for every dollar bet, on the average) does not depend on the amount of money being risked or on the player's strategy. While no mechanical wheel can be truly perfect, centuries of profits for casinos prove that, whatever small deviations may exist, they do not lead to a significant violation of the symmetry under permutations.

Not all gambling activities are symmetric under permutations of the players. Blackjack is a card game in which each gambler at the table plays against the dealer. Each number card has its face value, with all the picture cards having the value of ten, and the ace offering the option of being counted as one or eleven. The objective is to get the sum of the values of the dealt cards to be closer to twenty-one than the dealer's hand without exceeding twenty-one. What makes blackjack asymmetric with respect to permutations of the players is precisely the fact that strategy *is* important. In the 1960s casinos discovered the hard way the extent to which strategy counts. Mathematician Edward O. Thorp uncovered a flaw in the way casinos were calculating probabilities when the deck of cards was dwindling down. He used this information to develop an extremely profitable method of play. In case you wonder, the casinos have since taken corrective action. Nonetheless, it remains true that strategy does make a difference in blackjack. Indeed, six MIT students who communicated with card-count code words made millions in Las Vegas in the 1990s.

Permutation symmetry and some of its scientific close cousins have far-reaching consequences in the physics of the subatomic world, and we shall return to some of those in chapter 7. Here I will only mention briefly one simple example that explains an otherwise perplexing fact about atoms of different elements—they are all roughly the same size.

Atoms somewhat resemble miniature solar systems. The electrons in the atom are orbiting a central nucleus, just as the planets are revolving around the sun. The force that holds the electrons in their orbits, however, is electromagnetic, rather than gravitational. The nucleus contains protons that have positive electric charges (and neutrons, which are neutral), while the orbiting electrons (equal in number to the protons) are negatively charged. Opposite electric charges attract each other. Unlike planetary systems, which can have orbits of any size, atoms must obey the rules of the subatomic realm—*quantum mechanics.* The highest probability of finding the electrons is along certain specific, "quantized" orbits, restricted to a particular series of discrete sizes. The permitted orbits are characterized mainly by their energy. Broadly speaking, the higher the energy associated with the orbit, the larger its size. The situation is somewhat analogous to a flight of stairs, with the nucleus representing the bottom and the higher energy levels corresponding to increasingly higher steps. Here, however, comes the puzzle. Physics, and indeed everyday life, teach us that systems are most stable in their lowest possible energy state (e.g., a ball rolling down the steps reaches stability at the bottom). This would mean that whether we are dealing with the hydrogen atom, which has only one electron, the oxygen atom, with eight electrons, or uranium, ninety-two electrons, all the electrons would be

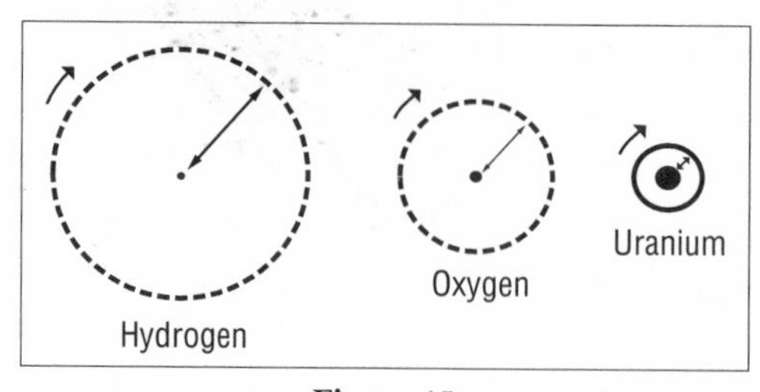

Figure 15

clustered in the smallest possible orbit. Because the more electron- and proton-rich atoms are, the stronger the electrical attraction between the nucleus and the electrons, we would expect that the oxygen atom would be smaller than the hydrogen atom, and the uranium atom much smaller still (as depicted schematically in figure 15). Experiments show, however, that this is far from being the case. Rather, irrespective of the number of electrons, the atoms are found to be roughly the same size. Why?

The explanation was given by the famous physicist Wolfgang Pauli (1900–58). He proposed in 1925 a powerful law of nature (which won him the Nobel Prize in 1945), known as the Pauli exclusion principle. The law refers to some elementary particles of the same type, such as electrons. All the electrons in the universe are precisely identical in terms of their intrinsic properties—there is no way to distinguish one from the other. In addition to their mass and electric charge, electrons have another fundamental property called *spin.* Spin could be thought of, for some purposes, as if the electron were a tiny ball spinning around its axis. Quantum mechanics—the theory that describes atoms, light, and subatomic particles—tells us that the electron spin can have only two states (loosely analogous to the ball spinning at a specific rate in one or the opposite direction). The Pauli exclusion principle asserts that no two electrons can be in precisely the same state; that is, having exactly the same orbit and direction of spin. How is this related to symmetry? To phrase the exclusion principle more accurately, we need to realize that quantum mechanics speaks in the language of probabilities. We can never determine precisely the location of an electron within the atom. Rather, we can only determine the different probabilities of finding it at various positions. The collection of all of these probabilities is known as the *probability function.* The probability function plays the role of a map, showing us where we are most likely to find the electron. Accordingly, Pauli also formulated his exclusion principle in terms of a property of the probability function describing the motion of electrons in the atom. He stated that the probability function is antisymmetrical with respect to interchanging any electron pair. Such a function is called antisymmetrical if transposing two electrons that move along the same orbit and have the same spin direction changes only the sign of the function (e.g., from plus to minus), but not its value. For instance, imagine that the letter a symbolizes the value of some property of the first electron, and the letter b the value of the same property for the second electron. A function that takes the value $a + b$ is symmetrical under the exchange of the two electrons since $a + b$ equals $b + a$. On the other hand, a function represented by $a - b$ is antisymmetrical, since changing a to b and b to a changes $a - b$ into $b - a$, and $b - a$ is precisely the negative of $a - b$ (e.g., $5 - 3 = 2$; $3 - 5 = -2$).

Pauli's statement is therefore the crux of the matter. On one hand, we know that if we interchange two identical electrons this should make no difference whatsoever, and the probability function should remain unchanged. On the other, the exclusion principle tells us that the probability function should change its sign (e.g., from positive to negative) under such a permutation. What kind of number is equal to the negative of itself? There is only one such number—zero. Changing the sign in front of a zero does not change the value by one iota; minus zero equals plus zero. In other words, the probability of finding two electrons with the same spin moving along the same orbit is zero—no such state exists.

Pauli's exclusion principle tells us that electrons with the same properties don't like to be bunched up in the same place. Consequently, no more than two electrons (one with each direction of spin) are allowed in any given orbit. Instead of all the electrons crowding the smallest (lowest energy) orbit, electrons are forced into successively higher-energy, larger-size orbits. The net result is that even though the sizes of all the quantized orbits are smaller in the heavier (more proton-rich) atoms, electrons have no choice but to occupy an increasing number of orbits. Amazingly, the behavior of the probability function under permutations of electrons provides the explanation why, unlike in figure 15, atoms are nearly equal in size.

Returning now to permutations in general, color transformation may be considered as near kin. For any pattern that has more than one color, such as a chessboard, the colors can be interchanged. Strictly speaking, actual patterns are usually not symmetrical under color transformation—they do change. A few of the imaginative designs of M. C. Escher come just about as close as one can expect to being color symmetrical (figure 16). Note that the image does not remain truly the same when black and white are transposed, and neither does a chessboard. However, the general visual impression remains the same.

Escher himself was never quite sure what had led him to his obsession with translation-symmetric, color-symmetric patterns. In his own words,

> I often wondered at my own mania of making periodic drawings. Once I asked a friend of mine, a psychologist, about the reason of my being so fascinated by them, but his answer: that I must be driven by a primitive, prototypical instinct, does not explain anything. What can be the reason of my being alone in this field? Why does none of my fellow-artists seem to be as fascinated as I am by these interlocking shapes? Yet their rules are purely objective ones, which every artist could apply in his own personal way!

Figure 16

Escher's retrospective musings touch upon two important topics: the role of symmetry in the "primitive" process of perception, and the rules that underlie symmetry. The latter topic will be the subject of several later chapters. However, since all the information we obtain about the world comes through our senses, the question of symmetry as a potential factor in perception becomes of immediate relevance.

– TWO –

eyE s'dniM eht ni yrtemmyS

Among all the human senses, vision is by far the most important vehicle for perception. However, the eyes are only optical devices; perception requires the participation of the brain. Visual perception is a complex of processes in the brain, combining sensations from the external world to produce an informative image. Our environment produces many more signals than we can possibly analyze. Consequently, perception involves sifting through the wealth of data and selecting the most useful features. When human chess players consider their next move, they do not mentally examine every possible move on the board. They focus on those few moves that appear to be most beneficial when viewed against the baseline of accumulated information—that thing we call memory. In the Woody Allen movie *The Curse of the Jade Scorpion,* Dan Aykroyd plays the boss of an insurance company. In one scene he tells one of his investigators, C. W. Briggs (played by Woody Allen), "You know, there's a word for people who think everyone is conspiring against them." To which Woody Allen replies, "Yeah—perceptive!" In reality, of course, paranoia represents a distortion of perception.

On the face of it, visual perception must accomplish an impossible task. It needs to transform the physical impinging of units of light energy (called *photons*) on receptors at the back of the eye into mental pictures of objects. As we shall soon see, symmetry provides an important aid toward this goal.

First, however, we must appreciate what types of difficulties have to be overcome. Astronomy can help to illustrate one of the many obstacles involved in this process—specifically, the perception of distance. Figure 17 shows a picture taken with the Hubble Space Telescope, peering through the spherical halo of stars surrounding the Andromeda galaxy (known to astronomers as M31). A galaxy is a vast sweep of a few hundred billion stars like the Sun. M31, at a distance of about 2.5 million light-years, is one of the nearest neighbors to our own Milky Way galaxy. (One light-year is about 6 trillion miles.) The picture in figure 17 contains about ten thousand stars in M31 and about a hundred other galaxies that are seen in the background (some of which appear as extended, fuzzy objects). Here, however, comes the problem. From simply looking at the picture, there is no way of telling that the stars are in our own backyard, relatively speaking (at a distance of 2.5 million light-years), while some of the galaxies are more than 10 *billion* light-years away!

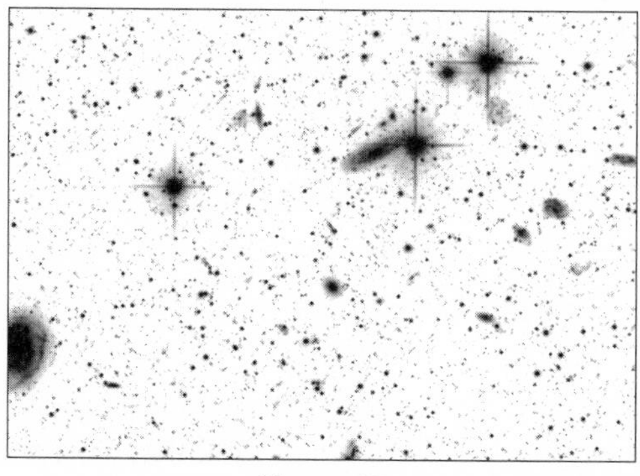

Figure 17

Similarly, when we gaze at the world around us, the eye recognizes only the direction of the light ray on which a photon traveled. Since the image is projected onto a two-dimensional surface (the retina), without some additional information the brain does not have a clue how far away the photon originated. In the case of a relatively nearby star, astronomers solve this problem of distance determination using a method known as *trigonometric parallax.* They view the star from two different spots along the Earth's orbit around the Sun (figure 18). During the course of a year, the nearby star appears to shift back and forth against the very distant (fixed) stars in the background. By measuring the angle associated with this apparent shift, knowing the diameter of the Earth's orbit, and using simple high-school trigonometry, one can calculate the distance to the star.

Humans use their two eyes in precisely the same way to produce spatial awareness. You can discover this mechanism, known as *stereoscopic vision,* by the following simple experiment. Extend your arm, hold up one finger, and look at the finger against some background. If you alternately close your right and left eyes, your finger will appear to shift back and forth relative to the background objects. Bring the finger closer to your eyes, and you will notice that the jump between the two positions increases. This apparent shifting (parallax) occurs because your two eyes view the finger from two different spots. Since the parallax depends on the distance of the object, by measuring the angle between the apparent positions and knowing the separation between the eyes, the brain "trigonometrizes" the distance to the object. If you are familiar with the relative loss of depth perception associated with closing one eye, you may think that the role of the two eyes in stereoscopic vision has been known since antiquity. Surprisingly, even some of the greatest researchers in perspective missed the concept of stereoscopic vision altogether. Mathematicians such as Euclid in ancient Greece, the Renaissance architects Brunelleschi and Alberti, the painters Piero della Francesca, Paolo Ucello, and Albrecht Dürer, and even the great Isaac Newton took the two eyes to be a mere manifestation of bilateral symmetry, with no other special function. The first to have noticed that two eyes can provide something one eye cannot was the quintessential Renaissance man—

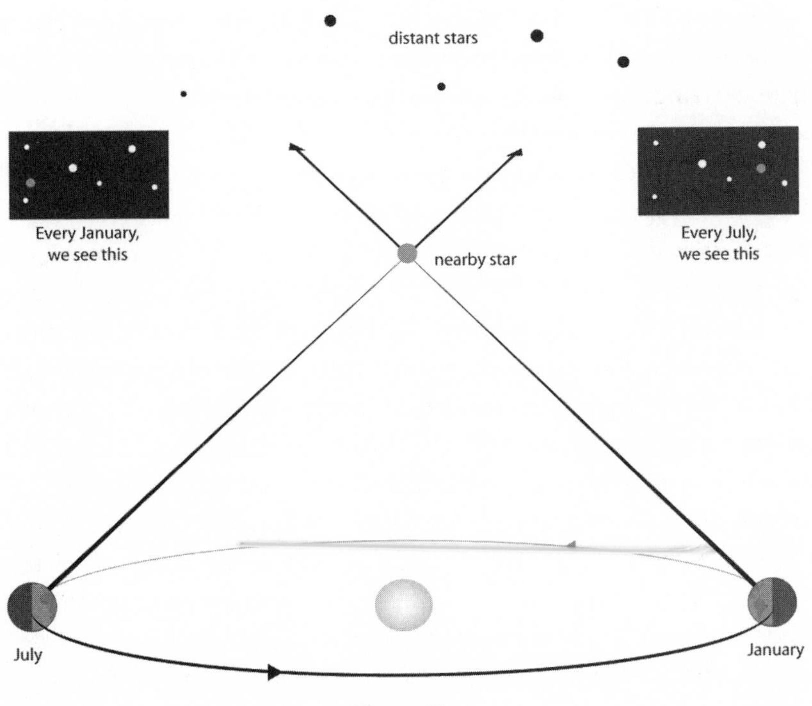

Figure 18

Leonardo da Vinci (1452–1519). Leonardo noted that when we look at an object with both eyes, the right eye manages to capture some of the space behind the object to its right, while the left eye sees around the object to its left. Leonardo therefore concluded that "the object . . . seen with both eyes becomes, as it were, transparent . . . but this cannot happen when an object . . . is viewed by a single eye." In spite of this foresight, by restricting his attention only to spheres, Leonardo missed the opportunity to discover that it was not only in the background, but also in the object itself, that the two eyes captured two different views. The person who established the importance of seeing with both eyes for the perception of distance was the German astronomer Johannes Kepler (1571–1630). In two remarkable books, *Astronomiae Pars Optica* (*The Optical Part of Astronomy*), published in 1604, and *Dioptrice* (*Dioptrics,* the part of optics that treats refraction), published in 1611, Kepler gave a detailed description of the optics of the eye, explained the operation of eyeglasses, and developed a theory of stereoscopic vision. Somehow, however, Kepler's work went relatively unnoticed, and even Charles Wheatstone, who rediscovered the mechanism of depth perception in 1838, appears to have been unaware of it.

Charles Wheatstone (1802–75) was born into a musical family, and his first investigations involved sound, vibrations of various devices such as strings and pipes, and musical instruments. In 1822 he set up a demonstration in his father's shop in Pall Mall in London that provided music not just to the ears, but also to the eyes. This "Enchanted Lyre" was suspended by a thin wire that passed through the ceiling to a room above and was connected to the soundboards of a piano, a harp, and a dulcimer. As Wheatstone played the instruments in the upper room, the Enchanted Lyre appeared to play by itself. A very imaginative experimentalist, Wheatstone invented the concertina (a musical instrument similar to a small accordion) and patented the electric telegraph in Britain.

Wheatstone started his experiments on stereoscopic vision in 1832 and presented his theory in a paper published on June 21, 1838. The title of the paper was "Contributions to the Physiology of Vision. Part the First. On some remarkable, and hitherto unobserved, Phenomena of Binocular Vision." The first paragraph of the paper describes the essence of the finding—that the incongruity of the images on the two retinas and the subsequent mental processing produce a spatial perception. In Wheatstone's words:

> When an object is viewed at so great a distance that the optic axes of both eyes are sensibly parallel when directed towards it, the perspective projections of it, seen by each eye separately, are similar, and the appearance to the two eyes is precisely the same as when the object is seen by one eye only. . . . But this similarity no longer exists when the object is placed so near the eyes that to view it the optic axes must converge; under these conditions a different perspective projection of it is seen by each eye. . . . This fact may be easily verified by placing any figure of three dimensions, an outline cube for instance, at a moderate distance before the eyes, and while the head is kept perfectly steady, viewing it with each eye successively while the other is closed.

I have gone to some length in describing the discovery of the processes involved in the perception of something as elementary as spatial depth because this story helps to exemplify the immense hurdles associated with developing a comprehensive understanding of perception. Theories of human perception can fill, and indeed have filled, entire volumes. Here I will focus solely on the function of symmetry in this process.

The role of symmetry in perception was thrust to center stage by the school of thought known as *Gestalt psychology.* Psychologists Max Wertheimer, Kurt Koffka, and Ivo Kohler, who initiated this doctrine, set up an influential laboratory for research in psychology at the University of Frankfurt in 1912. One of the key problems the Gestalt psychologists set out to address was that of perceptual organization—how the small bits of information received by the senses are organized into larger perceptual structures. How do we know which segments belong together to form an object? How do we separate objects from one another and how do we distinguish between object and background? The central "law" of perceptual organization of Gestalt psychology is known as the principle of Prägnanz, commonly referred to as the law of "good figure" (*Prägnanz* means "succinctness" in German). The law states: "Of several geometrically possible organizations, the one that is seen is the one which possesses the best, simplest and most stable shape."

To the Gestalt psychologists, therefore, symmetry was one of the key elements contributing significantly to the "goodness" of the figure. An arrangement of four dots as in figure 19 will be perceived as a square, because the "goodness" of the square as a symmetrical, closed, stable form is higher than that of, say, an arrangement of a triangle plus an extra dot. While the Gestalt psychologists never managed to formulate a precise theory of shape perception, later theorists, such as the Dutch psychologist Emanuel Leeuwenberg and the Americans Wendell Garner and Stephen Palmer, expanded on their basic principles. Garner and Palmer in particular recognized the role of symmetries of various types (such as symmetry under rotations and reflections) for the "goodness" of the figure.

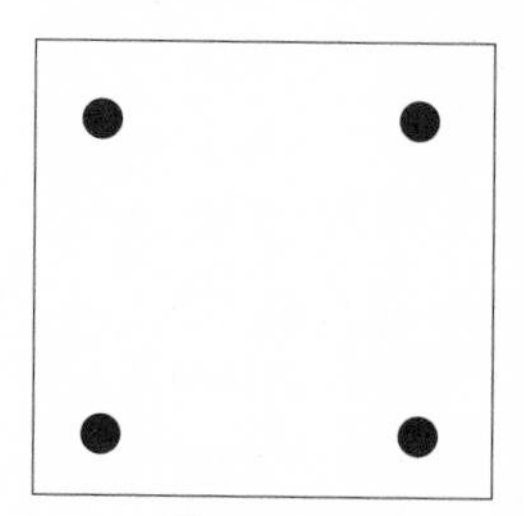

Figure 19

Leeuwenberg and his collaborators developed a theory of shape representation generally known as *structural information theory.* The two fundamental concepts of this theory are *codes* and *information loads.* Codes are simple perceptual descriptions that can generate an observed figure. For instance, to describe a rectangle, we can start from the upper left-hand corner and give the length of the segment that needs to be drawn (figure 20), followed by the angle adjustment that needs to be performed after that. We then give the next length, again followed by the angular adjustment. The final code to draw the rectangle would take the form *a* 90 *b* 90 *a* 90 *b* 90. You notice, however, that since the same instructions are repeated twice, we can simplify this code by writing 2*(*a* 90 *b* 90).

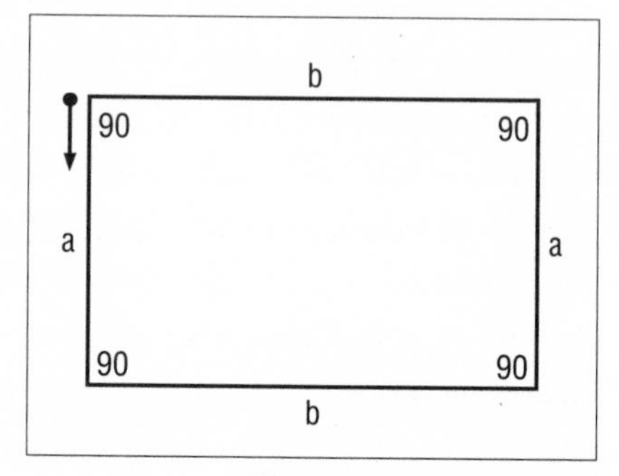

Figure 20

The information load measures the complexity of the simplest code that still gets the job done. Generally, you can compute the information load by simply counting the number of parameters in the code (such as *a, b,* and 90 in the previous example). The central idea of structural information theory is that the "goodness" of the figure is higher the lower the information load. Symmetric figures contain a lower information load and are therefore higher on the "goodness" scale. For the rectangle's code above, for instance,

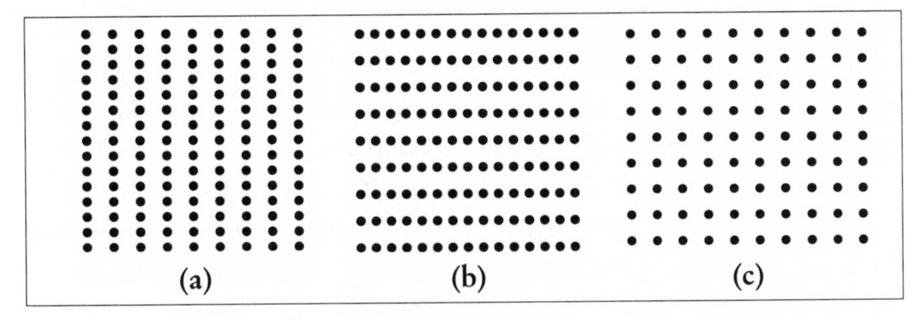

Figure 21

the information load is 4: the number of iterations (2); the two lengths (*a*, *b*); and the angle (90). For an arbitrary quadrilateral figure, on the other hand, the information load would be 8 (four lengths and four angles).

Two other important elements in the Gestalt principles of organization are *proximity* and *similarity.* The "law" of proximity expresses the fact that, generally, forms that are close together are grouped together mentally. In figure 21a we perceive columns, because the vertical spacings of the dots are smaller than the horizontal ones. The converse is true in figure 21b, resulting in the perception of rows. When the spacings are equal (as in figure 21c), we are left with an ambiguous impression.

Shapes that are similar also tend to be grouped in association, and similarity may sometimes be a more powerful organizational element than proximity. In figure 22 we tend to perceive columns because of the similarity of the dark circles, even though, because of their proximity alone, had all the circles been dark we would have seen rows.

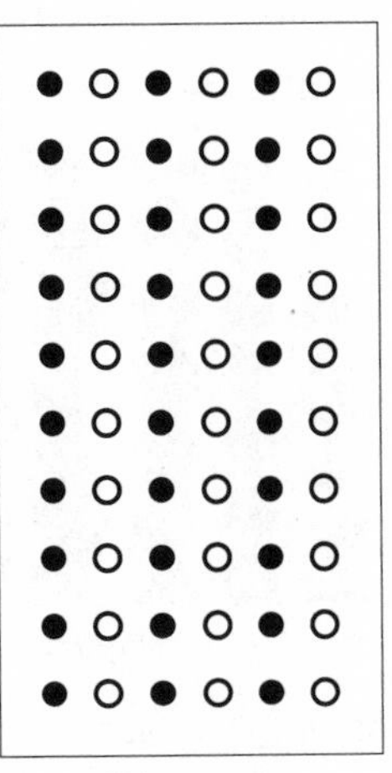

Figure 22

Symmetry plays an important role in the recognition of similarity because it represents a true *invariant*—an immunity to change. Consequently, symmetry is a particularly helpful feature for the perceptual system to use to determine whether the observed patterns are indeed similar or different.

Another Gestalt principle is *good continuation*—we perceive the symbol X as two lines crossing each other, not as one upright and one upside-down V connected at a vertex. *Common fate* is also a basis for grouping. We tend to group together things that are moving at the same speed, in the same direction. The biblical prophet Amos was already fully aware of this principle when he asked, "Do two walk together unless they have made an appointment?"

University of California–Berkeley psychologist Stephen Palmer and collaborators added to the principles of organization those of *common region, connectedness,* and *synchrony.* Figure 23 demonstrates these principles. Common region refers to the fact that elements are grouped together when they are enclosed within a region of space (23a). Connectedness means that we perceive as units elements that appear to be physically connected (23b). Finally, synchrony reflects the fact that simultaneous visual events are perceived as being associated (23c).

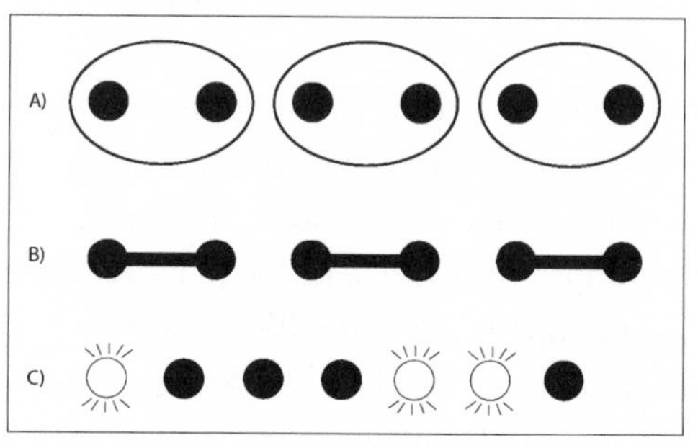

Figure 23

Symmetry, and in particular bilateral symmetry, is also one of the key elements in figure-ground segregation—the ability to see objects as figures that stand out from the background. Take a quick look at figure 24, both to the left and to the right, and decide which color is the figure and which one is the background. Bilaterally symmetric areas tend to be perceived as figures against asymmetrical backgrounds. Consequently, at the left of figure 24 we are inclined to identify the black areas as figures, while at the right, the white ones are the figures. Vertical and horizontal orientations are also more likely to be seen as figures than any other orientations. Finally, smaller areas surrounded by larger ones tend to be identified as figures, as are meaningful or familiar shapes.

Figure 24

You have probably noticed that the original Gestalt "laws" were no more than *heuristics*—best-guess principles that may work most of the time but not necessarily every time. They used rather vaguely defined concepts, such as "goodness" or "similarity." One may wonder why these principles work at all. The answer is that they probably represent a combination of learning and evolution. As Oscar Wilde once said, "Experience is the name everyone gives to their mistakes." Humans have been "practicing" perception for generations, and through their endless number of perceptual encounters they have learned what to expect. In spite of their shortcomings, the original Gestalt principles were useful because they provided a quick answer. When you want to find your keys, you go first to the two places where you normally leave them, and only after that has failed do you embark on a systematic search of the house.

Generally, recent psychological theories and experimental results confirm the important role of symmetry in perception. Many experiments show that bilateral symmetry about a vertical axis is the easiest to recognize (i.e., is recognized fastest) and that it is exploited as a diagnostic property for "same–different" judgment. Basically, symmetry is a property that catches the eye in the earliest stages of the vision process. Symmetry is also useful for discriminating living organisms (including potential predators) from inanimate articles and in the selection of desirable mates (I shall return to these topics in chapter 8). Other experiments have demonstrated that symmetrical figures are more easily reproduced than asymmetrical ones. In an interesting study, Stanford University psychologists Jennifer Freyd and Barbara Tversky found that in the first step, subjects quickly determined whether overall symmetry was present or absent. Then, if the form was perceived as having overall symmetry, some individuals mentally distorted the image and assumed it (sometimes incorrectly) to have symmetry in the details as well.

An intriguing suggestion that the preference for various types of symmetries may be a learned characteristic comes from experiments conducted by University of Illinois psychologist Ioannis Paraskevopoulos. His subjects were seventy-six elementary-school children. Paraskevopoulos found that double symmetry (vertical and horizontal reflection) was preferred at age six, bilateral symmetry (vertical reflection alone) at age seven, and horizontal symmetry (horizontal reflection) at age eleven.

Some of the most exciting recent studies are those attempting to use magnetic resonance imaging (MRI) to map the areas in the brain responding to symmetry. Psychologist Christopher W. Tyler of the Smith-Kettlewell Eye Research Institute in San Francisco presented subjects with a variety of translational- and reflection-symmetric patterns.

He found that these stimuli produced activation of a region of the occipital lobe, whose function is otherwise unknown. Surprisingly, very little or no activation was seen in other areas with known visual functions. Tyler concluded that this specialized region probably encodes the presence of symmetry in the visual field.

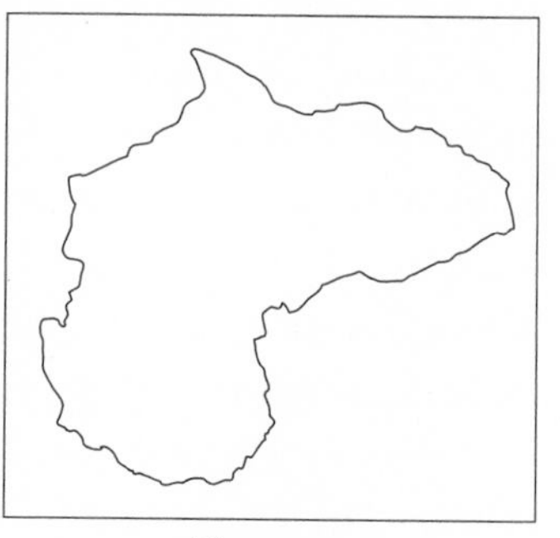
Figure 25

The interrelation between symmetry and orientation is also fascinating. Symmetrical figures do not change when rotated, reflected, or translated in certain ways. Many forms, however, are not symmetrical with respect to any transformation (except the identity, which leaves the form untouched), and how we perceive them is definitely affected, for instance, by their orientation. Take a quick look, for example, at figure 25. Did you recognize it as the map of Africa? Or, without turning the book upside down, do you recognize the person in figure 26?

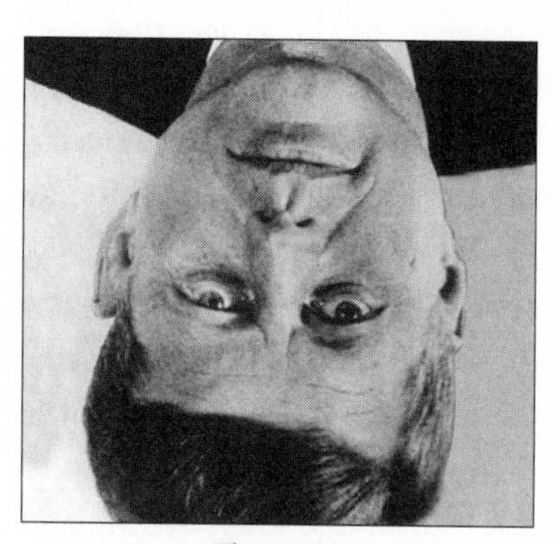
Figure 26

Even the perception of symmetry may be tricky. A shape may be reflection symmetric about some axis, as in figure 27a, but unless you turn it as in figure 27b, so that the axis of symmetry is vertical, you may not perceive the symmetry. Cognitive scientist Irvin Rock of Rutgers University and collaborators conducted a series of experiments designed to test the dependence of perception of form on orientation. In particular, they wanted to test if the perception of bilateral symmetry depends on whether the axis of symmetry is truly vertical in the retinal image or whether it is only perceived as being vertical. The researchers used a shape such as that shown in figure 28a as their standard form. This shape is symmetrical under both vertical and horizontal reflection. Subjects were asked to indicate

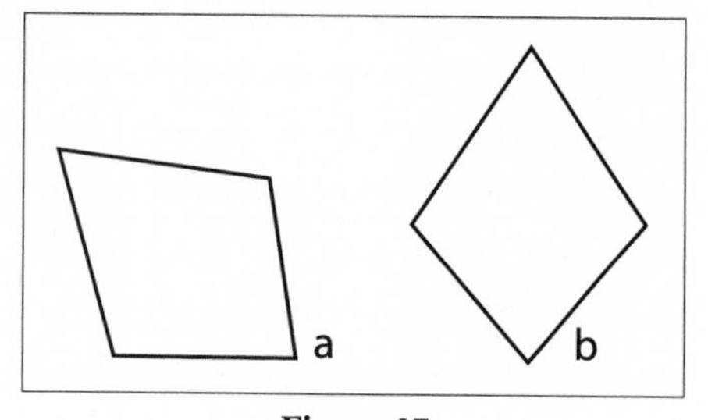

Figure 27

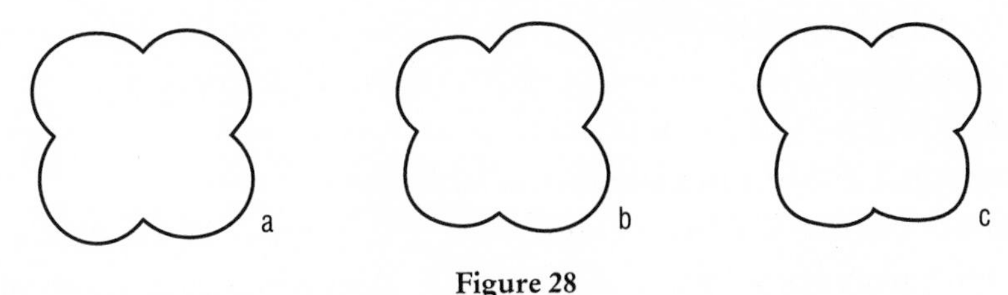

Figure 28

which of the two figures, 28b or 28c, they found to be more like 28a. Note that figure 28b was slightly changed so as not to be symmetric about a vertical axis, but still to preserve the symmetry about the horizontal axis. The converse was done for figure 28c. When the subjects observed the figures with their heads upright, most of them chose figure 28c. This was to be expected; the Moravian-Austrian physicist and philosopher Ernst Mach (1838–1916) had noted as early as 1914 that figures are perceived as symmetrical primarily as a result of reflection symmetry about a vertical axis. Here, however, came a surprise. When observers were tilted by 45 degrees, they still selected figure 28c as being more like Figure 28a, in spite of the fact that in this orientation, neither 28b nor 28c preserved vertical symmetry in the retinal image. From this and other experiments, Rock concluded that "for a novel figure there is little change in appearance when only the orientation of its retinal image is changed." Rock found that what really matters is not even so much the actual orientation of the figure in its environment, but the fact that we normally assign the directions top, bottom, left, and right to figures. These assignments typically depend on other visual cues, such as the direction of gravity or the environmental frame of reference. Disoriented figures with respect to the assigned directions are not easily recognized. Interestingly, Rock found that the effect on the perceived form is minimal when the only change performed is a left-right reversal. These results further confirm the primary importance of bilateral symmetry in perception. Rock did acknowledge, however, that some shapes, such as cursively written words or portraits, become very difficult to recognize even when only the orientation of the retinal image is altered.

While symmetry acts in most cases to facilitate perception, one type of symmetry may actually lure the eyes into a misinterpretation of what they see. Scottish physicist David Brewster (1781–1868), who also invented the kaleidoscope in 1816, noticed something strange when staring at wallpaper with translation-symmetric repeating patterns. The prolificacy of Morris and Company and their contemporaries ensured that such patterns were ubiquitous during the Victorian era. To his amazement, Brewster discovered that some of these designs literally "spring out" of the wall and become three-dimensional illusions, now known as *wallpaper* or *escalator illusions,* because both wallpapers and escalators have repetitive patterns. You may be familiar already with this phenomenon from the many Magic Eye books and posters. The fascination with these computer-generated *autostereograms*—patterns that leap into three-dimensionality when stared at with crossed eyes—reached the magnitude of a craze in the early 1990s. Figure 29 demonstrates the surprising effect. If you stare at it for about a minute as if you were focusing your gaze on an image behind the page, the surfers will miraculously materialize as three-dimensional entities. For reasons that are not entirely clear, some people cannot perceive the illusions created by autostereograms. So, if figure 29 did not suddenly gain depth for you, don't despair; you belong to an exclusive club. The idea behind the Magic Eye illusions stemmed from research in depth perception by the Hungarian-American psychologist Bella Julesz in 1959. Julesz's collaborator, psychologist Christopher

Figure 29

Tyler of the Smith-Kettlewell Eye Research Institute, discovered in 1979 that he could use an offset print technique to generate single-image stereograms. The basic explanation for the magic of the repeating patterns is rather simple. With each eye fixed on a different member of an adjacent pair in the repeating pattern, the brain erroneously perceives the two objects as a single one at a different distance (figure 30). The reason for the brain's "failure" is, of course, the fact that the repeating motifs create identical images on the two retinas, giving the impression that a single object is in focus.

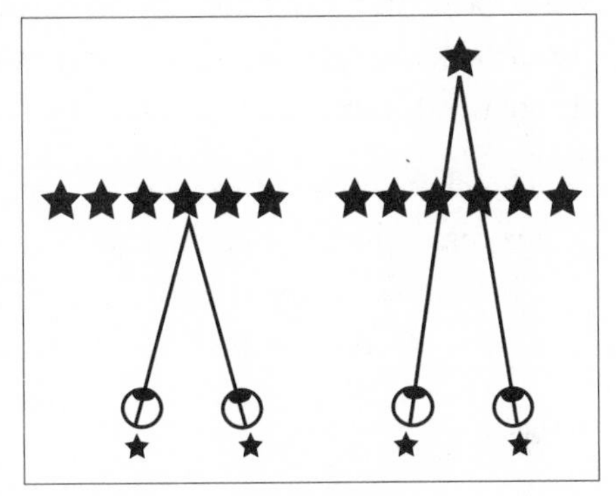

Figure 30

When the repetitive pattern is very closely spaced and consists of high-contrast motifs, it can induce a very powerful illusion of motion. British op artist Bridget Riley dazzled many observers with such hallucinatory patterns in her painting *Fall* (figure 31).

With the exception of permutations and of the Pauli exclusion principle, all the symmetries described so far were symmetries of shapes, forms, and configurations. They were symmetries of objects in space, imposed by the disposition of specific systems and perceived through the senses. We can *see* that a cathedral has bilateral symmetry, that a wallpaper design has translational symmetry, and that a circle has rotational symmetry. The symmetries underlying the fundamental laws of nature are close kin to the above symmetries, but rather than focusing on external form or figure, they concentrate on the question: What operations can be performed on the

Figure 31

world around us that would leave the laws that describe all the observed phenomena unchanged?

THE RULES OF THE GAME

What are the laws of nature? Biologist Thomas Henry Huxley (1825–95), the most passionate defender of Darwin's theory of evolution and natural selection, provided the following explanation:

> The chess board is the world, the pieces are the phenomena of the universe, the rules of the game are what we call the laws of Nature. The player on the other side is hidden from us. We know that his play is always fair, just, and patient. But also we know, to our cost, that he never overlooks a mistake, or makes the smallest allowance for ignorance.

This definition, by the man who was nicknamed "Darwin's Bulldog," lacks ambition by modern standards. Today's physicists would like the laws of nature not only to represent the rules of the game, but also to explain even the existence and properties of the chessboard and the pieces themselves!

Not until the seventeenth century did humans even dream of the possibility that a body of laws exists that would explain everything. Galileo Galilei (1564–1642), René Descartes (1596–1650), and in particular Isaac Newton (1642–1727) demonstrated for the first time that a handful of laws (such as the laws of motion and gravity) could explain a wealth of phenomena, ranging from falling apples and tides on the beach to the motion of planets.

Others followed in their giant footsteps. In 1873, Scottish physicist James Clerk Maxwell (1831–79) published his *Treatise on Electricity and Magnetism*—a monumental work unifying all the electric, magnetic, and light phenomena under the umbrella of only four mathematical equations. Building on the experimental results of the English physicist Michael Faraday (1791–1867), Maxwell was able to show that just as the force holding planets in their orbits and the one keeping objects on the Earth's surface are in fact one and the same, electricity and magnetism are simply different manifestations of a single physical essence. The twentieth century witnessed the birth of not one but two major scientific revolutions. First, Einstein's special and general theories of relativity changed forever the meaning of space and time. The latter two concepts have become inextricably linked in the entity now known as *spacetime.* General relativity also suggested that gravity is not some mysterious force that acts at a distance, but simply a manifestation of spacetime being warped by matter, like a rubber sheet sagging under the weight of a cannonball. Everything moving through this warped space—such as planets in their courses—travels not along straight lines, but on curved trajectories. Second, on a different front, all hope for a fully deterministic world was shattered with the introduction of *quantum mechanics.* In Newtonian mechanics, and even in general relativity, if you somehow knew the position of every single particle in the universe at a given moment and how fast and in which direction it was moving at that instant, you could both predict unambiguously the future of the universe and tell the entire story of the preceding cosmic history. The only limitations would have been associated with rare circumstances in which general relativity breaks down, as in the case of the collapsed objects known as black holes. Quantum mechanics changed all that. Even the position and velocity of a single particle cannot be determined precisely. The only things that are deterministic about the universe are the probabilities of various outcomes, not the outcomes themselves. Although for rather different reasons, the universe is a bit like the weather—the best we can do is predict the probability that it will rain tomorrow, not whether it will actually rain or not. God does play dice.

With every step toward the revolutions of relativity and quantum mechanics, the role of symmetry in the laws of nature has become increasingly appreciated. Physicists are no longer content with finding explanations for individual phenomena. Rather, they are now convinced more than ever that nature has an underlying design in which symmetry is the key ingredient. A symmetry of the laws means that when we observe natural phenomena from different points of view, we discover that the phenomena are governed by precisely the same laws of nature. For instance, whether we perform experiments in New York, Tokyo, or at the other edge of the Milky Way galaxy, the laws of nature that explain the results of those experiments will take the same form. Note that the symmetry of the laws does not imply that the results of the experiments themselves will necessarily remain unchanged. The strength of gravity on the Moon is different from that on Earth, and consequently astronauts on the Moon were seen to leap to greater heights than they would have on Earth. However, the dependence of the strength of the gravitational attraction on the Moon's mass and radius is the same as the dependence of the Earth's gravity on its own mass and radius. This symmetry of the laws—the immunity to changes when displaced from one place to another—is a translational symmetry. Without this symmetry under translation it would have been virtually impossible to understand the universe. The chief reason we can interpret relatively easily observations of galaxies ten billion light-years away is that we find that hydrogen atoms there obey precisely the same quantum mechanical laws they obey on Earth.

The laws of nature are also symmetrical under rotation. Physics has no preferred direction in space—we discover the same laws whether we perform the experiment standing upright or tilted whichever way, or whether we measure directions with respect to up, down, north, or southwest. This is less intuitive than you might think. Recall that for creatures that evolved on the surface of the Earth there is a clear distinction between up and down. Aristotle and his followers thought that objects fall downward because that is the natural place for heavy things. Newton, of course, made it clear that up and down seem different to us not because the laws of physics depend on these directions, but because we happen to feel the gravitational pull of this relatively large mass we call Earth underneath our feet. This is a change in the environment, not in the laws. In a way we are lucky—the symmetries under translation and rotation ensure that irrespective of where we are in space, or how we are oriented, we will discover the same laws.

Figure 32

A simple example can help to clarify further the difference between the symmetries of shapes and of laws. The ancient Greeks thought that the orbits of the planets must be circular, because this shape is symmetric under rotations by any angle. Instead, the symmetry of Newton's law of gravitation under rotation means that the orbits can have any orientation in space (figure 32). The orbits do not have to be circular; they can be, and indeed are, elliptical.

There exist other, more esoteric symmetries that leave the laws of nature invariant, and we shall return to a few of them and their important implications in chapter 7. The key point to keep in mind, however, is that symmetry is one of the most important tools in deciphering nature's design.

Until now, our swift survey of symmetries, whether of objects or of natural laws, has been like that of tourists in a foreign country. We have been able to admire the scenery, but to gain a deeper understanding of the culture we must learn to speak the language. So it's time for a crash Berlitz course.

THE MOTHER OF ALL SYMMETRIES

Even the brief glimpse of the world of symmetries we have caught so far makes it crystal clear that symmetry sits right at the intersection of science, art, and perceptual psychology. Symmetry represents the stubborn cores of forms, laws, and mathematical objects that remain unchanged under transformations. The language describing symmetries has to identify these invariant cores even when they are masquerading under different disciplinary disguises.

The language of the financial world, for instance, is the language of arithmetic operations. If you want to compare at a glance the economical strengths of two companies, you don't need to read entire volumes of prose; a comparison of some key numbers will do. When Isaac Newton formulated his celebrated laws of motion, he also developed the language of calculus to be able to express and manipulate them. One could argue that one of the achievements of abstract and nonobjective art in the twentieth century was the transformation of color into a language of meaning and emotion. Some painters abandoned the use of form and other visual elements almost entirely in favor of communicating exclusively by color.

To explore the labyrinths of symmetry, mathematicians, scientists, and artists light their way by the language of *group theory.* Like some exclusive clubs, a mathematical group is characterized by members that have to obey certain rules. A mathematical *set* is any collection of entities, whether the components of a dismantled airplane, the letters of the Hebrew alphabet, or a bizarre collection consisting of van Gogh's ear, the Easter bunny, all the Albanian newspapers, and the weather on Mars. A *group*, on the other hand, is a set that has to obey certain rules with respect to some operation. For instance, one of the most familiar groups is composed of all the integer numbers (positive, negative, and zero; i.e., . . . −4, −3, −2, −1, 0, 1, 2, 3, 4, . . .), in conjunction with the simple arithmetic operation of addition.

The properties that define a group are:

1. *Closure.* The offspring of any two members combined by the operation must itself be a member. In the group of integers, the sum of any two integers is also an integer (e.g., $3 + 5 = 8$).
2. *Associativity.* The operation must be associative—when combining (by the operation) three ordered members, you may combine any two of them first, and the result is the same, unaffected by the way they are bracketed. Addition, for instance, is associative: $(5 + 7) + 13 = 25$ and $5 + (7 + 13) = 25$, where the parentheses, the "punctuation marks" of mathematics, indicate which pair you add first.
3. *Identity element.* The group has to contain an identity element such that when combined with any member, it leaves the member unchanged. In the group of integers, the identity element is the number zero. For example, $0 + 3 = 3 + 0 = 3$.
4. *Inverse.* For every member in the group there must exist an inverse. When a member is combined with its inverse, it gives the identity element. For the integers, the inverse of any number is the number of the same absolute value, but with the opposite sign: e.g., the inverse of 4 is −4 and the inverse of −4 is 4; $4 + (-4) = 0$ and $(-4) + 4 = 0$.

The fact that this simple definition can lead to a theory that embraces and unifies all the symmetries of our world continues to amaze even mathematicians. As the great British geometer Henry Frederick Baker (1866–1956) once put it, "What a wealth, what a grandeur of thought may spring from what slight beginnings." Group theory has been called by the noted mathematics scholar James R. Newman "the supreme art of mathematical abstraction." It derives its incredible power from the intellectual flexibility afforded by its definition. As we shall see later in the book, the members of a group can be anything from the symmetries of the elementary particles of the universe or the different shuffles of a deck of cards to the symmetries of the equilateral triangle. The operation between the members may be something as mundane as arithmetic addition (as in the previous example), or more complicated, such as "followed by," for the operation of two symmetry transformations (as in a rotation by one angle followed by a rotation by another angle).

Group theory explains what happens when various transformations, such as rotation and reflection, are applied successively to a particular object, or when a particular operation (such as addition) scrambles different objects (such as numbers) together. This type of analysis exposes the most fundamental structures of mathematics. Consequently, when stock market analysts or elementary particle physicists encounter what appear to be insurmountable difficulties in the recognition of patterns, they can occasionally use the formalism of group theory to cross over into other disciplines and borrow tools developed there for similar problems.

To get an inkling of the relation between group theory and symmetries, let's start with the simple case of the symmetries of the human figure. Humans remain almost unchanged under only two symmetry transformations. One is the identity, which leaves everything as is and is therefore a precise symmetry. The second is reflection about a vertical plane—the (approximate) bilateral symmetry. Let us use the symbol I to denote the operation of the identity transformation and the symbol r to denote the reflection. The set of all the symmetry transformations of the human form therefore consists of just two members: I and r. What happens if we apply these transformations successively? A reflection followed by the identity is no different from performing a reflection alone. Symbolically we can express this as: $I \circ r = r$, where the symbol $\circ$ denotes "followed by." Note that the order is always such that the first symbol to the right is the transformation applied first, and the other follows. Therefore, $a \circ b \circ c$ means that c was applied first, followed by b and then by a.

The application of two reflections successively gets the human figure back to the original, since the first reflection interchanges left and right and the second interchanges them back. Applying r followed by r is therefore the same as applying the identity I: $r \circ r = I$.

We can now attempt to construct something like a multiplication table for the two symmetries, where the entry in row I and column r is

$I \circ r$, and so on. The word *multiplication* is used here loosely to represent the operation between the transformations (in this case, "followed by").

$\circ$	I	r
I	I	r
r	r	I

The multiplication table reveals an important truth: *The set of all the symmetry transformations of the human figure is a group!* Let us check that the defining properties of a group are indeed all satisfied:

1. *Closure.* The multiplication table demonstrates that the combination of any two symmetry transformations by the operation "followed by" is also a symmetry transformation. When you think about it, this is not surprising. Since any of the two transformations leaves the figure unchanged, so does their combined application.
2. *Associativity.* This is clearly satisfied, because it is true for any three transformations of this type combined by "followed by." Indeed, when we apply, say, $I \circ r \circ r$, it makes absolutely no difference how we bracket them.
3. *Identity element.* The identity is a symmetry transformation.
4. *Inverse.* The multiplication table shows that each one of the identity and the reflection transformations serves as its own inverse—applying either of them twice gives the identity: $I \circ I = I$ and $r \circ r = I$.

The group of symmetries of the human body contains only two elements, but the association we discovered between symmetries and groups is a powerful one. To choose a slightly richer example, examine the form of the three running legs in figure 33. This is the symbol of the British Isle of Man in the Irish Sea.

This shape has precisely three symmetry transformations: (1) rotation through 120 degrees about the center; (2) rotation by 240 degrees; (3) the identity (or rotation through 360 degrees). Note that the figure is not symmetric under reflection of any sort, because reflections make the feet point the wrong way. We can denote the rotation by 120 degrees by a, the rotation by 240 degrees by b, and the identity by I and examine again what happens when we combine symmetry transformations through the operation "followed by" (denoted by the symbol $\circ$). If we rotate by 120 degrees and again by 120 degrees we obtain a rotation by 240 degrees; meaning that $a \circ a = b$. Similarly, if we rotate twice by 240 degrees the result is the same as if we rotated through 120 degrees, because 480 degrees consists of one complete revolution (360 degrees = the identity) plus 120 degrees. We therefore have $b \circ b = a$. Finally, rotating through 120 degrees followed by a rotation through 240 degrees (or the other way around) results in a rotation through 360 degrees, or the identity: $b \circ a = a \circ b = I$. We are now in the position to complete the "multiplication table":

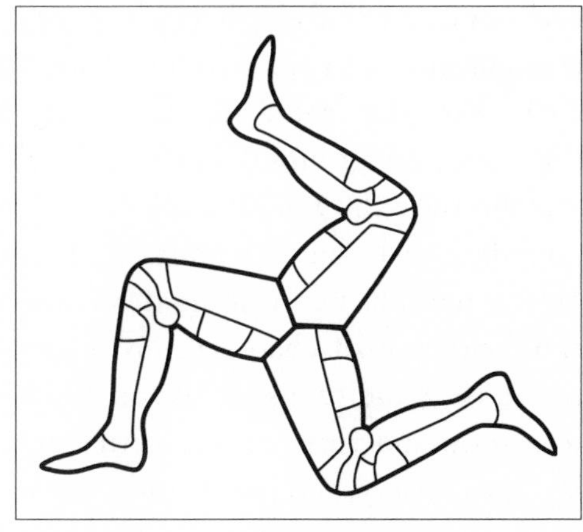

Figure 33

$\circ$	I	a	b
I	I	a	b
a	a	b	I
b	b	I	a

We find that the set of symmetry transformations of the three running legs symbol also forms a group. The table demonstrates closure, and the transformations a and b are each other's inverses—applying one after the other brings things back to the way they were, the identity.

You may begin to realize that groups will pop up wherever symmetries exist. In fact, *the collection of all the symmetry transformations of any system always forms a group.* This is easy to understand. If A is a symmetry transformation, that is, its application leaves the system unchanged, and B is another symmetry transformation, then clearly so is $A \circ B$ (B followed by A). Also, every transformation has an inverse, returning things to the original state. As we shall see throughout this book, the unifying powers of group theory are so colossal that historian of mathematics Eric Temple Bell (1883–1960) once commented, "Whenever groups disclosed themselves, or could be introduced, simplicity crystallized out of comparative chaos."

Unlike most mathematical discoveries, however, no one was looking for a theory of groups or even a theory of symmetries when the concept was discovered. Quite the contrary; group theory appeared somewhat serendipitously, out of a millennia-long search for a solution to an algebraic equation. Befitting its description as a concept that crystallized simplicity out of chaos, group theory was itself born out of one of the most tumultuous stories in the history of mathematics. Almost four thousand years of intellectual curiosity and struggle, spiced with intrigue, misery, and persecution, culminated in the creation of the theory in the nineteenth century. This amazing story, chronicled in the next three chapters, began with the dawn of mathematics on the banks of the Nile and Euphrates rivers.

– THREE –

Never Forget This in the Midst of Your Equations

In an address entitled "Science and Happiness" that was presented at the California Institute of Technology on February 16, 1931, Albert Einstein remarked, "Concern for man himself and his fate must always constitute the chief objective of all technological endeavors . . . in order that the creations of our mind shall be a blessing and not a curse to mankind. Never forget this in the midst of your diagrams and equations." Even Einstein himself could not have imagined how prophetic this admonition would become less than a decade later, during the dark days of World War II and the horrors of the Holocaust. The history of mathematical equations did start, however, solely with the benefit of humankind in mind. The first equation solvers attempted nothing more than to address specific everyday needs.

"US" AND "AHA"

Sometime in the fourth millennium BC the first Sumerian urban communities came into existence in Mesopotamia, the land between the Tigris and Euphrates rivers. Nearly half a million cuneiform tablets and other archeological artifacts found in this area tell the story of a society with organized agriculture, impressive architecture, and vibrant political and cultural history. Then, as today, this fertile land was prone to invasion from many directions, resulting in a frequent change of ruling populations. A few centuries after falling before the Akkadian king Sargon I (ca. 2276–2221 BC), Semitic Amorites took over the land of Sumer and established their capital in the commercial city of Babylon. Consequently, the culture of the entire region between roughly 2000 and 600 BC is conventionally referred to as "Babylonian." The rapidly evolving Babylonian society required massive records of supplies and distribution of goods. Computational tools were also needed for business transactions, for agricultural projects involving the partitioning of lots, and for the making of wills. To this end, the Babylonians developed the most sophisticated mathematics of that time. The texts of scores of cuneiform tablets demonstrate that the Babylonians not only mastered a variety of arithmetic manipulations, but literally anticipated more advanced algebra. Here I shall concentrate only on the emergence of "equations," since this is the most relevant part for the history of group theory. The reason I have put the word *equations* in quotation marks is that the Babylonians did not truly use the concept of algebraic equations in the same way we do today. Rather, they stated problems and solved them rhetorically, in the language of ordinary discourse. In other words, one problem after another was solved by precise verbal instructions, but no pattern or formula was ever identified as a general procedure.

There is little doubt that these mathematical problems first appeared in the context of the society's need to divvy up lots of land. The words used for the unknown quantities one needed to solve for were *us* (length), *sag* (width), and *asa* (area), even when no mensuration was involved.

The simplest equations one can formulate are the ones called *linear* (they are represented by straight lines when graphed). In modern notation these are equations of the type $2x + 3 = 7$, where x represents the unknown. To solve an equation means to find a value of x for which the equation holds true (in the above example, the solution is $x = 2$, since $2 \times 2 + 3 = 7$). Several tablets contain problems that need to be solved using linear equations.

Sometimes to find the answer, one needed to solve for the value of two unknowns. For instance, in one problem the values of the width and length are called for if one-quarter of the width plus the length are equal to 7 hands (a unit of length), and the length plus the width are equal to 10 hands. Using the algebra we learn in school, if we denote the length by x and the width by y, this problem translates into the system of two linear equations: $\frac{1}{4}y + x = 7$, $x + y = 10$. The Babylonian scribe notes correctly that a length of 6 hands (or 30 fingers, one hand being equal to

5 fingers) and a width of 4 hands (20 fingers) satisfy both equations (in appendix 2, I present for the interested reader a brief reminder of how one solves such systems of equations).

Linear equations featured even more prominently in the mathematics of ancient Egypt. Apparently the Babylonians found them too elementary to deserve detailed documentation. Much of our knowledge of Egyptian mathematics comes from the fascinating Ahmes Papyrus. This large papyrus (about eighteen feet long) currently resides in the British Museum (except for a few fragments, discovered unexpectedly in a collection of medical papers, that are in the Brooklyn Museum). The papyrus was purchased by Scottish Egyptologist Alexander Henry Rhind in 1858 and is consequently often referred to as the Rhind Papyrus (figure 34). According to the scribe Ahmes's own testimony, he copied the papyrus around 1650 BC from an original document that had been written a couple of hundred years earlier (during the rule of King Ammenemes III of the Twelfth Dynasty). The papyrus, described by British scientist D'Arcy Thompson as "one of the ancient monuments of learning," contains eighty-seven problems. These are preceded by a table of "recipes" for divisions and an introduction. The introduction describes the document somewhat grandiloquently as "The entrance into the knowledge of all existing things and all obscure secrets." The problems that Ahmes presents and solves, on the other hand, deal mostly with a variety of practical issues, from the fair partition of loaves of bread to the slope of pyramids. The unknown is called *aha,* meaning "heap." For example, problem 26 calls for the value of *aha*

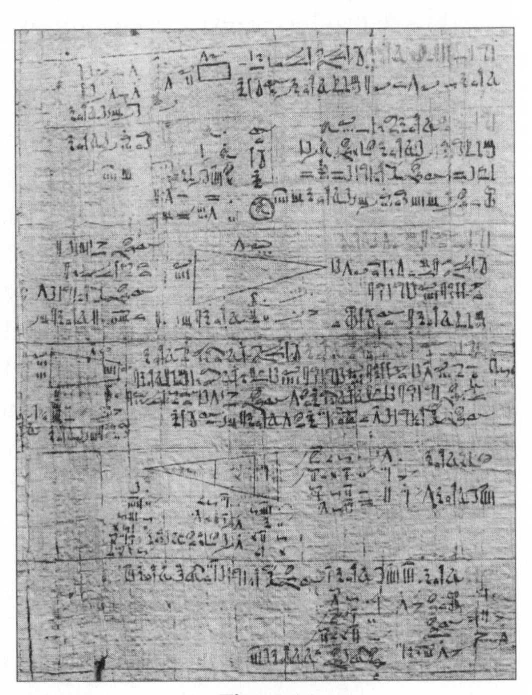

Figure 34

if *aha* and its quarter are added to become 15. In modern notation we would formulate the equation $x + \frac{1}{4}x = 15$, to which the answer is, as Ahmes correctly finds, $x = 12$.

Not all the mathematical problems in the Ahmes Papyrus address the pressing questions of the time. Some were clearly introduced as exercises for students, and at least one was chosen purely for its charm. Problem 79 reads: "Houses 7, Cats 49, Mice 343, Spelt 2,401, *Hekats* 16,807, Total 19,607." Evidently, a playful Ahmes describes here a puzzle, in which in each of seven houses there are seven cats, each of which ate seven mice, each of which would have eaten seven ears of wheat, each of which would have produced seven *hekats* (measures) of grain. The unknown called for in this problem is the total, which, being the sum of all houses, cats, mice, spelts, and *hekats,* is clearly of no practical worth. Many have speculated that this ancient brain twister metamorphosed over the centuries into two other known puzzles. In 1202, the famous Italian mathematician Leonardo of Pisa (nicknamed Fibonacci; lived ca. 1170–1240), published a book entitled *Liber abaci* (*Book of the Abacus*). In this book, he poses a problem that reads, "Seven old women are traveling to Rome, and each has seven mules. On each mule there are seven sacks, in each sack there are seven loaves of bread, in each loaf there are seven knives, and each knife has seven sheaths. Find the total of all of them."

Half a millennium later still, in the eighteenth-century *Mother Goose* collection of nursery rhymes, we find:

As I was going to St. Ives,
I met a man with seven wives.
Every wife had seven sacks,
Every sack had seven cats,
Every cat had seven kits;
Kits, cats, sacks, and wives,
How many were going to St. Ives?

Was this nursery rhyme truly inspired by the Ahmes Papyrus of more than three thousand years earlier? Hard to believe. Note, incidentally, that depending on the interpretation, the correct answer to the nursery rhyme puzzle is either one (the narrator; all others were *coming from* St. Ives) or none (the narrator does not belong in the group of "kits, cats, sacks, and wives"). Geometrical series of this type, in which every successive number is increased by the same multiplier, have always fascinated humans. Furthermore, spiritual qualities have been associated with the number seven in both Eastern and Western traditions (e.g., seven days of the week, seven gods of luck in Japan, seven deadly sins). The three puzzles might therefore have been the independent creations of three imaginative brains, separated by centuries.

The knowledge of how to solve linear equations was not exclusive to the Middle East. The impressive Chinese collection *Nine Chapters on Mathematical Art* (*Jiu zhang suan shu*) was composed sometime between 206 BC and AD 221 and was based on a yet earlier collection. In chapter 8 of *Nine Chapters,* we find problems that involve no fewer than three linear equations with three unknowns, all solved brilliantly.

The next level up, in terms of the intricacy of algebraic equations, is represented by *quadratic equations.* The extra complication is introduced by the fact that in such equations the unknown, x, appears squared, as in $3x^2 + x = 4$. To the novice this may not look like a dramatic change, yet quadratic equations are actually more difficult to solve than linear ones. As incredible as this may sound, the topic of equations in general and of quadratic equations in particular became the subject of a heated debate in the British parliament in 2003. In a brilliant speech on the curriculum at school, member of Parliament Tony McWalter explained:

> Why should anyone feel passionate about the *x*s and *y*s in a system of equations? One answer is this: because if one does not make the effort to see what those *x*s and *y*s conceal, one will be cut off from having any real understanding of science. . . . Why should anyone try to understand quadratic equations and the principles that lie behind solving them? They underpin modern science as surely as the smelting methods of the Romans were the key to their building culture.

However, you may wonder, who were the first to have encountered the need to formulate and solve such equations?

THE PROTECTORS OF THE PUBLIC

In the Jewish code of civil and canon law—the Talmud—we find a story of an exilarch upon whom a huge fine had been imposed. He had to fill a granary having a base surface of 40 upon 40 with wheat. The distressed man went to Rabbi Huna (ca. AD 212–97), the head of the Academy of

Sura in Babylonia, for advice. The scholar told him, "Persuade them to take from you [two installments:] now a surface of 20 upon 20 and after some time another installment of 20 upon 20 and you will profit the half." Of course the area of a square with a side of 40 units is $40 \times 40 = 1{,}600$ square units, while the combined area of two 20×20 squares is only 800 square units. Rabbi Huna takes advantage here of a common mistake in ancient times—the notion that the area of a figure depends entirely on its perimeter. The Greek historian Polybius (ca. 207–125 BC), for instance, tells us that many people of his time refused to believe that Sparta, with a surrounding wall of 48 stadia, could have double the capacity of Megalopolis, with a perimeter of 50 stadia. Figure 35 presents a simple demonstration of how a figure with a smaller perimeter can have a larger area. The elongated rectangle has a perimeter of $2 \times (100 + 10) = 220$ units and an area of $100 \times 10 = 1{,}000$ square units. The shorter rectangle has a smaller perimeter, $2 \times (50 + 40) = 180$ units, yet it has twice the area, $50 \times 40 = 2{,}000$ square units. The Greek mathematician Proclus (AD 410–85) noted that even as late as the fifth century, members of certain communities used to cheat their fellow citizens by giving them land of larger perimeter but smaller area than they selected for themselves. To add insult to injury, these scoundrels used this scheme to earn a reputation for generosity.

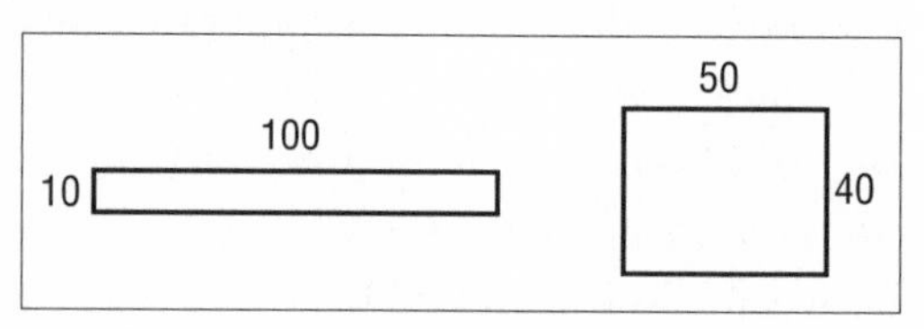

Figure 35

Let us examine for a moment what is involved in resolving the perimeter-area confusion. Suppose we have a rectangle with a perimeter of 18 units. If we denote its length by x and its width by y, then $x + y = 9$ (since the perimeter is composed of twice the length and twice the width). Assume further that the area is given as 20 square units. This means that $xy = 20$ (the area is the product of the length and width). We therefore have the system of two equations with two unknowns:

$$x + y = 9$$
$$xy = 20$$

A straightforward way to solve this problem would be to isolate the unknown y from the first equation (by subtracting x from both sides), $y = 9 - x$, and to substitute this expression for y in the second equation:

$x(9 - x) = 20$. If we now multiply through, on the left-hand side, we obtain the quadratic equation $9x - x^2 = 20$. Many Babylonian problems leading to quadratic equations are broadly of this general form. For instance, problem 2 in tablet 13901 in the British Museum reads, "I subtracted the side from the area of my square. 870." This corresponds to the quadratic equation $x^2 - x = 870$. One speculation is, therefore, that quadratic equations came to light as an attempt by conscientious Babylonian mathematicians to protect the public from manipulators and scheming land thieves. How these mathematicians discovered the solution to the quadratic equation remains a mystery, since, while the Babylonians always spell out in great detail the steps of the procedure leading to a solution, they never tell us how they derived that procedure.

The ancient Egyptians could handle only the simplest of the quadratic equations, of the type $x^2 = 4$, but not "mixed" equations that included both x^2 and x. What is the solution to $x^2 = 4$? It is the square root of 4, denoted as $\sqrt{4}$. One obvious answer is 2, since $2 \times 2 = 4$. That was all the Egyptians cared about, since the number was supposed to represent quantities such as length or loaves of bread, which have to be positive. However, the equation $x^2 = 4$ actually admits a second, less obvious solution: -2. When one negative number is multiplied by a second negative number, the result is a positive number. In other words, $(-2) \times (-2) = 4$, and therefore the equation $x^2 = 4$ has two solutions: $x = 2$ and $x = -2$. This is the first indication that quadratic equations may have two different solutions, not just one. While the Babylonians knew how to solve mixed quadratic equations, they were still interested only in positive solutions, since the unknowns typically represented lengths. They also avoided those cases in which two different positive solutions could be found, since those must have struck them as illogical absurdities.

In spite of their superb mathematical abilities, the very early Greek mathematicians concentrated primarily on geometry and logic and paid relatively little attention to algebra. The clear perception of form and number as two aspects of one mathematics had to await the brilliant mathematical minds of the seventeenth century. The great Euclid of Alexandria, whose monumental work *The Elements* (published ca. 300 BC) laid the foundations of geometry, addresses quadratic equations only obliquely. He solves the equations geometrically, by formulating methods for finding lengths, which are in fact solutions to quadratic equations. Arab mathematicians were to further expand upon this type of geometric algebra centuries later.

THE FATHERS OF ALGEBRA

The great Greek school of Alexandria produced many outstanding mathematicians during two golden ages. Notwithstanding many ups and downs, the city of Alexandria, its school (known as the Museum), and the associated library, reputed to hold some seven hundred thousand books (many confiscated from ill-omened tourists), endured for almost seven hundred years. One of the most original thinkers of the Alexandrian school was Diophantus, a man sometimes called the "father of algebra." Details of the life of Diophantus are so veiled in obscurity that we don't know with certainty even in which century he lived, except that it has to be later than about 150 BC (since he quotes the mathematician Hypsicles, who lived ca. 180 BC to 120 BC) and earlier than about AD 270 (since he is mentioned by Anatolius, bishop of Laodicea, who took office around that time). Generally, Diophantus is assumed to have flourished around AD 250, although the possibility that he had lived a century earlier cannot be ruled out. We know of Diophantus's ingenious work mostly through his major treatise, *Arithmetica,* which originally contained thirteen books. Only six books in Greek have survived the onslaught of the Muslims on the Alexandrian library in the seventh century. An Arabic translation of what may be four more books (attributed to the ninth-century mathematician Qusta Ibn Luqa) was miraculously discovered in 1969.

Despite the honorific "father of algebra," most of *Arithmetica* actually deals with problems from the theory of numbers. Nevertheless, Diophantus certainly represents a crucial stage in the evolution of algebra that is intermediate between the purely rhetorical style of the Babylonians and the symbolic forms of equations (e.g., $2x^2 + x = 3$) we use today. The German mathematician and astronomer Johannes Regiomontanus could not curb his admiration for *Arithmetica* in 1463: "In these old books the very flower of the whole of arithmetic lies hid, the *ars rei et census* [art of the "thing" and enumeration; referring to equations with unknowns and arithmetic] which today we call by the Arabic name of algebra." Diophantus demonstrated incredible creativity and skill in his solutions to many problems. Yet he only considered positive answers, and even among those, only the ones that could be expressed either as whole numbers (such as 1, 2, 3, . . .) or as fractions (such as 2/3, 4/9, 5/13; collectively, the whole numbers and the fractions are known as *rational numbers*). As an example of Diophantus's ingenuity, consider problem 28 from the first book: "To find two numbers such that their sum and the sum of their squares are given numbers." Clearly, this is a problem with two unknowns (the two numbers). Yet, Diophantus succeeds by a brilliant trick to reduce the number of unknowns from two to one, and to obtain for it a simple equation. (For the interested reader I present Diophantus's solution in appendix 3.) The *Arithmetica* makes it abundantly clear that Diophantus knew how to solve quadratic equations of the three types: $ax^2 + bx = c$ (where a, b, c are given positive numbers, as in $2x^2 + 3x = 14$); $ax^2 = bx + c$; and $ax^2 + c = bx$. These were precisely the types of equations revisited by Arab mathematicians more than five centuries later.

Diophantus is best known today for a special class of equations that bears his name—Diophantine equations—and also because of his very unusual epitaph. Diophantine equations are truly bizarre in that, on the face of it, they appear to admit any number as a solution. Consider, for instance, the equation: $29x + 4 = 8y$. For what values of x and y does the equality hold true? If we choose, say, $y = 5$, we obtain $x = 36/29$. If we choose $y = 1$, we obtain $x = 4/29$, and so on. We have an infinity of values to choose from for y, and for any value we happen to choose, we can find a corresponding x that satisfies the equation. What makes Diophantine equations special is that we are actually supposed to be seeking only solutions for x and y that are both whole numbers (such as 1, 2, 3, . . .). This immediately limits the possible solutions and makes them much harder to find. Can you discover a solution for the Diophantine equation above? (If not, I present it in appendix 4.)

The most famous Diophantine equation in history is the one known as Fermat's Last Theorem, the celebrated statement by Pierre de Fermat (1601–55) that there are no whole number solutions to the equation $x^n + y^n = z^n$, where n is any number greater than 2. When $n = 2$, there are many solutions (in fact an infinite number). For instance, $3^2 + 4^2 = 5^2$ ($9 + 16 = 25$); or $12^2 + 5^2 = 13^2$ ($144 + 25 = 169$). Miraculously, when we go from $n = 2$ to $n = 3$, there are no whole numbers x, y, z that satisfy $x^3 + y^3 = z^3$, and the same is true for any other value of n that is greater than 2. Appropriately, it was in the margin of the second book of

Diophantus's *Arithmetica,* which Fermat was eagerly reading, that he wrote his extraordinary claim—one that took no fewer than 356 years to prove.

A sixth-century collection known as *The Greek Anthology* contains some six thousand epigrams. One of these supposedly gives us a scanty record of the life of Diophantus:

> God granted him to be a boy for the sixth part of his life, and adding a twelfth part to this, He clothed his cheeks with down; He lit him the light of wedlock after a seventh part, and five years after his marriage He granted him a son. Alas! late-born wretched child; after attaining the measure of half his father's life, chill Fate took him. After consoling his grief by this science of numbers for four years he ended his life.

Diophantus himself would have probably been somewhat offended by the fact that his life story has been reduced to a mere linear equation, of the type that had never really interested him. If the description is correct, he lived to be eighty-four years old.

Recognizing that the Babylonians, Greeks, and in particular Hindu mathematicians of the seventh century already knew how to solve quadratic equations of various types, we should not be surprised that the solution of such equations is considered today to be part of elementary algebra. The most general form of a quadratic equation is: $ax^2 + bx + c = 0$, where a, b, c can be any given numbers (a cannot be zero, or the equation is not quadratic). The real question is whether there exists some universal recipe or formula that can be relied upon to give the solutions every time. You may have at least dim memories from high-school algebra that such a formula indeed exists. It reads:

$$\frac{-b \pm \sqrt{b^2 - 4ac}}{2a}$$

In spite of its somewhat disconcerting appearance, this is really a simple formula, which when the given values of the numbers a, b, c are substituted into it, yields immediately the values of x for which the equation holds true. For instance, suppose we need to solve the equation: $x^2 - 6x + 8 = 0$, where $a = 1$, $b = -6$, $c = 8$. All we need to do is put these values of a, b, and c into the formula above and we find the two possible solutions: $x = 2$ or $x = 4$ (the symbol $\pm$ means that we choose plus to obtain one solution and minus to obtain the other).

Following the decline and fall of the Alexandrian school, European mathematics seems to have gone into hibernation for almost a millennium. The baton of keeping mathematics, and indeed science in general, alive was passed on to India and the Arab world. Accordingly, the path from Diophantus to the modern solution of the quadratic equation passes through non-European mathematicians. The Indian mathematician and astronomer Brahmagupta (598–670) managed to solve a few impressive Diophantine equations, as well as quadratic equations that for the first time involved negative numbers. He referred to such numbers as "debts," realizing that negative numbers appear most frequently in monetary transactions. In the same spirit, he called positive numbers "fortunes." The rules for multiplying or dividing positive and negative numbers were therefore stated as, "The product or ratio of two debts is a fortune; the product or ratio of a debt and a fortune is a debt."

The man who literally gave algebra its name was Muhammad ibn Musa al-Khwarizmi (ca. 780–850; figure 36 shows him as he appeared on a Soviet stamp). The book he composed in Baghdad—*Kitab al-jabr wa al-muqabalah* (*The Condensed Book on Restoration and Balancing*)—became synonymous with the theory of equations for centuries. From one of the words in the title of this book (*al-jabr*) comes the word "algebra." Even the word "algorithm," used today to describe any special method for solving a problem by following a succession of procedural steps, comes from a distortion of al-Khwarizmi's name. While al-Khwarizmi's book was not particularly groundbreaking in terms of its contents, it was the first to expose in a systematic way the solutions of quadratic equations. The word *al-jabr*, meaning "restoration" or "completion," referred to moving negative terms from one side of the equation to the other, as in transforming $x^2 = 40x - 4x^2$ (by adding $4x^2$ to both

Figure 36

sides) into $5x^2 = 40x$. So great was the influence of al-Khwarizmi's book that even eight centuries later, in the masterful burlesque of the popular romance of chivalry *Don Quixote de la Mancha,* we find that the bonesetter is called "algebrista," because of his job of restoration.

The first book to include the full solution to the most general quadratic equation appeared in Europe only in the twelfth century. The author was the eclectic Spanish Jewish mathematician Abraham bar Hiyya Ha-nasi (1070–1136; "Ha-nasi" means "the leader"). As if to remind us of the early origins of quadratic equations, the book was entitled: *Hibbur ha-meshihah ve-ha-tishboret* (*Treatise on Measurement and Calculation*). Abraham bar Hiyya explains:

> Who wishes correctly to learn the ways to measure areas and to divide them, must necessarily thoroughly understand the general theorems of geometry and arithmetic, on which the teaching of measurement . . . rests. If he has completely mastered these ideas, he . . . can never deviate from the truth.

This brought to an end a long era during which Arab mathematicians acted as the safe custodians of mathematics. Progress during the three thousand years that followed the Old Babylonian period has been only incremental. With the tremendous intellectual awakening of the Renaissance, however, the center of gravity was about to move to northern Italy, with other western European countries soon to follow. Humanists discovered ancient Greek works and encouraged a process of delving into all the Greeks' accumulated knowledge, including mathematics. As the copying of manuscripts became a major industry (according to one report the influential Florentine banker Cosimo de Medici employed forty-five scribes), the invention of printing with moveable type was only to be anticipated, with the ensuing proliferation of scientific knowledge.

There was nothing in the relatively tranquil and rather sluggish history of the quadratic equation to indicate that the next stage in the solution of equations was going to be particularly dramatic. This was, however, only the calm before the storm. The next chapter was about to begin.

THE CUBIC

In the same way that problems dealing with areas result in quadratic equations (because one length is multiplied by the other, producing a length squared), the calculation of volumes of solids such as the cube (where one multiplies length by width and by height) leads to *cubic equations*. The most general cubic equation has the form $ax^3 + bx^2 + cx + d = 0$, where a, b, c, d are any given numbers (a has to be different from zero). The goal of all the aspiring equation solvers was clear: to find a formula, similar to the one for the quadratic equation, that upon substitution of a, b, c, d would give the desired solutions. The ancient Babylonians did generate some tables that allowed them to solve a few very specific cubics, and the Persian poet-mathematician Omar Khayyam presented a geometric solution to a few more in the twelfth century. However, the solution to the general cubic equation defied mathematicians until the sixteenth century. This was not for lack of trying. Three famous Florentine algebraists, Maestro Benedetto in the fifteenth century and his two fourteenth-century predecessors, Maestro Biaggio and Antonio Mazzinghi, had put considerable toil into the understanding of equations and their solutions. Their efforts, however, proved insufficient for the cubic. The fourteenth-century mathematician Maestro Dardi of Pisa also presented ingenious solutions to no fewer than 198 different types of equations—but not to the general cubic. Even the famous Renaissance painter Piero della Francesca, who was also a gifted mathematician, contributed his part to the attempts to find a solution. In spite of these and other valiant efforts, the answer remained elusive. No wonder that mathematician and author Luca Pacioli (1445–1517) concluded his 1494 influential book *Summa de arithmetica, geometria, proportioni et proportionalità* (*The Collected Knowledge of Arithmetic, Geometry, Proportion and Proportionality*) in a defeatist mood. "For the cubic and quartic [involving x^4] equations," he said, "it has not been possible until now to form general rules." The good news was that Pacioli's encyclopedic six-hundred-page work was written in the accessible Italian. Consequently, the book promoted algebraic studies even among those not versed in Latin. At that point, practicality gave way to ambition. No one was searching for a solution to the cubic for some practical purposes. Solving the cubic equation had become an intellectual challenge worthy of consideration by the best mathematical minds. Enter a modest hero—

a mathematician from Bologna named Scipione dal Ferro (1465–1526) who unknowingly becomes part of an unfolding drama.

Scipione dal Ferro was the son of a paper maker, Floriano, and his wife, Filippa. In the century that witnessed the invention of printing, the production of paper became a desirable profession. Little is known of Scipione's youth or of what motivated him to study mathematics. He probably completed his education at the University of Bologna. This prestigious institution, the oldest university still operating today (figure 37 shows a spectacular corridor in the oldest building, which currently houses a library), was established in 1088, and by the fifteenth century it had gained the reputation of being one of Europe's finest. Mathematics (beyond Euclid's basic geometry) had been made part of the regular curriculum at Bologna by the late fourteenth century, and in 1450 Pope Nicholas V added to the teaching staff four positions in mathematics. In 1496, dal Ferro became one of the five joint holders of the chair of mathematics at the university, and except for a short leave that he spent in Venice, he continued in this post for the rest of his life. Although several sources describe him as a great algebraist, no original manuscripts of his work in either script or printed form have survived. A collection of lecture notes from the University of Bologna dated to 1554–68 may include a copy of some of dal Ferro's writings (figure 38). The passage is headed by, "From the Cavaliere Bolognetti, who had it from the Bolognese master of former days, Scipion dal Ferro." Scipione probably met Luca Pacioli in 1501, when the latter was lecturing in Bologna. Pacioli was not exactly a mathematical powerhouse, but he was a great communicator of mathematical knowledge. Being frustrated by his inability to solve the cubic equation, Pacioli may have convinced Scipione, who had great dexterity in manipulating expressions involving cubic roots and square roots, to try it himself. Around 1515, dal Ferro's efforts finally bore fruit. He created a major mathematical breakthrough by managing to solve the cubic equa-

Figure 37

tion of the form $ax^3 + bx = c$. In the mathematical language of the sixteenth century such equations were described as "unknowns and cubes equal to numbers." While this was not the most general form, it opened the door for the discoveries to follow. Scipione dal Ferro did not rush to publish his curtain-raising result. Keeping mathematical discoveries secret was quite common until the eighteenth century (what a difference from today's scientific paper chase!). Nevertheless, he did divulge the solution to his student and son-in-law, Annibale della Nave, and to at least one other student, the Venetian Antonio Maria Fiore. He also expounded his method in a manuscript that came into the possession of his son-in-law after Scipione's death.

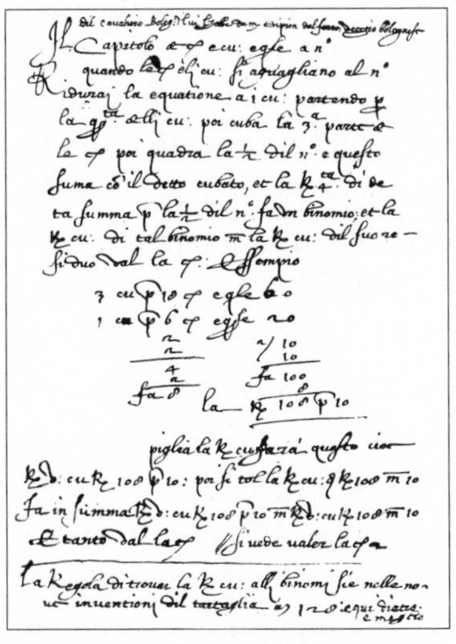

Figure 38

Bologna of the sixteenth century experienced a surge of interest in mathematics. Mathematicians and other scholars were sometimes involved in public debates and oral disputations that attracted large crowds. In attendance were not only university officials and appointed judges, but also students, supporters of the contestants, and spectators who came for entertainment and for a betting opportunity. Often, the disputants themselves would wager considerable amounts of money on their anticipated victory. According to one description by a nineteenth-century historian of mathematics, mathematicians were interested in such confrontations of wits, because on their results

> depended not only their reputation in the city or in the University, but also tenure of appointment and increase in salary. Disputations took place in public squares, in churches, and in the courts kept by noblemen and princes, who esteemed it an honor to count among their retinue scholars skilled not only in the casting of astrological predictions, but also in disputation on difficult and rare mathematical problems.

Antonio Maria Fiore, who was brought into the secret of dal Ferro's solution, was a rather mediocre mathematician. Upon dal Ferro's death, he also did not publish the solution immediately, even though he treated it as if it were his to be exploited. Rather, he decided to wait for the right moment—one that would allow him to make a name for himself. In a society in which the renewal of university appointments largely depended on success in debates, possession of a secret weapon could mean the difference between survival and perishing. The opportunity finally presented itself in 1535, and Fiore challenged the mathematician Niccolò Tartaglia to a public problem-solving contest. Who was this Tartaglia, and why did Fiore select him from a long list of potential candidates as his opponent?

Niccolò Tartaglia (figure 39) was born in Brescia in 1499 or 1500. His original surname was probably Fontana, but he was nicknamed Tartaglia (meaning "the stammerer") because of a saber cut to his mouth that he had received at age twelve from a French soldier. The young boy was left for dead in the cathedral in which he sought sanctuary, and he was slowly nursed back to health by his mother. As an adult, he always wore a beard to hide the disfiguring scars. Tartaglia came from a very poor family. His father Michele, a postal courier, died when Niccolò was about six years old, leaving the widow and her children in heartbreaking misery. Tartaglia had to stop his studies of reading and writing of the alphabet upon reaching the letter *k* because the family ran out of money to pay for the tutor. In his later retrospection, Tartaglia described the completion of his education: "I never returned to a tutor, but continued to labor by myself over the works of dead men, accompanied only by the daughter of poverty that is called industry." In spite of these ill-starred circumstances, Tartaglia proved to be a talented mathematician. Eventually he moved to Venice in 1534 as a teacher of mathematics, after having spent some time in Verona. In his mathematical memoirs

Figure 39

Tartaglia states that in 1530 he managed after considerable effort to solve the cubic equation $x^3 + 3x^2 = 5$. This was a challenge posed to him by a fellow Brescian, Zuanne de Tonini da Coi. Rumors of Tartaglia's claim that he was able to solve cubics must have reached the ears of Antonio Maria Fiore, but the latter greeted the information with skepticism, thinking that Tartaglia was bluffing. Confident in his ability to defeat Tartaglia due to his secret knowledge of dal Ferro's solution, Fiore issued the challenge. Shortly afterward, Fiore and Tartaglia reached an agreement on the precise conditions of the contest. Each side was to propose thirty problems for his opponent to solve. The problems were then to be sealed and deposited with the notary Master Per Iacomo di Zambelli. The two contestants fixed a term of forty to fifty days for each to attempt to solve the problems, once the seals were opened. They agreed that whoever solved more problems would be considered the winner, and in addition to honors, would receive a handsome reward suggested for each problem (according to some sources the loser was supposed to pick up the tab for a feast attended by the winner and thirty of his friends). As it turned out, Fiore indeed had only one arrow to his bow—all the problems he put forward were of the form for which he knew the solution from dal Ferro, $ax^3 + bx = c$. Tartaglia's list, on the other hand, contained thirty diverse problems, each one of a different kind, in his words, "to show that I thought little of him and had no cause whatever to fear him."

The date of the contest was set for February 12, 1535. Various university dignitaries and some of the Venetian intellectual high society must have been in attendance. As the problems were presented to the two adversaries, something totally unexpected happened. To the spectators' amazement, Tartaglia blasted through all the problems thrown at him in the space of two hours! Fiore failed to solve even one of Tartaglia's problems. In his account of the events some twenty years later, Tartaglia recalled:

> The reason why I was able to solve his 30 [problems] in so short a time is that all 30 concerned work involving the algebra of unknowns and cubes equaling numbers [equations of the form $ax^3 + bx = c$]. [He did this] believing that I would be unable to solve any of them, because Fra Luca [Pacioli] asserts in his treatise that it is impossible to solve such problems by any general rule. However, by good fortune, only eight days before the time fixed for collecting from the notary the two sets of 30 sealed problems, I had discovered the general rule for such expressions.

In fact, a day after discovering the solution to $ax^3 + bx = c$, Tartaglia also discovered the solution to $ax + b = x^3$. Since he also already knew how to solve $x^3 + ax^2 = b$ (the challenge posed to him by da Coi), Tartaglia became overnight literally the world expert on the solution of cubic equations. Nevertheless, he waved aside a suggestion from da Coi to publish his solution immediately, explaining that he intended to write a book on the subject. The formulae Tartaglia discovered were so complicated that he found it hard to remember his own rules for the three cases. To help himself memorize them, he composed some verses that started with:

In cases where the cube and the unknown
Together equal some whole number, known:
Find first two numbers diff'ring by that same;
Their product, then, as is the common fame . . .

Tartaglia's complete verses and his formula are presented in appendix 5.

Tartaglia was now no longer an anonymous math teacher—he was a mathematical celebrity. But in Renaissance Italy no story, even a story of mathematics, comes without its operatic moments.

THE PLOT THICKENS

Word of the contest between Tartaglia and Fiore spread throughout Italy like wildfire and reached one of the most brilliant and controversial figures of the sixteenth century—the physician, mathematician, astrologer, gambler, and philosopher Gerolamo Cardano (1501–76; figure 40).

Even compared to the many colorful geniuses of the Renaissance, Cardano's life readily catches the imagination. He was the illegitimate son of the Milanese lawyer Fazio Cardano and the much younger widow Chiara Micheri. In his later autobiography, *De vita propria liber* (*The Book of My Life*), Cardano delights in describing in great and unnecessary detail all the medical problems from which he suffered early in life, including his sexual impotence between the ages of twenty-one and thirty-one. Encouraged by his educated father, who advised Leonardo da Vinci in geometry on several occasions, Gerolamo studied mathematics, the classics, and medicine at the universities of Pavia and Padua. During his student days, gambling became his chief source of

Figure 40

financial support. He played cards, dice, and chess, turning his knowledge of probability theory into profits. Later in life he would transform his addiction to gambling into an interesting book: *Liber de ludo aleae* (*The Book on Games of Chance*), the first book on the calculation of probabilities. Having a very loud voice and a rude attitude, Cardano managed to alienate many of his professors, and at the end of his studies, the first ballot denied him the doctorate of medicine with the overwhelming vote of 47 against 9. Only after two more rounds of votes did he finally get the degree. While Cardano's first attempts to obtain a position as a physician in Milan failed miserably, his luck soon changed drastically. In 1534 he was appointed, through the influence of his father's acquaintances, lecturer of mathematics at the Piatti Foundation. Simultaneously, he started a clandestine practice of medicine, in which he was extremely effective. His success did not, however, gain him the support of the College of Physicians in Milan. In 1536 Cardano decided to bring his dispute with the college to a showdown. He published a viciously aggressive book entitled *De malo recentiorum medicorum medendi usu libellus* (*On Bad Practices of Medicine in Common Use*). In particular, Cardano ridiculed the grandiloquent manners of the physicians of his time: "The things which give most reputation to a physician nowadays are his manners, servants, carriage, clothes, smartness and caginess, all displayed in a sort of artificial and insipid way; learning and experience seem to count for nothing." Incredibly enough, not only did Cardano's offensive get him a physician's position, but by the middle of the century he was to become one of Europe's best-known medical practitioners, second only to the legendary anatomist Andreas Vesalius.

Cardano appears to have thrived on controversy and competition. This may have stemmed from his passion for gambling. He once noted, "Even if gambling were altogether an evil, still, on account of the very large number of people who play, it would seem to be a natural evil. For that very reason it ought to be discussed by a medical doctor like one of the incurable diseases." Quick-witted and sharp-tongued, Cardano won many disputes, both during his student days and as a mature scholar. No wonder then that the news of the Tartaglia-Fiore contest kindled his curiosity. At the time, he was completing his second mathematical book, *Practica arithmeticae generalis et mensurandi singularis* (*The Practice of Arithmetic and Simple Mensuration*), and he found the idea of including the solution to the cubic in the book very attractive. In the few years that followed, Cardano must have tried in vain to discover the solution by himself. Having failed, he decided to send the bookseller Zuan Antonio da Bassano to Tartaglia to convince the latter to reveal his formula. Tartaglia later described his own reply in no uncertain terms: "Tell his Excellency that he must pardon me, that when I publish my invention it will be in my own work and not in that of others, so that his Excellency must hold me excused." After a few rather lengthy and quite acrimonious exchanges, in which Tartaglia brushed aside all of Cardano's overtures, he finally was lured into accepting an invitation to visit Cardano in Milan. The bait that did the trick was a promise by Cardano to introduce Tartaglia to the Spanish viceroy and commander in chief in Milan, Alfonso d'Avalos. Tartaglia had written a book on artillery and such a contact could guarantee him a nice income.

In Milan, Cardano subjected Tartaglia to a heavy dose of charming hospitality, still attempting to schmooze the solution out of him. But Tartaglia's lips remained sealed, at least for a while. He even rejected a proposal that Cardano would include a special chapter in the book heralding Tartaglia as the discoverer of the solution.

Unfortunately, from this point on, our information of the subsequent events relies almost exclusively on Tartaglia's far from objective testimony. According to Tartaglia, he eventually did agree to divulge the secret to Cardano, but only after the latter had taken the following solemn oath: "I swear to you by the Sacred Gospel, and on my faith as a gentleman, not only never to publish your discoveries, if you tell them to me, but I also promise and pledge my faith as a true Christian to put them down in cipher so that after my death no one shall be able to understand them." This weighty conversation took place on March 25, 1539. Ludovico Ferrari, then a young secretary in Cardano's household, tells a rather different story. According to Ferrari, Cardano took no oath of secrecy. Ferrari claimed to have been present at the conversation and said that Tartaglia revealed the secret simply in return for Cardano's hospitality. However, as we shall soon see, Ferrari's own objectivity is at least as questionable as that of Tartaglia. The fact remains, nevertheless, that the *Practica arithmeticae generalis* appeared in May 1539 without Tartaglia's solution.

Ludovico Ferrari (1522–65) takes center stage as the next character in this tragicomic drama. He first arrived at Cardano's house from Bologna as a boy of fourteen. Cardano soon recognized the youngster's exceptional talents and took full responsibility for his education. Ferrari, however, was as irritable as he was sharp. In one brawl, at age seventeen, he lost the fingers on his right hand. As soon as Cardano learned Tartaglia's solution, he succeeded not only in providing a proof for it, but he also started to work on more general cubic equations. Recall that Tartaglia actually managed to solve only particular forms of the cubic, such as $x^3 + ax = b$ or $x^3 = ax + b$. The realization that these are only special cases of the general equation $ax^3 + bx^2 + cx + d = 0$ had not yet sunk in with the sixteenth-century mathematicians. Rather, they treated each one of thirteen different forms of cubic equations separately. At the same time, with Cardano's encouragement, the brilliant Ferrari managed in 1540 to find a beautiful solution to the *quartic equation,* such as $x^4 + 6x^2 + 36 = 60x$. The master and his student were now really on a roll. A rumor that dal Ferro had left his original formula with his son-in-law reached Cardano. In 1543 Cardano and Ferrari took a special trip to Bologna to meet with Annibale della Nave, to whom Scipione dal Ferro's original paper had been entrusted. There they were able to confirm firsthand that dal Ferro had indeed, twenty years earlier, discovered the same solution as Tartaglia. Even if Cardano had truly given an oath to Tartaglia, this was probably all he felt he needed to free himself of the obligation. The oath was formally, after all, not to reveal Tartaglia's formula, not dal Ferro's. In 1545 Cardano published the book regarded by many mathematicians as marking the beginning of modern algebra—*Artis magnae sive de regulis algebraicis liber unus* (*The Great Art or the Rules of Algebra Book One*), commonly known as *Ars magna* (*The Great Art;* figure 41 shows the frontispiece of the book). In this book, Cardano explores in great detail the cubic and quartic equations and

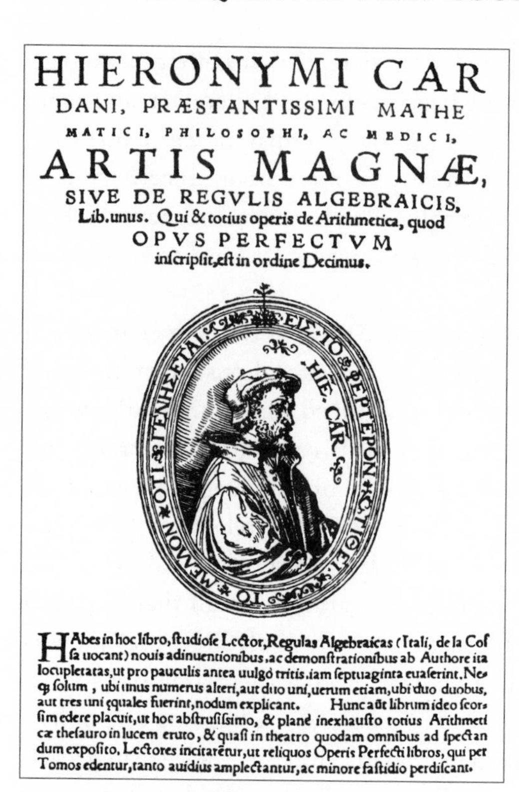

HIERONYMI CAR
DANI, PRÆSTANTISSIMI MATHE
MATICI, PHILOSOPHI, AC MEDICI,
ARTIS MAGNÆ,
SIVE DE REGVLIS ALGEBRAICIS,
Lib. unus. Qui & totius operis de Arithmetica, quod
OPVS PERFECTVM
inſcripſit, eſt in ordine Decimus.

HAbes in hoc libro, ſtudioſe Lector, Regulas Algebraicas (Itali, de la Coſſa uocant) nouis adinuentionibus, ac demonſtrationibus ab Authore ita locupletatas, ut pro pauculis antea uulgò tritis, iam ſeptuaginta euaſerint. Neq; ſolum, ubi unus numerus alteri, aut duo uni, uerum etiam, ubi duo duobus, aut tres uni æquales fuerint, nodum explicant. Hunc aũt librum ideo ſeorſim edere placuit, ut hoc abſtruſiſſimo, & planè inexhauſto totius Arithmeticæ theſauro in lucem eruto, & quaſi in theatro quodam omnibus ad ſpectandum expoſito, Lectores incitarẽtur, ut reliquos Operis Perfecti libros, qui per Tomos edentur, tanto auidius amplectantur, ac minore faſtidio perdiſcant.

Figure 41

their solutions. He demonstrates for the first time that solutions can be negative, irrational, and in some cases may even involve square roots of negative numbers—quantities he refers to as "sophistic"—to be dubbed "imaginary numbers" in the seventeenth century. The printer Johannes Petreius of Nürnberg published the first edition of *Ars magna,* and it swept mathematical Europe, winning immediate acclaim. One mathematician was, needless to say, less respectful. Tartaglia's fury was unimaginable. In less than a year he published a book, *Quesiti et inventioni diverse* (*New Problems and Inventions*), in which he directly accused Cardano of perjury. Presenting what was supposed to be a word-for-word account of all of their exchanges (even though those had taken place a full seven years earlier), Tartaglia used the most offensive language against Cardano. His justification: "I truly do not know of any greater infamy than to break an oath." But was Cardano a mathematical plagiarist? By standard scientific ethics, certainly not. The second paragraph in the opening chapter of *Ars magna* reads:

> In our own days Scipione dal Ferro of Bologna has solved the case of the cube and the first power equal to a constant, a very elegant and admirable accomplishment. Since this art surpasses all human subtlety and the perspicuity of mortal talent and is a truly celestial gift and a very clear test of the capacity of men's minds, whoever applies himself to it will believe that there is nothing that he cannot understand. In emulation of him, my friend Niccolò Tartaglia of Brescia,

> wanting not to be outdone, solved the same case when he got into a contest with his [Scipione's] pupil, Antonia Maria Fior, and, moved by my many entreaties, gave it to me. For I had been deceived by the words of Luca Pacioli, who denied that any more general rule could be discovered than his own. Notwithstanding the many things which I already discovered, as is well known, I had despaired and had not attempted to look any further. Then however, having received Tartaglia's solution and seeking for the proof of it, I came to understand that there were a great many other things that could also be had. Pursuing this thought and with increased confidence, I discovered these others, partly by myself and partly through Ludovico Ferrari, formerly my pupil.

In chapter XI ("On the Cube and First Power Equal to the Number") Cardano repeats briefly the same credit:

> Scipio Ferro of Bologna well-nigh thirty years ago discovered this rule and handed it on to Antonia Maria Fior of Venice, whose contest with Niccolò Tartaglia of Brescia gave Niccolò occasion to discover it. He [Tartaglia] gave it to me in response to my entreaties, though withholding the demonstration. Armed with this assistance, I sought out its demonstration in [various] forms. This was very difficult. My version of it follows.

Tartaglia was far from being appeased by the acknowledgment Cardano granted him. In fact, the battle of offenses had not only heated up, but turned into an ugly show of insults played with great ferocity before the entire Italian public. While Cardano himself stayed clear of the feud, his ill-tempered collaborator, Ludovico Ferrari, gladly jumped into the role of intellectual gladiator to defend his (in his words) "creator." In response to Tartaglia's book, Ferrari issued a *cartello*—a letter of challenge—that he distributed to fifty-three scholars and dignitaries across Italy. Ferrari adopted a viciously degrading style: "By reading your nonsense one has the impression of reading the jokes of Piovano Arlotto [a priest who lived in the fifteenth century, known for his practical jokes]." He then goes on contemptuously and accuses Tartaglia himself of plagiarism: "Among the more than one thousand errors in your book I note first that in section eight you give a result by Giordano [referring to the thirteenth-century German mathematician Jordanus

Nemorarius, also known as Jordanus de Nemore] as your own, without mentioning him, and this is theft." The first *cartello* was sent on February 10, 1547. Tartaglia received it on the thirteenth and took only six days for a counterattack. He first complained about the fact that Cardano himself did not bother to answer:

> This I advise you again that in case the said Signor Gerolamo Cardano does not intend to write to me, acknowledging wisely that he was wrong, then he has no reason for complaint against me. . . . You should at least make certain that he also signs your cartel in his own hand as your associate in this dispute.

In response to Ferrari's invitation for a public dispute on mathematics, Tartaglia declared that he would gladly dispute with Cardano himself. Clearly, Tartaglia saw no point in entering a contest with a youngster of no particular distinction, where even a victory would not mean much, and he preferred to battle with Cardano, whose reputation on the continent was on a spectacular rise. Cardano, however, was at a stage in life where he was anxious to promote a more balanced temperament (he advocated that scholars adopt a lifestyle of "reading love stories"), and he remained silent.

Between February 10, 1547, and July 24, 1548, Tartaglia and Ferrari exchanged no fewer than twelve *cartelli* (six challenges and six responses), all circulated to the entire intellectual high society. In spite of the generally disparaging style, the *cartelli* also serve as an interesting documentation of the knowledge of two leading Renaissance mathematicians. Tartaglia's continuing attempts to drag Cardano into the dispute failed miserably. In 1548, Tartaglia was offered the position of lecturer in geometry in his hometown, Brescia. Due to the high profile of his exchanges with Ferrari, however, the appointment was most probably made on the condition of him defeating Ferrari in a public contest. Consequently, Tartaglia was forced, reluctantly, to commit to a debate. The agreed-upon topics of the debate were sixty-two problems proposed by the two disputants (thirty-one by each)—the ones presented in the exchanged *cartelli.* Most of the problems were in mathematics, but in the Renaissance spirit, there were also questions in other areas, such as architecture, astronomy, geography, and optics.

The debate took place on August 10, 1548, in a church within the garden of the Frati Zoccolanti, in Milan. All the Milanese who's who showed up, including the governor, Don Ferrante di Gonzaga, who was supposed to be the ultimate arbiter. Ferrari turned up with a large entourage of supporters, while Tartaglia may have been accompanied only by his own brother. Cardano made sure he was out of town during the debate. Unfortunately, no official record of either the debate itself or the final verdict exists. In two later books, Tartaglia presents rather confused accounts of the proceedings. In particular, he blames the audience for loudly interfering and preventing him from presenting his arguments in full. The dry facts, however, paint a rather different picture. Tartaglia left the dispute before its conclusion, immediately after the end of the first day. We also know that Tartaglia was denied his salary after a year of lecturing in the position in Brescia, and he was obliged to return to his modest teaching job in Venice. All signs point, therefore, to Tartaglia's having suffered an agonizing and humiliating defeat in Milan. Cardano also mentions briefly in his writings that Ferrari was more than a match for Tartaglia.

As for the triumphant Ludovico Ferrari, his career skyrocketed. Following his victory, offers for positions started to pour in. Ferrari even declined the opportunity to tutor the emperor's son for the more lucrative appointment as a tax assessor for the governor of Milan. His life, however, was to end unexpectedly, providing the final act to this drama.

Upon his return to Bologna sometime after 1556, Ferrari was accompanied by his sister Maddalena, a poor widow. While no direct proof of her poisoning him in 1565 exists, her subsequent behavior and the ensuing circumstances raise serious suspicion. Maddalena married two weeks after Ferrari's death, and she transferred to her husband all the money and property she had inherited from her brother. When Cardano came to Bologna to retrieve some of his own books and notes, he found nothing. Maddalena's husband took possession of everything, apparently intending to publish some material in the name of his son from a previous marriage.

The history of the solutions to the cubic and quartic equations raises interesting questions beyond the realm of mathematics. This story would be incomplete without some contemplation on the questions of intellectual property and proprietary rights on scientific information. During the bitter Tartaglia-Ferrari exchanges, Ferrari claimed that Cardano had actually done Tartaglia a service by rescuing his formula from oblivion and planting it in a "fertile garden"—the *Ars magna.* But was this true? Or was Tartaglia right in replying that without his formula Cardano's garden would have remained an obscure, weedy field? There is no question that from Tartaglia's perspective Cardano was the devil. Not only had he broken an oath, but by doing so, he had denied Tartaglia the recognition and fame the latter felt were rightfully his. No credit lines in Cardano's book could have healed this wound. The fact remained that all references from that point on were to "Cardano's formula," and to his book. Worse yet, since Cardano added many solutions and proofs of his own to all the forms of the cubic and quartic equations, the breakthrough nature of Tartaglia's formula was lost in the shuffle.

But what about Cardano's viewpoint? Solemn oath or not, surely he felt that he was entitled at the very least to publish his own seminal work on the subject. Cardano's standpoint is even more understandable once we realize (as did he) that Tartaglia was not the original discoverer of the formula—Scipione dal Ferro was. What right did Tartaglia have to suppress the publication of a formula that dal Ferro himself had left for posterity? Tartaglia's claim that he was about to publish a book on new algebra himself also does not hold water. In fact, in spite of the substantial head start that Tartaglia had on Cardano, he got distracted by the pursuit of other projects and the book on new algebra never got off the ground.

A couple of present-day examples of common scientific practices concerning publication of discoveries can help to demonstrate that the issue of ownership of discoveries is not simple. Astronomers propose on a yearly basis for observations to be performed by the Hubble Space Telescope. After a very detailed process of evaluation of the proposals by panels of experts, only about one out of seven proposals is actually selected for the observations to be executed. The data collected are made available to the proposer within a few days after the observation takes place. Following that, there is a proprietary period of one year, during which only the proposer has access to the data. The proposer can use this time to analyze the data and publish the results. After one year the data become public, for all the astronomers in the world to use. This process has been established first and foremost in recognition of the fact that scientific discoveries (especially those made with taxpayers' funding) belong to the community at large and should not be treated as private property. Second, the procedures have been designed so as to discourage scientific procrastinators from merely sitting on important data.

At the same time, private companies that deal with, say, mathematical modeling of stock market behavior are extremely secretive about their findings, but not more so perhaps than some chefs about their secret recipes.

From a purely scientific point of view it would make the most sense to refer to the formula for solving the cubic as "dal Ferro's formula," since there is no doubt that he was the first to discover it. This is neither the first nor the last case, however, where scientific innovations are not named after the true discoverer. Tartaglia's attitude with regard to intellectual property appears somewhat hypocritical when one considers his own practices. For instance, Tartaglia produced a translation of some of Archimedes' works under his own name, when in fact he merely published a thirteenth-century Latin translation by the Flemish scholar William of Moerbeke. Similarly, he presented a solution to the mechanics of a heavy body on an inclined plane without crediting the originator of that solution, the German mathematician Jordanus de Nemore.

The entire dal Ferro-Tartaglia-Cardano-Ferrari sequence of events remains one of the most controversial affairs in the history of mathematics. No wonder that many historians of science have enjoyed sinking their teeth into it. From the point of view of the present book, what is important is that as the curtain fell on this drama, mathematicians knew how to solve cubic and quartic equations, even if a general theory of equations was still missing.

Cardano never denied his good fortune. In *The Book of My Life* he writes:

> Although happiness suggests a state quite contrary to my nature, I can truthfully say that I was privileged from time to time to attain and share a certain measure of felicity. If there is anything good at all in life with which we can adorn this comedy's stage, I have not been cheated of such gifts.

Given the role that the solution of equations was to play centuries later in the formulation of group theory as the "official" language of symmetry in nature and in the arts, the following historical fact stands out as an amusing curiosity. Cardano published horoscopes of one hundred prominent men of his century. Only one of those, the German painter Albrecht Dürer, was an artist.

To conclude this story I must add a personal note. In the summer of

Figure 42

2003 I decided that I had to find the birthplace of the true hero of the cubic equation—Scipione dal Ferro. After some effort I discovered the place. Today it is located at the corner of via Guerrazzi and via S. Petronio Vecchio in Bologna. An easy-to-miss plaque on the side wall marks the house as dal Ferro's birthplace (figure 42). I rang the entrance buzzer at a few apartments randomly, and an old lady showed up at the window of a third-floor apartment. I explained to her in my pathetic Italian that I was researching the life of Scipione dal Ferro. She told me to wait for her husband to descend. The pleasant old gentleman explained to me in a broken mixture of Italian and English that there was nothing else in the building to indicate the fact that the man responsible for one of the major breakthroughs in algebra had lived there. We both stared silently at the plaque for a few minutes and then parted.

After the brilliant dal Ferro-Cardano-Ferrari work, it was only natural to believe that the *quintic equation*, of the form $ax^5 + bx^4 + cx^3 + dx^2 + ex + f = 0$, could also be solved by a formula. In fact, with the confidence gained from the *Ars magna*, the expectation was that the solution was right around the corner, and it prompted some of the sharpest mathematical minds to hunt for this treasure.

WILL TELL ALOUD YOUR GREATEST FAILING

The satirical author Jonathan Swift (1667–1745), best known for *Gulliver's Travels*, wrote an amusing poem in 1727 entitled "The Furniture of a Woman's Mind." A few lines read:

For conversation well endu'd,
She calls it witty to be rude;
And, placing raillery in railing,
Will tell aloud your greatest failing.

The story of the search for a formulaic solution to the quintic equation in the 250 years following Cardano is one of great failing. It started with another Bolognese, Rafael Bombelli (1526–72). By a historical coincidence, Bombelli was born precisely in the year that dal Ferro died. Having studied with great admiration the *Ars magna,* Bombelli felt that Cardano's exposition had not been sufficiently clear and self-contained; in Bombelli's words, "In what he said it was obscure." Consequently, he spent two decades writing an influential book called *L'algebra.* Unlike the other Italian mathematicians, Bombelli was not a university professor, but rather a hydraulic engineer. Bombelli's greatest original contribution was the realization that one cannot avoid having to deal with square roots of negative numbers. This truly required a mental leap. After all, what is the square root of -1? Clearly, no ordinary (real) number multiplied by itself gives -1, since even the multiplication of a negative number by itself gives a positive result. Nevertheless, the solution to the cubic equation (see appendix 5) sometimes produced a square root of a negative number as an intermediate step, even when the final solution was a real number. Cardano, who was puzzled by these "sophistic" numbers, concluded that they were "so subtle that they were useless," and when he needed to calculate with them he said he was doing so by "dismissing mental torture." Bombelli, on the other hand, had the remarkable insight to understand that these new numbers, which he called "plus of minus," were a necessary vehicle that could bridge the gap between the cubic equation (which was expressed in real numbers) and the final solutions (which were also real numbers). In other words, while both the beginning and the end involve real numbers, the solution has to traverse the new world of "imaginary" numbers. The square root of -1 was denoted by i in 1777 by the great Swiss mathematician Leonhard Euler. The numbers in the new vistas revealed by Bombelli's work are now called *complex numbers*—these are sums of real numbers (all the ordinary numbers) and imaginary numbers (which involve square roots of negative numbers).

There was also an important historical lesson to be learned here. The study of equations had afforded mathematicians a first glimpse of new kinds of numbers several times throughout history. There were the negative numbers, such as -1 and -2; the irrational numbers, such as $\sqrt{2}$, that could not be expressed as fractions; and through Bombelli's work, even imaginary numbers, such as $\sqrt{-1}$. Who knew what insights might emerge out of the solution of the quintic?

In the centuries that followed, cracking the enigma of the quintic became one of the most intriguing challenges in mathematics. Unfortunately, the solutions discovered by dal Ferro and Ferrari (for the cubic and quartic, respectively) did not offer much help. These represented brilliant but ad hoc tricks rather than methodical studies that could be extended to equations of higher degrees. What was badly needed was a more comprehensive theory of equations in general, rather than experiments with isolated cases. To use a medical metaphor, mathematics had to move on from the treatment of the symptoms to the understanding of the causes and the associated side effects.

The French lawyer François Viète (1540–1603) and the English astronomer Thomas Harriot (1560–1621) took steps in the right direction. They introduced improvements to both the notation used to describe algebraic equations (which was extremely cumbersome in Cardano's work) and to the methods of solution themselves. Viète was also the person responsible for the word *coefficients,* used to define the numbers that describe an equation (e.g., a, b, c in $ax^2 + bx + c = 0$). Although not a mathematician by profession, Viète came on one occasion to the rescue of the honor of the entire French mathematical society. In 1593, at the end of the preface of his book *Ideae mathematicae,* the Belgian mathematician Adriaan van Roomen (1561–1615) challenged all the mathematicians of his time to decipher a problem that involved no less than solving an intimidating equation of degree 45 (see appendix 6). The ambassador to Paris from the Netherlands was only too delighted to remark mockingly to King Henry IV that there was no French mathematician who could solve the problem. The embarrassed king called upon Viète for help and was pleasantly surprised when the latter was able (according to legend) to find the positive solutions within a few minutes, upon discovering that a trigonometric relation was underlying the problem. In fact, Viète did much more—he showed that the equation has twenty-three positive solutions and twenty-two negative ones.

The first serious, but alas unsuccessful, attempt at a solution of the quintic was made by the Scot James Gregory (1638–75). Gregory is known primarily for a reflecting telescope (the Gregorian telescope) that he invented. During the year before he died (at the young age of thirty-six), he had begun to doubt whether a formula for the quintic could be found at all. Nevertheless, he did discover relations between the solutions of various equations and their coefficients. The next step was taken by the German Count Ehrenfried Walther von Tschirnhaus (1651–1708). A man of many accomplishments, from glasswork to algebra, Tschirnhaus elaborated on an interesting method that for a while gave hope that there was light at the end of the tunnel. The basic idea was simple. If one could somehow reduce the quintic equation to equations of a lower degree (such as the quartic or cubic), then one could use the known solutions to those equations. In particular, Tschirnhaus was able by some clever substitutions to get rid of the x^4 and x^3 terms in the quintic. Unfortunately, there was still a major obstacle in Tschirnhaus's method, which was soon noticed by the mathematician Gottfried Wilhelm Leibniz (1646–1716), and after much effort in this direction Tschirnhaus conceded defeat.

The eighteenth century brought about a renewed interest and a vigorous series of attacks on the problem. The Frenchman Étienne Bézout (1730–83), who published several works on the theory of algebraic equations, adopted methods somewhat similar to those of Tschirnhaus, but again to no avail. At that point, the most prolific mathematician of all time entered the race.

Leonhard Euler (figure 43) was so productive that an entire volume is needed merely to reproduce the list of his publications. Euler's body of published works in mathematics and mathematical physics constitutes about one third of all the work published in these areas during the last three-quarters of the eighteenth century. Euler conjectured that the solution to the quintic could be expressed in terms of some four quantities, and he concluded in a hopeful tone: "One might suspect that if the elimination were done carefully, it might possibly lead to an equation of degree 4." In other words, he also optimistically believed that the problem could be reduced to one that had already been solved. This general philosophy is characteristic of advances in mathematics. In an old joke, a physicist and a mathematician are asked what they would do if they needed to iron their pants, but although they are in possession of an iron, the electric outlet

Figure 43

is in the adjacent room. Both answer that they would take the iron to the second room and plug it in there. Now they are asked what they would do if they were already in the room in which the outlet is located. The physicist answers that he would plug the iron into the outlet directly. The mathematician, on the other hand, says that he would take the iron to the room without the outlet, since that problem has already been solved.

In spite of Euler's optimism, he failed to solve the general quintic. He did manage to show, however, that a few special quintics, such as $x^5 - 5px^3 + 5p^2x - q = 0$ (where p and q are given numbers), were solvable by a formula. This left the door open for potential future endeavors. Next in line was the Swede Erland Samuel Bring (1736–98). A teacher of history at Lund University by profession, Bring's favorite pastime was mathematics. And what better riddle to solve than the quintic? Bring achieved what appeared to be a huge step toward a solution. He found a mathematical transformation that could reduce the general quintic ($ax^5 + bx^4 + cx^3 + dx^2 + ex + f = 0$) to the much simpler form $x^5 + px + q = 0$. Unfortunately, not only did even this shorter and seemingly much more tractable form still present an insurmountable obstacle, but Bring's remarkable transformation went entirely unnoticed, only to be independently rediscovered by the English mathematician George Birch Jerrard in the nineteenth century.

Three further undertakings, by mathematicians working nearly simultaneously in three different countries, also fell short of producing a solution. Still, the profound works of these mathematicians introduced an exciting new idea into the search. In particular, they showed that properties of the permutations of the putative solutions of equations might have something to do with whether the equations are solvable by a formula or not. Since this was historically the first point of connection between the solutions of equations and the concept of symmetry, let me give a brief explanation of the basic principle. Examine, for instance, the quadratic equation $ax^2 + bx + c = 0$ (where a, b, c are known numbers). One can easily show that if the two solutions of the equation (given by the formula on page 60) are denoted by x_1 and x_2, then both the sum of the solutions, $x_1 + x_2$, and their product, x_1x_2, can be expressed in terms of the coefficients of the equation, a, b, c (see appendix 7). In fact, $x_1 + x_2 = -b/a$ and $x_1x_2 = c/a$. In other words, in the equation $x^2 - 9x + 20$

= 0, the sum of the two solutions is equal to 9, and their product is equal to 20. The formula for the solutions on page 60 can in itself be expressed (appendix 7) as a combination of $(x_1 + x_2)$ and $x_1 x_2$:

$$\frac{1}{2}\left[(x_1 + x_2) \pm \sqrt{(x_1 + x_2)^2 - 4x_1 x_2}\right].$$

The important point to notice here is that this expression is symmetric under the interchange of the two solutions x_1 and x_2—the formula remains unchanged when x_1 and x_2 are transposed. The question raised by the Frenchman Alexandre-Théophile Vandermonde (1735–96) and the Englishman Edward Waring (1736–98) was whether the solution to the quintic, and indeed equations of any degree, could not be represented by a similar, symmetric expression. This could, in principle, lead to a formula for the solutions. The idea was picked up by the person considered by Napoleon Bonaparte to be "the lofty pyramid of the mathematical sciences"—Joseph-Louis Lagrange (1736–1813).

Lagrange (figure 44) was born in Turin (now Italy), but his family was partly of French ancestry on his father's side, and he considered himself "more" French than Italian. His father, who was originally wealthy, managed to squander all the family's fortune in speculations, leaving his son with no inheritance. Later in life, Lagrange described this economic catastrophe as the best thing that had ever happened to him: "Had I inherited a fortune I would probably not have cast my lot with mathematics."

In his outstanding treatise (published in Berlin) *Reflections on the Resolution of Algebraic Equations,* Lagrange first reviewed carefully the contributions of Bézout, Tschirnhaus, and Euler. He then showed that all the tricks by which solutions had been obtained for the linear, quadratic, cubic, and quartic equations could be replaced by a uniform procedure. Here, however, came a nasty surprise. For degrees 2, 3, and 4, the equations had been solved by reducing the equation to one of a lower degree than the one being discussed (i.e., reducing the quartic to a cubic, and so on). When precisely the same process was attempted on the quintic, something unexpected happened. The resulting equation, instead of being a quartic, turned out to be one of degree 6! The method that had worked beautifully for degrees 2, 3, 4 failed utterly at the quintic.

Figure 44

Disappointed, Lagrange concluded that "it is therefore unlikely that these methods will lead to the solution of the quintic—one of the most celebrated and important problems of algebra."

As a way out of the impasse, Lagrange introduced a more general discussion of permutations. Recall that permutations are the operations that produce different arrangements of objects, such as the transformations of *ABC* into *BAC* or *CBA*. Lagrange made the important discovery that the properties of equations and their solubility depend on certain symmetries of the solutions under permutations.

Even Lagrange's new insights, as groundbreaking as they were, proved insufficient for a solution to the quintic. Remaining optimistic that his analysis would generate the necessary breakthrough, he wrote, "We hope to return to this question at another time and we are content here in having given the fundamentals of a theory which appears to us new and general." As history would have it, Lagrange never returned to the quintic. Two days before his death he summarized his life thus: "My career has come to an end; I have acquired a modicum of renown in mathematics. I have not hated anyone, nor have I done ill by anyone; it is good to come to the end."

There was another algebraic problem that was being debated in mathematical circles around the same time, and it had implications for the attempts to solve the quintic. The question was: Do all the equations (of any order) have at least one solution? For example, how do we know if there is any value of x for which the equation $x^4 + 3x^3 - 2x^2 + 19x + 253 = 0$ holds true? Even more acutely, if we have an equation of degree n (where n can be any whole number 1, 2, 3, 4, . . .) and we allow for the solutions to be either real or complex numbers (involving $i = \sqrt{-1}$), how many solutions are there? We already know the answer in the case of the quadratic equation—there are always precisely two solutions. But what about $n = 5$ or $n = 17$? Although many mathematicians, including Leibniz, Euler, and Lagrange, attempted to give an answer, the definitive statement was left to the Swiss accountant Jean-Robert Argand (1768–1822) and to the man acknowledged as the "prince of mathematicians"—Johann Carl Friedrich Gauss (1777–1855; figure 45).

Gauss's genius was recognized already at age seven, when he was able to sum the whole numbers from 1 to 100 instantly in his head, simply by noticing that the sum consists of fifty pairs of numbers, each totaling 101. In his doctoral dissertation in 1799, Gauss gave his first proof of what has become known as the *fundamental theorem of algebra*—the statement that every equation of degree *n* has precisely *n* solutions (which can be real or complex numbers). Gauss's first proof had some logical gaps in it, but he would end up giving three more proofs during his life, all rigorous. Argand's proof, published in 1814, was actually the first correct proof.

Figure 45

The fundamental theorem demonstrated unambiguously that the general quintic equation must have five solutions. But could those be found by a formula? In the same year that Gauss published his first proof for the fundamental theorem he also expressed his skepticism about a formulaic solution to the quintic: "After the labors of many geometers left little hope of ever arriving at the resolution of the general equation algebraically, it appears more and more likely that this resolution is impossible and contradictory." He then added an intriguing note: "Perhaps it will not be so difficult to prove, with all rigor, the impossibility for the fifth degree." Gauss would never publish another word on this topic.

The repeated frustrations of the quintic hunters for more than two centuries prompted the French historian of mathematics Jean Étienne Montucla (1725–99) to use military metaphors in describing the attack on the quintic: "The ramparts are raised all around, but, enclosed in its last redoubt, the problem defends itself desperately. Who will be the fortunate genius who will lead the assault upon it or force it to capitulate?"

By another historical coincidence, the final and conclusive series of offensives on the quintic was about to begin in the year Montucla died. As in the case of the cubic and the quartic, this phase started with another Italian.

Paolo Ruffini (1765–1822; figure 46) was born in Valentano, Italy. He was the son of Basilio Ruffini, a medical doctor, and Maria Francesca Ippoliti. The family moved to Reggio, near Modena, during Ruffini's teens, and it was in Modena that he studied mathematics, medicine, literature, and philosophy, graduating in 1788. Extraordinarily versatile, Ruffini started practicing medicine and teaching mathematics at the same time. In the wake of the French Revolution, these were extremely uncertain times. The French army, under the command of Napoleon Bonaparte, took one Italian town after another, capturing Modena in 1796. Ruffini was initially appointed as a representative to the Junior Council of the Cisalpine Republic set up by Napoleon, only to lose his teaching appointment upon refusing to pledge allegiance to the new republic. Curiously, it was during this period of upheaval that Ruffini did his most important work. He claimed to have proven that the general quintic equation cannot be solved by a formula that involves only the simple operations of addition, subtraction, multiplication, division, and the extraction of roots.

We have to pause here for a moment to appreciate the magnitude of Ruffini's claim. The formula for the solutions of the quadratic equation had been known essentially since Babylonian times. The formula for the solutions of the cubic was discovered by dal Ferro, Tartaglia, and Cardano. Ferrari came up with the solutions to the quartic. All of these formulae were expressed by simple arithmetic operations and the taking of roots. Then came two and a half centuries of failed expectations, during which some of the most brilliant mathematicians tried in vain to find such a formula for the quintic. Now Ruffini was claiming that he could prove that the quintic equation could not be solved by a formula of this type, no matter how hard one tried. This represented a dramatic revolution in the thinking about equations. Mathematicians have grown accustomed to the fact that some equations are very difficult to solve, but here Ruffini's proof was supposed to show that in the case of the quintic, the effort was doomed from the start.

Figure 46

Ruffini published his proof in a two-volume treatise entitled *Teoria generale delle equazioni* (*General Theory of Equations*), which appeared in 1799. However, the proof was extremely intricate, and the tortuous reasoning made it difficult to follow through the 516 pages of the book. Not surprisingly, the reaction from the mathematical world was one of skepticism and suspicion at best. Ruffini sent a copy of *Teoria* to Lagrange around 1801 but received no reply. Still not discouraged, he sent a second copy, noting:

> Because of the uncertainty that you may have received my book, I send you another copy. If I have erred in my proof, or if I have said something which I believed new, and which is really not new, finally if I have written a useless book, I pray you point out to me sincerely.

Lagrange did not respond to this letter either. Ruffini tried one last time in 1802, starting with praise for Lagrange's work:

> No one has more right . . . to receive the book which I take the liberty of sending to you. . . . In writing this book, I had principally in mind to give proof of the impossibility of solving equations of degree higher than 4.

Still no reply.

Thwarted by the reception his work had received, Ruffini attempted to publish more rigorous and somewhat less abstruse proofs in 1803 and 1806. He also discussed the proof with fellow mathematicians Gianfrancesco Malfatti (who published a treatise about the quintic in 1771) and Pietro Paoli. The latter conversations led to a final version of the proof that was published in 1813, in a paper entitled "Reflections on the Solution of General Algebraic Equations." Unfortunately, even this supposedly more transparent proof did not make headlines in the mathematical community.

In a report to the king entitled "Historical Report on the Progress of the Mathematical Sciences since 1709," the French mathematician and astronomer Jean-Baptiste Joseph Delambre (1749–1822) did mention Ruffini's work briefly. He used, however, rather tentative language: "Ruffini proposes to prove that it is impossible." The exasperated Ruffini was quick to protest: "I not only proposed to prove but in reality did prove." Even this exchange did not result in a general acceptance of Ruffini's proof by his contemporaries and successors. Worse yet,

Delambre explained to Ruffini that it was hopeless to expect a definitive answer, because "whatever decision your Referees [mathematicians Lagrange, Lacroix, and Legendre] would have reached [concerning the validity of the proof], they had to work considerably either to motivate their approval or to refute your proof." From some comments the elderly Lagrange made to the scientist and pharmacist Gaultier de Claubry, we may gather that while he was generally impressed with Ruffini's work, even he was not quite intellectually inclined to accept such a revolutionary concept as the impossibility of solving the quintic by a formula. Consequently, Lagrange never made any public statements concerning Ruffini's proof.

In desperation, Ruffini sent his proof to the Royal Society in London. He received a polite reply stating that while a few members who had read his work found it satisfactory, it was not the society's policy to publish official approvals of proofs. The one distinguished mathematician who accorded credence to Ruffini's result was Augustin-Louis Cauchy (1789–1857). Cauchy's productivity was so prodigious (he published a staggering 789 mathematical papers) that at one point he had to found his own journal. In a letter received about six months before Ruffini's death, Cauchy, generally reserved with compliments, writes:

> Your memoir on the general resolution of equations is a work which has always seemed to me worthy of the attention of mathematicians and which, in my judgment, proves completely the insolvability of the general equation of degree greater than 4. . . . I add moreover, that your work on the insolvability is precisely the title of a lecture which I gave to several members of the academy.

Even with Cauchy's appreciation, Ruffini's proof became neither widely known nor accepted. Most mathematicians still found his arguments so convoluted that they were unable to ascertain their soundness.

But did Ruffini truly prove that the quintic cannot be solved by a formula that involves simple operations? With the wisdom of hindsight, we can say that he did not quite prove it. There was still a significant gap in the proof, where Ruffini made an assumption, without realizing that it was necessary to prove his assumption. Instead, he was satisfied to note that any other assumption would lead to a more complicated situation, so that "we can completely abandon it." This imperfection, however, takes nothing away from the originality of his discovery. In fact, none of Ruffini's contemporaries located the gap in his proof. Ruffini was the person responsible for a revolutionary change in the approach to equations. Instead of trying to solve the quintic, the effort was soon to turn to attempts to prove that it cannot be solved.

When we come today to evaluate Ruffini's work, we realize that he actually did much more than merely change ideas about the quintic equation. He took the relations between solutions of the cubic and quartic and certain permutations one step further. This marked the beginning of the transition from the traditional algebra, which deals only with numbers, to the roots of group theory, which involves operations between elements of any sort. Recall that members of groups can be anything from integer numbers to the symmetries of the human body. The birth of abstract algebra was on the horizon.

Ruffini was always conscientious to a fault. He once refused a chair in mathematics at Padua because he did not want to forsake all the families he was treating as a physician. Infinitely devoted to his patients, Ruffini contracted severe typhoid fever during the 1817–18 epidemic. He used that traumatic experience to write *Memoir of Contagious Typhus.* Although greatly weakened, he continued to visit patients and did not abandon his mathematical research. In April 1822, he was struck by chronic pericarditis and passed away the following month. Strangely, after his death, his work was all but forgotten, and with the exception of Cauchy, the mathematicians who followed him essentially had to rediscover his ideas.

This was the setting into which two young men, perhaps the most tragic figures in the history of science, appeared. The Norwegian Niels Henrik Abel and the Frenchman Évariste Galois were about to change the course of algebra forever. The life stories of these two remarkable individuals are so heartrending that I feel compelled to describe them in some detail in the next two chapters.

– FOUR –

The Poverty-Stricken Mathematician

The first lines in Erich Segal's celebrated novel *Love Story* read, "What can you say about a twenty-five-year-old girl who died? That she was beautiful, and brilliant, that she loved Mozart and Bach. And the Beatles and me." One can easily paraphrase this sad summary for Évariste Galois (1811–32) and Niels Henrik Abel (1802–29). For Galois it would probably read something like, "What can you say about a twenty-year-old boy who died? That he was a romantic, and a genius, that he loved mathematics. And he succumbed to misunderstanding and self-destruction." Or, for Abel: "What can you say about a twenty-six-year-old boy who died? That he was shy, and a genius, that he loved mathematics and the theater. And he was condemned to death by poverty." The Swedish mathematician Gösta Mittag-Leffler (1846–1927) described Abel's mathematical achievements with the words, "The best works of Abel are truly lyric poems of sublime beauty . . . raised farther above life's common-place and emanating more directly from the very soul than any poet, in the ordinary sense of the word, could produce." The great Austrian mathematician Emil Artin (1898–1962) wrote about Galois, "Since my mathematical youth, I have been under the spell of the classical theory of Galois. This charm has forced me to return to it again and again." Indeed, the genius of Abel and Galois could be compared only to a supernova—an exploding star that for a short while outshines all the billions of stars in its host galaxy.

ABEL—THE EARLY YEARS

Niels Henrik Abel was born on August 5, 1802. He was the second son of a Lutheran pastor, Søren Georg Abel, and Anne Marie Simonsen, the daughter of a shipping merchant (figure 47 shows silhouettes of Niels Henrik's parents). Some years after Abel's birth, his mother reported that she had actually given birth three months prematurely and that the newborn showed signs of life only after being washed in red wine. The unlikely combination of a father from a long line of men of the cloth and a remarkably beautiful woman known for her passion for earthly delights did not hold promise for a successful marriage. Before Niels Henrik was two, his father took a post in the village of Gjerstad, replacing his own father as minister. Norway, which was a part of Denmark during those years, was constantly in the shadow of war, first by sea, with the English fleet, then by land with Sweden. The results of the blockade of Norway's shipping routes by British warships were devastating. All timber exports came to a halt by mid-1808, and the trade in grain from Denmark had become so dangerous that it was also reduced to a trickle. Hunger and starvation spread over Norway in 1809. Pastor Abel barely managed to fight famine in his own parish by convincing the people of Gjerstad to eat the previously taboo horse meat.

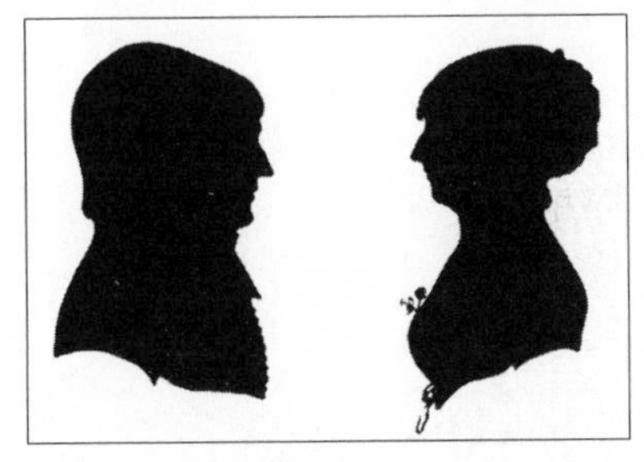

Figure 47

Niels Henrik was taught by his father, at the vicarage, till age thirteen. The pastor did not take the responsibility for this early education lightly. He had actually prepared a handwritten textbook from which he catechized his children. The book included grammar, geography, history, and mathematics. Amazingly, the first page on the topic of arithmetic addition (figure 48) contains a glaring error: 1 + 0 = 0! Fortunately, the world of mathematics did not lose one of its brightest stars because of this early misinformation. In 1815 Niels Henrik was sent to the Cathedral School in Christiania (today's Oslo). The deteriorating family life in a house in which both parents were increasingly indulging in alcoholism and the mother was rather free with her sexual favors probably hastened the young boy's departure. His father wrote, "May

God protect him! But it is without anxiety that I send him out into this depraved world."

Niels Henrik entered the Cathedral School at a rather low point in this institution's history. The opening of the Christiania University a few years earlier had robbed the Cathedral School of all of its best teachers, leaving mostly the unqualified behind. The mathematics teacher in particular, one Hans Peter Bader, was a heartless brute who terrorized the children and often beat pupils black and blue. Niels Henrik's grades were at first satisfactory, even though he showed little interest in the rather long and dreary school days. He tended to fall into depression when not in the company of friends, correctly diagnosing himself later: "As it happens, I am so constituted that I absolutely cannot, or at least only with the greatest difficulty, be alone. Then I become quite melancholy and not in the mood for work." Then, as in later years, his great escape from the burden of life's inevitable chores was the theater. There he could lose himself in the lives of fictional characters, instead of having to struggle with problems for the solution of which he never had the chance to receive the proper guidance. Niels Hendrik was shy and insecure, and his relations with the opposite sex remained very limited, not only during his student days, but in fact until his death. By the end of 1816 Niels Henrik's performance at school was on a slippery slope, and after having been beaten several times by Bader he had to quit for a short period. His grades fell so low that in 1817 he was allowed to pass the year only provisionally. In November of 1817, however, a fateful event at the Cathedral School was to mark a dramatic turning point in Abel's life. On November 16, a student, Henrik Stoltenberg, was taken ill with nervous fever accompanied by typhus. He died a week later. Eight of Stoltenberg's classmates signed statements to the effect that the hated mathematics teacher Bader had not only violently beaten the student with his fists, but that he had continued to kick

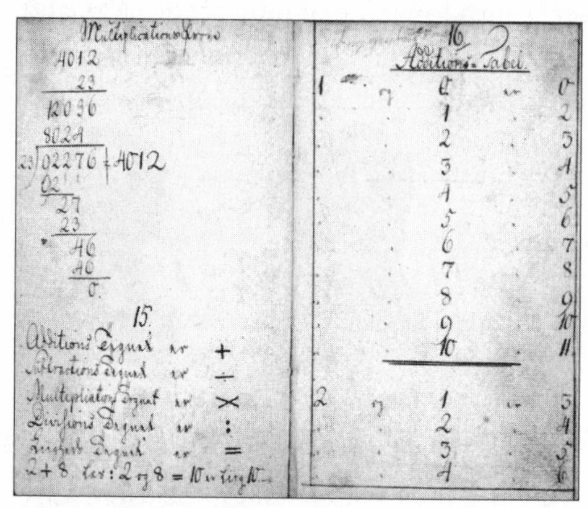

Figure 48

him even after poor Stoltenberg was lying helpless on the floor. While beating as a cause of death was never confirmed by the medical examiner, Bader was discharged.

As a substitute teacher the school hired Bernt Michael Holmboe (1795–1850), a graduate of the Cathedral School himself, who was only seven years older than Abel. Holmboe introduced a new syllabus that started with the training of students to understand mathematical symbols fully. It did not take him long to discover that the dream of every teacher of mathematics had come true in his class—he had a genius on his hands. After whizzing through the standard curriculum, Abel started, with Holmboe's enthusiastic and inspiring encouragement, to immerse himself in the original works of the great mathematicians Euler, Newton, Laplace, Gauss, and, in particular, Lagrange. Holmboe could not restrain his admiration. In Abel's report card for 1819 he unabashedly exclaimed, "A remarkable mathematical genius." In the following year he went even further. Holmboe's assessment is written right across all the school subjects: "With the most incredible genius he unites an insatiable interest in and ardour for mathematics, such that quite probably, if he lives, he will become one of the great mathematicians." Underlying the last sentence a few words have been scratched out, which can still be made out to read "the world's greatest mathematician." Apparently the school board insisted on Holmboe tempering his praise. The words "if he lives" turned out to be tragically prophetic.

A STRUGGLING GENIUS

During his last year in school Abel made his first attempt at spreading his own wings, and what an undertaking that was. With the chutzpah characterizing only youngsters in their first venture into unfamiliar territory, Abel tried no less than to solve the quintic equation. Here was a mathematical problem with which the best mathematicians of Europe had struggled for nearly three centuries, and now a high-school kid was claiming that he had solved it. Abel showed his solution to Holmboe, who found nothing wrong with it. Lacking the confidence of a seasoned mathematician, however, Holmboe presented the solution to the two mathematicians at Christiania University, Christopher Hansteen and Søren Rasmussen. They also did not find any errors in the solution. Realizing the magnitude of the discovery, Hansteen decided to forward the work to the leading Scandinavian mathematician of the time—Ferdinand Degen of Copenhagen—for publication by the Danish Academy.

Degen was a pragmatic person who preferred to err on the side of caution. Even though he found no fault with Abel's solution, he asked that Abel send him "a more detailed deduction of his result and also a numerical illustration" of the method—for instance, a solution to the equation $x^5 + 2x^4 + 3x^2 - 4x + 5 = 0$. After all, the *a priori* chances that a disciple from Cathedral School would solve one of the most celebrated problems in mathematics were not very high. While attempting to produce specific examples, Abel discovered, to his great dismay, that his solution was in fact incorrect. Far from signaling the end of his quest, however, this temporary setback was about to lead Abel to a monumental breakthrough. Degen, in any case, was sufficiently impressed to offer Abel a piece of advice. To Degen, the study of equations appeared to be "sterile subject matter." He suggested that Abel instead concentrate his efforts in the new field of elliptical integrals (special types of mathematical entities in calculus, called such because one can use them to calculate the arc length of an ellipse). There, Degen said, "a serious investigator with the proper approach . . . could discover a Strait of Magellan leading into wide expanses of an immense analytic ocean."

Even as Abel's mathematical genius was starting to shine through, the skies were darkening on the family front. The years 1818–20 were to become drainingly grievous for Abel. His father the pastor managed to get elected to the Storting (parliament) on December 10, 1817, but this seemingly prestigious event turned into utter disaster. At first, the newly elected and very energetic parliamentarian brought forward a few successful bills on education. In particular, he was instrumental in the establishment of a veterinary school. However, due perhaps to a judgment impaired by excessive drinking and an insatiable craving for self-promotion, he was driven to commit what was tantamount to political suicide. During one calamitous session, on April 2, 1818, he unexpectedly accused two representatives of unjustly imprisoning a former constable of one of the ironworks. The charges turned out to be totally unfounded, thus marking the beginning of Pastor Abel's downfall. The political and public furor that erupted led to threats of impeachment. Søren Georg Abel was given a last chance to apologize, which he hardheadedly declined. In the fall of 1818, the disgraced and disillusioned pastor returned to Gjerstad. He was increasingly inclined to drown his troubles in alcohol, which only brought about a rapid deterioration of his health. When he died in 1820, no one in Gjerstad expressed much sorrow. The morally weak widow was said to have received the consoling visitors in bed, with a servant whose services went beyond household chores.

Anne Marie and Niels Henrik's five siblings were left with a tiny pension that was far from sufficient even to support their own needs. The question of money to allow Abel to complete his education could not even be raised, let alone addressed. Through circumstances that were nothing short of miraculous, Abel managed, nevertheless, to enter the university in 1821. In an environment in which personal contact between students and professors was generally discouraged and professors adopted a distant and aloof attitude, no fewer than three professors volunteered to support Abel out of their own, rather shallow pockets. This generosity persisted until 1824, when Abel finally received a stipend to live on. During his first years at the university Abel became a frequent and welcome guest in the house of Professor Christopher Hansteen, and it was in a periodical started by Hansteen that Abel published his first mathematical paper, in 1823. This was not exactly an earthshaking article (nor was it, or Abel's second paper, comprehensible to most of the magazine's readership). Abel's third publication, however, "Solution of a Pair of Propositions by Means of Definite Integrals," addressed what was much later to become the mathematical basis of modern radiology (for which physicist Allan Cormack and electrical engineer Godfrey Hounsfield received the 1979 Nobel Prize in Medicine).

In the meantime, Professors Hansteen and Rasmussen continued relentlessly to search for ways to support Abel's work, and in particular to allow him to travel abroad to expand his horizons. When one such plea to the Academic Collegium had totally disappeared within the university's bureaucracy, Rasmussen gave Abel a personal gift of one hundred speciedaler so that he would be able to travel to Denmark to meet Degen and other Danish mathematicians. Against all odds, therefore, Abel spent the summer holiday of 1823 in Copenhagen. There, he discovered that "the men of science think that Norway is pure barbarity" and he did everything in his power "to convince them to the contrary." The trip to Copenhagen had another unexpected outcome—Abel met his future fiancée, Christine (nicknamed "Crelly") Kemp. The first meeting between the two took place at a party in Abel's uncle's house. Abel asked Christine to dance, but to his and her embarrassment the orchestra started playing what was then the new sensation—the waltz—which neither of them knew. Disconcerted, they stared at each other for a few minutes and then quietly left the dance floor. Abel's entire relationship with Crelly is somewhat of a mystery. After spending Christmas with her in 1824, Abel shocked his friends at the university with the announcement that he was engaged to be married. Apparently Abel never participated either verbally or physically in any of the erotic experiences that were quite typical for student life in the capital. The engagement at a very young age generated a protective wall that allowed him to avoid the need for any further explanations when the topic of women came up. Niels Henrik never married Christine. It was unthinkable at that time that anyone would marry before he had the means to support a household. Sadly, Abel never reached that position. Five years after the engagement, on his deathbed, overcome with guilt and responsibility, Abel would ask his good friend, Baltazar Mathias Keilhau, to take care of Crelly. "She is not pretty," he would be heard to say, "has red hair and freckles, but is a splendid human being." Keilhau, who up to that point had not even seen Crelly, indeed married Kemp in 1830, and the two spent the rest of their lives together.

THE QUINTIC

Ever since his unsuccessful attempt to solve the quintic by a formula, this subject had not left Abel's mind. While he did not ignore Degen's advice to embark on pioneering studies in two other areas of mathematics, the obsession with the quintic persisted. Upon his return from Copenhagen, he therefore decided to revisit this topic with fresh eyes. Instead of attacking the problem again with the goal of finding a solution, he was now determined to show that a formulaic solution did not exist. Recall that this was precisely what Ruffini had claimed to have proven in a series of works in the period 1799–1813, without realizing that his "proof" contained a serious gap. Since Ruffini's work had not been widely publicized, Abel was unaware of it in 1823. After a few months of intensive work, the twenty-one-year-old student from remote Norway brought a centuries-old quest to an end. He succeeded in proving, rigorously and unambiguously, that it is impossible to find a solution of the quintic equation that can be expressed as a simple formula of the coefficients that involves only the four arithmetic operations and the extraction of roots.

Let me first clarify briefly what Abel's proof means, and equally important, what it does not mean. Abel proved that in the case of the general quintic equation, and equations of higher degrees, one cannot repeat what had been achieved for the quadratic, cubic, and quartic equations. In other words, a solution to the quintic in the form of an algebraic formula that involves only the coefficients simply does not exist. All the toil put in by scores of brilliant mathematicians amounted to no more than a Sisyphean effort. Abel's proof does not imply that quintic equations cannot be solved. The quintic equation $x^5 - 243 = 0$, for instance, has the obvious solution $x = 3$, because $3^5 = 243$. Furthermore, even the general quintic equation can be solved, either numerically, using computers, or by introducing more advanced mathematical tools, such as elliptic functions. What Abel discovered was a fundamental shortcoming of basic algebra when it comes to the taming of the quintic. The familiar operations of addition, subtraction, multiplication, and the extraction of roots simply reach the limit of their usefulness when faced with the quintic. This was a monumental realization in the history of mathematics. It changed the entire approach to equations from mere attempts to find solutions to the necessity to prove whether solutions of certain types exist at all.

Abel's proof is too technical to be reproduced in detail in a popular text. I refer the more mathematically inclined readers to a clear exposition in Peter Pesic's book *Abel's Proof.* Here let me only note that the proof relied on the logical tool known as *reductio ad absurdum.* The idea behind this method is that you prove a proposition by proving the falsity of its contradictory. In other words, Abel assumed that the quintic is solvable and showed that this assumption leads to a logical contradiction.

Abel was not oblivious to the significance of his discovery. Unlike his previous papers, which were written in the inaccessible Norwegian, he wrote the proof for the insolvability of the quintic in French, hoping to attract the attention of the leading mathematicians of his time. He also decided to use the proof as his "business card," thinking that this "would be the best introduction I could have." Accordingly, he paid the printer Grøndahl from his own pocket (probably by skipping quite a few meals) to produce the article in the form of a pamphlet. In order to save on the printing expenses, however, he condensed the article *"Mémoire sur les équations algébriques où l'on démontre l'impossibilité de la résolution de l'équation générale du cinquième degré"* ("Essay on Algebraic Equations, Where the Impossibility of Solving the General Equation of Fifth Degree Is Demonstrated") to only a scant six pages. This parsimony proved to be costly in other ways. The greatly abbreviated, almost telegraphic version was so opaque to most mathematicians that even though Abel sent copies of the pamphlet to his friends in Copenhagen and to the great Carl Friedrich Gauss, the paper received little attention. Gauss apparently did not even bother to open Abel's pamphlet—after his death the article was found, uncut, among his papers. One of the greatest masterpieces of mathematical literature found no readership.

Around that time, Abel's guardian angels, Professors Hansteen and Rasmussen, concluded that, for him to realize his full potential, it was no longer practical for them to continue to support him out of their own meager means. Consequently, in 1824 they applied to the Norwegian government for a travel grant for Abel. They justified the unusual request by noting that for this extraordinary talent "a stay abroad in those places where the most remarkable mathematicians are, would most excellently contribute to his scientific and scholarly education." After the usual bureaucratic delays, the finance department did approve a modest grant for Abel. This was a truly remarkable achievement given the country's dire financial situation at the time. The approval did introduce, however, two important modifications to the original request. First, Abel was required to stay in Norway for another eighteen months to "further his scholarly scientific education, particularly, perhaps in further study of the learned languages," in order to be prepared for travel. Second, and as it turned out more important, no money was allocated to support Abel upon his return home. This latter omission would prove to have devastating consequences.

A EUROPEAN EXPERIENCE

In September of 1825 Abel finally took his farewells from Crelly, who was by then a governess for the children of a family in the small town of Son, near Christiania, and left for the Continent accompanied by three friends. Two of them were later to become geologists and the third a veterinarian. Originally, upon Hansteen's advice, Abel planned to spend his time in Paris, after a brief stay in Copenhagen. However, when his friends decided to go to Berlin, the horror of being left alone in Paris convinced Abel to take a detour through Berlin. In this particular case, Abel's mortal fear of isolation produced a fortunate outcome. In Berlin he met an influential construction engineer who had a great passion for mathematics and was about to become Abel's greatest admirer, fatherly friend, and benefactor. August Leopold Crelle (1780–1855) was at first rather unclear about the purpose of the visit of the young Norwegian, who barely spoke German. In a letter to Hansteen, Abel described the event:

> A considerable time elapsed before I could make clear to him what the purpose of my visit was, and it all appeared to be heading for a melancholy end when he asked me what I had already read in mathematics. I took courage, and mentioned to him the works of a couple of the foremost mathematicians. He then became very amiable and, as it appeared, really happy. He began an extensive conversation with me about various difficult problems which were still not resolved. When we came to the solution of the quintic equation, and I told him that I had demonstrated the impossibility of giving a general algebraic solution, he would not believe it and said he would dispute it. I therefore gave him a copy, but he said he could not see the reason for several of my conclusions. Others have said the same, and I consequently have made a revision of it.

Following this meeting, Crelle founded a mathematical journal, usually referred to as *Crelle's Journal* (the official name was *Journal for Pure and Applied Mathematics*), which became the premier German mathematical publication of the nineteenth century. The first volume of *Crelle's Journal* appeared in 1826, and it included the astonishing number of six papers by Abel (written in French and translated by Crelle). One of those papers was a more detailed and elaborated exposition of the proof for the insolvability of the quintic by a simple formula. Abel was apparently still unaware of Ruffini's proof at the beginning of 1826, but he probably discovered it around the summer of that year, via a summary of Ruffini's ideas by an anonymous author. In a manuscript dated 1828 that was published posthumously, Abel notes, "The first, and if I am not mistaken, the only one before me who tried to prove the impossibility of the resolution of algebraic equations is the geometer Ruffini. But his memoir is so complicated that it is difficult to judge the validity of his reasoning. It seems to me that his reasoning is not always satisfying."

In the midst of these impressive scientific endeavors, the harsh reality of his financial situation continued to haunt Abel. From his extremely modest means he was also partially supporting his siblings. In a letter to Mrs. Hansteen he wrote:

> God bless you for not forgetting my brother [referring to his troubled brother Peder]. I am so worried that things might be going badly for him. If he should need more money than he already has received, may I ask you to give him still a little more. When the 50 daler have been used up, I shall make arrangements for you to receive more.

A more serious affair was about to cast a dark shadow over Abel's expectations and future prospects. Professor Rasmussen found it no longer possible to juggle his teaching responsibilities and his public duties, and he resigned from the university to take a position with the Bank of Norway. This opened up what seemed to be a golden opportunity for Abel, one he had always dreamt about—a university appointment. There were, however, two other potential candidates for the position: Abel's former teacher Holmboe and the young Niels Henrik. When the news about the opening reached the young travelers in Berlin, one of them, Christian Peter Boeck (himself an aspiring veterinarian), was quick to write to Hansteen:

> My cousin Johan Collett writes to me about Rasmussen's bank appointment. What will happen to his position? Is there any hope that Abel might obtain it upon his return, or perhaps Holmboe is ahead of him? However reasonable the latter may be in certain respects, it does not appear quite just, since Abel presumably ranks a head above Holmboe.

The letter was written on October 25, 1825. The faculty met on December 16 to discuss and approve the recommendation for the new appointment. They recommended Holmboe to fill the vacancy. The chief reason given for preferring Holmboe to Abel was that the latter "cannot as easily adjust himself to the comprehension of the younger students as a more experienced teacher, and thus would not be able to present so fruitfully the elementary parts of mathematics, which is the principal object of the above mentioned position." This type of tension between talent in teaching and aptitude for research as qualifications for an appointment is not uncommon. In fact, I can testify from firsthand experience (having served on scores of search committees) that discussions of this sort continue to characterize university appointments to the present day. In this particular case, however, with one candidate being head, shoulders, chest, and knees above the other, there is no question that the shortsighted faculty had committed a serious error. Not entirely unaware of their problematic decision, the Norwegian faculty concluded, "We also consider it a duty to point out how important it is for science in general and our university in particular that student Abel not be lost from sight."

Even with his hopes pulverized and the realization of his uncertain future starting to sink in, the generous-spirited Abel made every effort to keep his friendship with Holmboe intact. In a warm letter to Holmboe he wrote, "Among other news he told me that you, my friend, had been recommended to be lecturer to take the place of Rasmussen. Receive my most sincere congratulations and be assured that none of your friends is as pleased over it as I. Believe me, I have often wished for a change in your position." The comradeship between Abel and Holmboe indeed stood this test, and they remained devoted friends for the rest of Abel's life. The disappointed Abel did feel compelled, however, to inform Crelly that their marriage plans had to be put on hold.

Despite these troublesome circumstances, that winter in Berlin proved to be one of Abel's happiest times. He was extremely productive, contributing seminal papers in integral calculus and on the theory of sums of various infinite series. The young scientists did not miss any opportunity to go to the theater—Abel's passion—and the students were occasionally invited to balls or threw parties of their own. The latter events, which were quite noisy, sometimes annoyed the famous philosopher Georg Hegel (1770–1831), who happened to live in the same house. He was once heard to refer to his clamorous neighbors as "Russian bears."

As spring was approaching, Abel started to make travel plans that would bring him to his original destination—Paris. However, the thought of being separated from his friends was again such a deterrent that he ended up first traveling with Keilhau to Freiberg, and then with two more friends through Dresden, Bohemia, Vienna, northern Italy, and Switzerland, reaching Paris only in July of 1826.

PARIS

Anyone who arrives in Paris during July or August knows what it's like. As Abel soon found out, everybody was on vacation and away from the capital. Still, Paris was the indisputable mathematical capital of the world, and Abel eagerly awaited the opportunity to meet with mathematical giants he revered. After all, the works of Cauchy, Laplace, and Legendre constituted the bulk of Abel's bedtime reading. In his first letter to Hansteen from Paris he exclaimed exuberantly, "I have finally arrived at the focus of all my mathematical desires, in Paris." Little did Abel know that the Paris visit would cause only disappointment and disillusion.

Abel took up lodging with the Cotte family at 41, rue Ste. Marguerite, across from the famous St. Germain-des-Prés quarter. For the somewhat outrageous sum of 120 francs a month he had an "extremely plain" room, clean clothes, and two meals a day. The landlord, who was a "rascally dilettante in mathematics" according to Abel, took him on his first attempt to meet with the famous mathematician Adrien-Marie Legendre (1752–1833). Unfortunately, the latter was stepping into his carriage just as Abel arrived, and the exchange between the two was limited to a few polite greetings. A few years later Legendre would come to regret not having talked more with Abel while the young mathematician was still in Paris. (They did, in fact, have a quite fruitful exchange in 1829, but this was 1826, and the aging Legendre had no idea who Abel was.)

During his first few months in Paris, Abel worked ceaselessly on what was to become a true tour de force, now known as Abel's theorem. Although this theorem is not directly related to the quintic or to group theory, it played such a major role in Abel's life that no biography of him would be complete without it. The theorem dealt with a special class of functions known as *transcendental functions,* and it vastly generalized a relation previously obtained by Euler. It would not be an exaggeration to say that Abel's theorem literally afforded the world of mathematics new perspectives. The clarity and intrinsic simplicity of Abel's proof has been likened to the classical statues of the Greek sculptor Phidias. Abel's originality was revealed, in particular, by his ability to turn problems inside out. Let me give a nonmathematical example for this type of inversion logic.

Imagine that someone suggests that one of the reasons that firearms are so common in some inner cities is that the number of homicides is so high—people acquire weapons in order to protect themselves. One could, however, turn this problem on its head and submit that one of the reasons for the frequent homicides is the unrestricted availability of guns. In mathematics, examine for instance the relation $x = \sqrt[3]{y}$ (read as "x equals the cubic root of y"). It implies that to calculate x we need to extract the cube root of y, as in $2 = \sqrt[3]{8}$. However, the inverted relation $y = x^3$ is precisely equivalent to the previous one (e.g., $8 = 2^3$), yet most people would agree that calculating the third power is much easier and more convenient than manipulating cube roots. This was precisely the type of insight that Abel provided in his theorem, one that had escaped Legendre in almost forty years of work.

Abel's paper turned out to be one of his longest (it fills sixty-seven pages in his collected works). This remarkable treatise, entitled "*Mémoire sur une propriété générale d'une classe très étendue des fonctions transcendantes*" (Memoir on a General Property of a Very Extensive Class of Transcendental Functions), included both the theory and its applications. When it was completed, Abel could barely contain his excitement. He submitted the paper with great anticipation to the French Academy of Sciences on October 30, 1826. Here was the work, he thought, that would be his passport to recognition. Abel was actually present at the session at the French Institute when the paper was introduced. He listened with a great sense of accomplishment as the secretary of the academy, mathematical physicist Joseph Fourier (1768–1830), read the introduction to the work. Cauchy and Legendre were immediately appointed as referees, and Cauchy was put in charge of communicating a report to the academy.

Abel spent the next two months in Paris eagerly awaiting the verdict. He felt increasingly lonely, gloomy, and anxious: "Though I am in the most boisterous and lively place in the continent," he wrote to Holmboe, "I feel as though I were in a desert. I know almost nobody." Partly perhaps because of his own melancholic mood when not surrounded by friends, he found it difficult to communicate:

> On the whole, I do not like the French as well as the Germans; the French are extremely reserved toward strangers. It is very difficult to become more closely associated with them, and I dare not hope for it. Everybody works for himself without concern for others. All want to instruct, and nobody wants to learn. The most absolute egotism reigns supreme.

As always, the theater remained Abel's principal source of amusement and joy: "I know of no greater enjoyment than to see a play by Molière in which Mademoiselle Mars [the best known actress of the time, Anne-Françoise-Hippolyte Boutet, known as Mars] plays. I am really quite enthused," he wrote. The other "attractions" of nineteenth-century Paris left him cold:

> Occasionally I visit Palais Royal [figure 49], which the French call un lieu de perdition [a den of vice]. There one sees des femmes de bonne volonté [women of "goodwill"] in considerable numbers, and they are not at all intrusive. The only thing one hears is: "Voulez-vous monter avec moi? Mon petit ami, petit méchant." [Do you want to come up with me? My little friend, little bad boy.] Being an engaged man I never listen to them and leave Palais Royal without the least temptation.

One compatriot that Abel did meet in Paris was the painter Johan Gørbitz. Gørbitz was working in the atelier of the famous historical painter Jean-Antoine Gros, and he had lived in Paris since 1809. Gørbitz produced during that winter the only authentic portrait of Abel painted during his lifetime (figure 50). The portrait depicts a handsome young man of delicate features. While Abel's mother was a woman of great beauty, none of Abel's contemporaries mention him as being particularly good looking. The flattering portrait may therefore represent the beautifying tendency of painters at the time.

Figure 49

Figure 50

Perhaps the best glimpse into the complex workings of Abel's creatively wandering mind can be gleaned from the pages of his Paris notebook. Among formulae of various integrals and expressions involving complex numbers, we find a menagerie of doodles and various fragments of sentences that hop relentlessly from one stream of thought to the next. The page shown in figure 51, for instance, contains (in no particular order) the following snippets of phrases: "complete solution to the equations in which the . . . goddamn . . . goddamn, my ∞ [infinity sign]"; "Our Father who art in Heaven, give me my bread and beer. Listen for once," referring perhaps to his rapidly deteriorating financial situation; "Come to me in God's name"; "My friend, my beloved"; "Tell me, my dear Eliza . . . listen . . . listen," referring either to his beloved sister Elizabeth, to whom he sent a gift from Paris, or to some sexual encounter implied by the language of the last phrase below; "Suleiman the Second," referring to the Ottoman sultan of the seventeenth century—Abel read extensively on European history before his trip; "Come to me my friend"; "now, for once my"; "solutions to algebraic equations"; "Come to me in all your lewdness."

Abel was extremely optimistic about the paper he had submitted to the academy, and he was absolutely convinced that a laudatory report was forthcoming. After all, he surmised, surely those great mathematicians would recognize the value of the work. What he did not realize, however, was the fact that the two mathematicians appointed as evaluators were, for different reasons, totally unsuited to the task. Legendre was seventy-four at the time, and he lacked the patience to go through a lengthy manuscript that was (in his own words) "barely legible . . . written in very thin ink, the letters badly formed." Cauchy, on the other hand, was at the peak of his egotistical phase, or, in the words of historian of mathematics Eric Temple Bell, "so busy laying eggs of his own and cackling about them that he had no time to examine the veritable roc's egg which the modest Abel had deposited in his nest." The net result of these unfortunate circumstances was that Legendre couldn't be

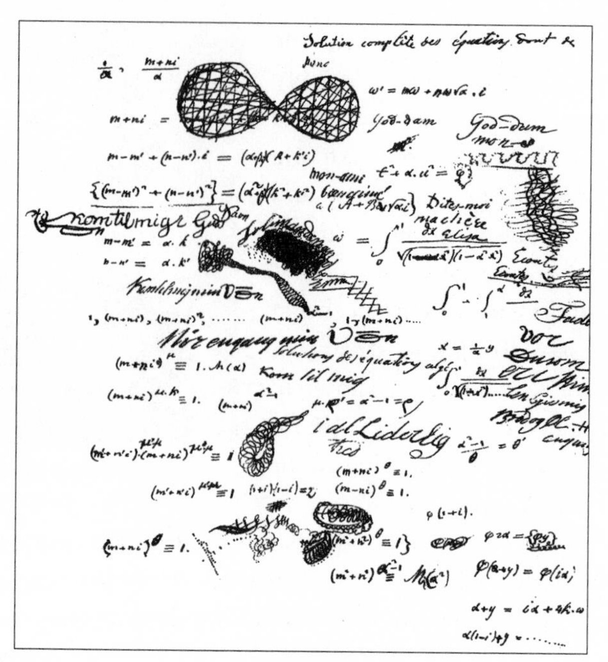

Figure 51

bothered and Cauchy mislaid the memoir somewhere among his piles of papers and forgot about it. Imagine this, a genuine *chef d'oeuvre*—as seminal perhaps as Claude Monet's painting *Impression: Sunrise* for the development of impressionism—being mislaid and lost. Only two years later would Legendre learn of the contents of the manuscript through a correspondence with Abel, who by then was back in Norway.

Another person who in 1829 became familiar with Abel's paper was the great German mathematician Carl Gustav Jacob Jacobi (1804–51). He wrote with unconcealed excitement to Legendre on March 14, 1829:

> What a discovery by Herr Abel, this generalization of Euler's integral! Has anything like it ever been seen? But how is it possible that this discovery, perhaps the most important in our century, could have avoided the attention of yourself and your colleagues after having been communicated to the Academy more than two years ago?

It was in response to this perplexed inquiry that Legendre gave the lame excuse about the paper being "barely legible."

Abel spent two more months in Paris, with dwindling resources, an increasingly gloomy mood, and deteriorating health. He made only two new acquaintances of note. One was the mathematician Johann Dirichlet (1805–59), who although younger than Abel had already made a name for himself by proving (with Legendre) Fermat's Last Theorem for the case $n = 5$. That is, he proved that there are no whole numbers x, y, z such that $x^5 + y^5 = z^5$. The other was Jacques Frédéric Saigey, editor of the mathematics and astronomy review *Ferrusac's Bulletin*, for which Abel wrote a few articles, basically summarizing his papers in *Crelle's Journal.*

What Abel thought was a nagging cold started to bother him, and he must have consulted a few physicians. Two years later, on his deathbed, he would be heard to exclaim, "There, you can see that it was not true what they said in Paris—I certainly do not have consumption." From this we can conclude that the diagnosis of the French doctors was alarming—tuberculosis. Refusing at the time to acknowledge his medical condition, even with his hopes dashed and his funds running dry, Abel decided to leave Paris on December 26 for Berlin.

Shortly after his arrival in Berlin he became ill. These were probably the first signs of his rapidly declining health. Crelle did his best to help Abel financially, and Abel also received a loan from Holmboe. Miraculously, neither his economic worries nor his worsening health prevented Abel from completing his most extensive publication yet—"Research on Elliptic Functions" is 125 pages long in his *Complete Works.* This treatise presented an immense generalization of the familiar trigonometric functions (e.g., sine, cosine), and it had important ramifications even into number theory. Crelle attempted to persuade Abel to remain in Berlin until he could secure a position for him there. Abel, however, was tired and painfully homesick. On May 20, 1827, heavily in debt and with no prospects for a position, he returned to Christiania.

COMING HOME

The situation in Christiania in 1827 confirmed Abel's worst fears. Recall that the conditions of his grant were such that no provisions were made to sponsor him in Norway. After the Department of Finance turned down his application for an extension on his fellowship, the university managed to come up with a small stipend for him to live on (not before the Department of Finance reserved the right to deduct this award from Abel's future earnings). Even with this allowance, Abel had no choice but to tutor schoolboys to make ends meet. Crelly, his fiancée, took a new governess position with the Smith family, who owned ironworks in Froland in southern Norway.

The beginning of 1828 brought with it significant financial improvement. Professor Hansteen succeeded in obtaining a large grant for studying the Earth's magnetic field, thereby enabling Abel to become his temporary substitute at both the university and the military academy. At the same time, Abel suddenly found himself involved in a scientific race to publish, of a sort he had never experienced before. September of 1827 witnessed the appearance of not one but two papers on elliptic functions. One was the first part of Abel's massive treatise "Research on Elliptic Functions," and the other was a paper announcing related results by the young German mathematician Jacob Jacobi. Not to be scooped, Abel frantically rushed to press with the second part of his manuscript, to which he added a note that showed how Jacobi's results could be obtained from his own. More important from the perspective of the present book, he stopped working on what was supposed to be his definitive answer to the question of which equations can be solved by a formula. This left the door open for another young genius—Évariste Galois—to provide the answer, and in the process to introduce group theory.

The recognition of Abel's genius was by now spreading throughout Europe. Legendre, who began corresponding with both Abel and Jacobi on the theory of elliptic functions, declared that "through these works you two [Abel and Jacobi] will be placed in the class of the foremost analysts of our time." Besides the mathematical fame, the reality of Abel's precarious economic situation started to reach the ears of some European mathematicians, especially through the efforts of the tireless Crelle. In an unprecedented act of support, four eminent members of the French Academy of Sciences wrote to King Charles XIV of Norway and Sweden and urged him to see to it that a position commensurate with Abel's talents would be created. The effort was to no avail.

Abel spent the summer of 1828 in Froland with Crelly and the Smith family. This was a place where, in his words, he felt "among all the angels." Hanna Smith, one of the daughters, then twenty years old, described Abel later in her memoirs as being generally lively and playful. She gives a moving description of how he used to sit with his mathematical papers, surrounded by the ladies of the house, writing his new manuscripts on the thinnest paper to save on the postage cost.

The disastrous terms imposed by the stipend allocated to Abel two years earlier came back to haunt him. The minister of finance insisted that the collegium would "take care that the aforementioned advance, in adequate installments be deducted from Mr. Abel's salary." While the university refused to follow these outrageous instructions, Abel's finances were sinking faster than a lump of lead. He concluded one of his notes to Mrs. Hansteen that summer with, "Your most poverty-stricken creature," and another note with, "I am as poor as a church-mouse, . . . Yours destroyed."

By the fall of 1828 Abel was back in Christiania, preparing for the school year to begin. For a few weeks in September he was so ill that he had to be confined to bed. Nevertheless, in mid-December, during a particularly cold winter, he ignored his sister's advice and left again for Froland to spend Christmas with his fiancée. He became ill shortly after Christmas, starting to cough to exhaustion. In spite of his debilitating condition, he managed to produce a very brief summary of his Paris treatise (one he feared had been permanently lost), which he sent to *Crelle's Journal.* On January 9, as it became clear that Abel was spitting blood, the district doctor was called for help. The doctor hesitated to use the dreaded words "tuberculosis" or "consumptive sickness," which were effectively a death sentence, and diagnosed Abel's illness as pneumonia. The next few months became a horrid nightmare for everyone involved. Crelly and the two eldest daughters of the Smith family took turns at his bedside night and day. During the painful, sleepless nights, Abel would be heard cursing the entire medical profession for not having made sufficient progress to be able to help him. The days were slightly better. Abel would repeat several times that the mathematician Jacob Jacobi was the person who could best understand the value of his work. Sometimes Abel would collapse into self-pity and would complain bitterly about the poverty that had been his most constant companion. As the winter progressed, Abel's speech became more and more hoarse by the day, to the point where his words became barely intelligible to the people around him. As April arrived, his condition was visibly deteriorating. After an excruciating night on April 5, the young Norwegian genius passed away on April 6 at 4:00 p.m., with Crelly and one of the Smith sisters at his bedside. He was twenty-six years old. The devastated Crelly wrote to Mrs. Hansteen on April 11: "My Abel is dead! I have lost everything on Earth! Nothing, I have nothing left."

On April 8, still unaware of Abel's death, Crelle wrote to him from Berlin, in ecstatic jubilation: "Now my dear, precious friend, I can bring you good news. The Ministry of Education has decided to call you to Berlin and employ you here."

Abel was buried in Froland on April 13, 1829, a day after a violent blizzard. His friends paid for the gravestone. In his obituary, Crelle wrote:

> All of Abel's work is shaped by an exceptional brilliance and force of thought . . . difficulties seem to vanish in front of the victorious onslaught of his genius. But it was not only his great talent which . . . made his loss infinitely regrettable. He distinguished himself equally by the purity and nobility of his character and by the exceptional modesty which made his person cherished to the same unusual degree as was his genius.

On June 28, 1830, the French Academy of Sciences announced that the Grand Prix for mathematical achievements would be awarded jointly to Abel and Jacobi.

But what was the fate of Abel's Paris memoir? Following Jacobi's exchanges with Legendre and an intervention by the Norwegian consul to Paris, Cauchy eventually managed to uncover the manuscript in 1830. It would take eleven more years for the manuscript to get to print. Finally, as an almost comical conclusion to this saga of neglect, the manuscript vanished again during the printing process, only to resurface in Florence in 1952.

In 2002, the Norwegian government established a fund of $22 million to award the Abel Prize in Mathematics. This prize is presented Nobel-style by the king of Norway. The first prize, in the amount of $816,000, was awarded on June 3, 2003, to the famous French mathematician Jean-Pierre Serre; the second prize was awarded on May 25, 2004, jointly to two other remarkable mathematicians—Sir Michael Francis Atiyah of the University of Edinburgh and Isadore M. Singer of MIT. This prize has finally brought the name of the mathematician who proved that a certain equation cannot be solved by a formula to the attention of the general public. Ironically, the brilliant work of this poorest of mathematicians is celebrated with a huge monetary award.

There was one meeting that never happened during that bleak autumn of 1826 in Paris. Unbeknownst to Abel, a young French mathematician who lived only a couple of miles away was starting to be obsessed with precisely the same problems that had intrigued the young Norwegian. Could the quintic be solved by a formula? Or even more generally, which equations can be solved by a formula? Évariste Galois was only fifteen when Abel was in Paris, but he was already devouring books on mathematics as if they were adventure stories. Sadly, we shall never know how a meeting between these two star-crossed individuals might have changed their lives. One thing is certain: If there could conceivably be an even more tragic story than that of Abel, it was that of Galois.

– FIVE –

The Romantic Mathematician

On the morning of May 30, 1832, a single shot fired from twenty-five paces hit Évariste Galois in the stomach. Although fatally wounded, Galois did not die on the spot. He remained lying on the ground until an anonymous good Samaritan, perhaps a former army officer, perhaps a peasant, picked him up and brought him to the Cochin Hospital in Paris. The following day, with his younger brother Alfred at his side, Galois died of peritonitis. His last known words were, "Don't cry, I need all my courage to die at twenty."

This was the grim end of the life of one of the most visionary of all mathematicians—the unlikely combination of a genius like Mozart and a romantic like Lord Byron, all swathed in a tale that rivals in its woe that of Romeo and Juliet.

GALOIS—THE EARLY YEARS

Évariste Galois was born during the night of October 25, 1811, and was named after the saint celebrated on October 26 (figure 52 shows the birth certificate and appendix 8 gives the extended family tree). His father, Nicolas-Gabriel Galois (figure 53), was an educated man, who was managing at the time a quite reputable school for boys in Bourg-la-Reine (today a Paris suburb)—a position he inherited from Évariste's grandfather. In his spare time, Nicolas-Gabriel composed witty verses and entertaining plays, both of which made him a popular guest at house parties of the time. Évariste's mother, Adéläide Marie Demante, the daughter of a jurisconsult in the Paris Faculty of Law, was herself well

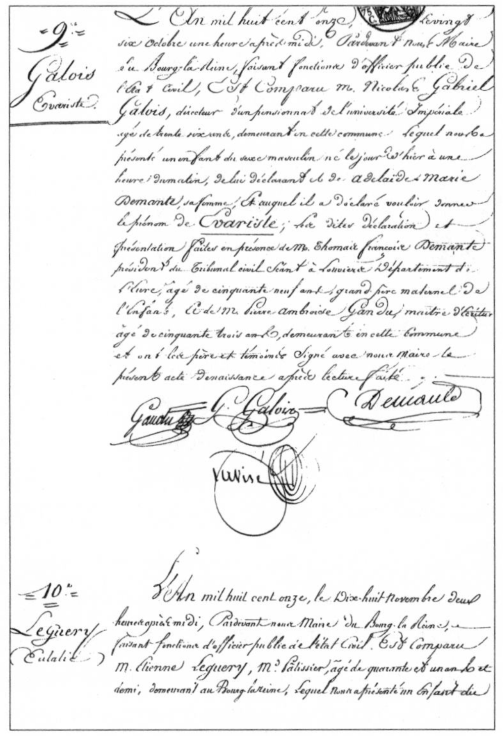

9e
Galois
Evariste

L'An mil huit cent onze, le vingt six Octobre une heure après midi, Pardevant nous Maire de Bourg-la-Reine, faisant fonctions d'officier public de l'état civil, Est Comparu M. Nicolas Gabriel Galois, directeur d'un pensionnat de l'université Impériale, agé de trente six ans, demeurant en cette commune. Lequel nous a présenté un enfant du sexe masculin né le jour d'hier à une heure du matin, de lui déclarant et de adelaïde marie Demante, sa femme, Et auquel il a déclaré vouloir donner le prénom de Evariste; les dites déclaration et présentation faites en présence de M. Thomas françois Demante président du Tribunal civil séant à Louviers Département de l'Eure, agé de cinquante neuf ans, grand père maternel de l'enfant, Et de M. Pierre ambroise Gandu, maître d'écriture agé de cinquante trois ans, demeurant en cette Commune et ont les père et témoins signé avec nous Maire le présent acte de naissance après lecture faite.

Gandu G. Galois Demante

10e
Leguery
Eulalie

L'An mil huit cent onze, le Dix-huit Novembre deux heures après midi, Pardevant nous Maire de Bourg-la-Reine, faisant fonctions d'officier public de l'état Civil. Est Comparu M. Etienne Leguery, Md Pâtissier, agé de quarante et un an et demi, demeurant au Bourg-la-Reine, Lequel nous a présenté un Enfant du

Figure 52

versed in classical studies. The Demante family lived almost right across from number 54, Grand Rue—the Galois home (figure 54 shows Galois's home, when it still existed).

In the midst of the Napoleonic era, Nicolas-Gabriel was a loyal subject to the emperor. His brother went even further to become an officer in the Imperial Guard. The postrevolutionary times were extraordinarily turbulent, however, and following his colossal defeat in Russia, Napoleon was forced to abdicate in 1814 in favor of the Bourbon king Louis XVIII. The megalomaniacal practices of this king, which were accompanied by a gradual restoration of the power of the church, were sufficient to rekindle the liberal movement, with Nicolas-Gabriel as a vocal proponent. Riding the wave of public dissatisfaction, Napoleon seized the opportunity to return to power in March of 1815, only to fall again one hundred days later, this time permanently. Still, during Napoleon's brief return, Nicolas-Gabriel was appointed mayor of Bourg-la-Reine, a position he continued to hold even after Napoleon met his Waterloo (figure 55 shows Nicolas-Gabriel's equivalent of a passport). The frequent changes of power and the chameleon-like nature of the political climate helped to polarize French society into two rather distinct camps. On the left were the liberals and the republicans, broadly inspired by the sweeping ideals of the French Revolution. On the right were the "legitimists" or "ultras" (short for ultraroyalists), whose model state was a church-dominated monarchy.

Figure 53

Like Abel, Évariste received his early education at home. Adéläide Marie offered her children a strong background in the classics and in religious studies while also infusing them with liberal ideas. Even after the boy's tenth birthday, Évariste's mother regretted her original intention to send him to a school in Reims and decided to keep him at home instead for two more years.

In October of 1823, Évariste finally did leave home, for the Parisian boarding school Lycée Louis-le-Grand. This prestigious institution had existed since the sixteenth century, and it counted among its illustrious graduates people such as the revolutionary Robespierre and, later, the novelist Victor Hugo. Prior to Galois's enrollment, the school had the distinction of having stayed open even during the tumultuous times of the French Revolution. In spite of its academic preeminence, the school was housed in a prisonlike building that was in desperate need of repair. The student body provided an excellent representation of the entire political spectrum in the French society of the time, which was a sure recipe for unrest. Rebellion, quarrels among students, and riots were the norm at Louis-le-Grand. Disobedience was further bred by the more-stringent-than-

Figure 54

military discipline enforced on the pupils. The spartan daily program, which started at 5:30 a.m. and ended at 8:30 p.m. sharp, was meticulously structured and allowed for very little recreation time. Silence was imposed even during meals, which themselves were extremely lean. Breakfast, for instance, consisted of dry bread and water.

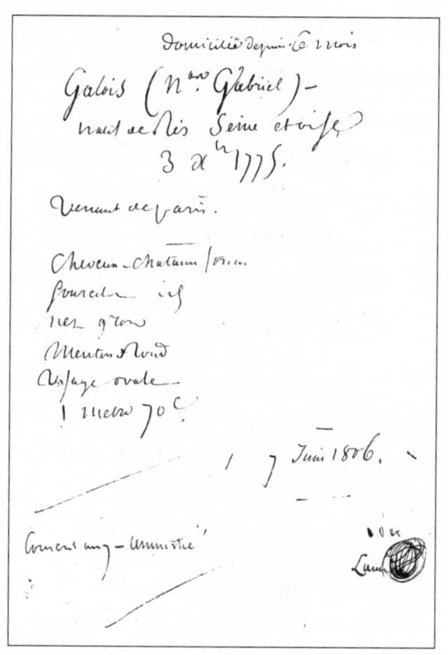
Galois (Nre Gabriel) -
Venant de Paris.
Nez gros
Menton rond
Visage ovale
7 Juin 1806.

Figure 55

In the classrooms, students sat in pairs on bare steps, with lighting provided by candles, one for each pair. The sight of rats crossing the classroom floor during lessons was so common that it attracted no attention. The slightest deviation from the mandatory routines—even a mere refusal of food during meals—resulted in solitary confinement in one of twelve special cells. Overall, the transition from the peaceful, happy atmosphere at home to the violent and confining milieu at school must have been quite shocking for Galois.

Évariste arrived at Louis-le-Grand shortly after the conservative Nicolas Berthot was appointed as school principal. The students suspected that this assignment was but the first step in an attempt by the right to drag the school back to its Jesuit roots. Students expressed their discontent by refusing to chant at the chapel service and by ignoring the customary toasts to King Louis XVIII and other dignitaries at a school banquet on January 28, 1824. The reaction was swift and harsh—117 students were immediately expelled. Galois, who was then just in his first term, was not involved, but was undoubtedly emotionally affected.

In spite of the humiliating conditions and the inhumanely strict discipline, Évariste's first two years at Louis-le-Grand were characterized by considerable successes. His mother's superior preparation in the classics soon translated into distinctions for narration in Latin and for translation from Greek. In the comprehensive competitive examination, he also received the prize for mathematics. Nevertheless, the dismal environment took its toll. The damp winter of 1825–26 brought about a painful earache that persisted for many months, which did not help to improve Galois's generally downbeat mood. The separation from his father, with whom he used to enjoy the exchange of witty couplets, was particularly hard on the young boy. Consequently, his schoolwork began to deteriorate.

Outside the school walls, events were progressing rapidly. Louis XVIII died in September of 1824 and was succeeded by his brother, who assumed the title King Charles X. The transition was marked by a dramatic growth in the influence of the clergy and of the extreme-right ultras. Conviction for the dubiously defined "crimes against religion" could now carry the death penalty.

THE BIRTH OF A MATHEMATICIAN

The fall of 1826 witnessed Galois's first humiliating setback. This was in the rhetoric class. While Galois's diligent if unenthusiastic efforts in this subject had generally been appreciated by his teacher, the new ultraconservative school headmaster, Pierre-Laurent Laborie, had rather different ideas. In his rigid opinion Galois was too young for this advanced class, which required "judgment that only comes with maturity." In January, therefore, Galois was forced, to his and his father's dismay, to repeat the third-year classes. Phrases such as "original and bizarre" and "good but singular" started appearing in the report card describing his character. The unpleasant experience with rhetoric, however, turned out to be a blessing in disguise—Galois discovered mathematics. Figure 56 shows a portrait of Galois at about this time, drawn by a classmate.

Figure 56

The new teacher for Preparatory Mathematics, Mr. Hippolyte Vernier, decided to introduce a new book for the study of geometry. This was Legendre's *Elements of Geometry,* which first appeared in 1794 and had rapidly become the book of choice all over Europe. This by now classic text broke with the somewhat tedious Euclidean tradition of high-school geometry. Legend has it that the mathematics-hungry Galois swallowed all of Legendre's book, originally intended for a full two-year course, in just two days. While it is impossible to verify the validity of this (probably exaggerated) story, there is no question that by the fall of 1827 Galois had lost all interest in any other subject and had become passionate about mathematics. His rhetoric teacher, who at first misunderstood Galois's indifference in class and described his uninspired performance by saying, "There is nothing in his work except strange fantasies and negligence," concluded correctly after the second term that "he is under the spell of the excitement of mathematics. I think it would be best for him if his parents would allow him to study nothing but this." The third trimester only confirmed the verdict: "Dominated by his passion for mathematics, he has totally neglected everything else."

Galois was indeed bewitched. He tossed aside the conventional textbooks and went straight to the original research papers. Galloping from one professional mathematics article to the next as now a more ordinary youth would with successive volumes of the Harry Potter stories, he now immersed himself totally in Lagrange's memoirs, *Resolution of Algebraic Equations* and *Theory of Analytic Functions.* This mind-opening experience led to an ambitious endeavor. Totally unaware of Ruffini's and Abel's work, Galois tried for two months to solve the quintic. And just like the young Norwegian before him, he also thought at first that he had found the formula, only to be disappointed later, as he discovered an error in his solution. Figure 57 shows a later editorial footnote referring to the fact that Abel's error (of thinking that he had solved the quintic) had been repeated by Galois, and that "it is not the only striking analogy between the Norwegian geometer who starved to death and the French geometer condemned to live or die . . . behind the lock of a prison." As in Abel's case, this minor setback only impelled Galois on the course to bigger things concerning the solvability of algebraic equations.

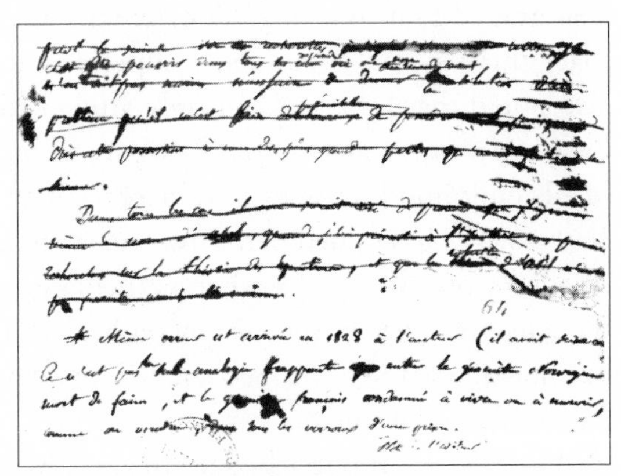

Figure 57

More serious hindrances were still to present themselves, some of

Galois's own making. As Mr. Vernier correctly diagnosed, in spite of his genius and creative imagination, Galois was never able to study methodically and work systematically. Extremely advanced in some subjects, he lacked some of the most fundamental basics in others. Unaware of his own deficiencies and turning a deaf ear to Vernier's advice, Galois tried boldly in June of 1828 to take the entrance examination to the legendary École polytechnique a year early. The École polytechnique had been founded in 1794 as the main school for the training of engineers and scientists. Lagrange, Legendre, Laplace, and other famous scientists were, at one time or another, on the teaching staff of this institution. The school had also been known for its liberal atmosphere. Had Galois passed the examination, the Polytechnique would have been the perfect breeding ground for his soaring spirits. Given his inadequate preparation, however, Galois predictably failed the exam. This blasted expectation may have been the seed for his later feeling of persecution that grew to clear paranoiac dimensions.

Forced to continue at Lycée Louis-le-Grand, Galois enrolled in the Special Mathematics class of Louis-Paul-Émile Richard (1795–1849). Richard proved to be for Galois what Holmboe had been for Abel—an inspiring and motivating teacher and supporter. Richard was not a brilliant mathematician himself, but was well read in the latest mathematical developments. He immediately recognized Galois's unusual abilities and encouraged him to engage in original research, stating enthusiastically that "this student is markedly superior to all his school fellows." He also noted that "this student only studies higher mathematics." Just as Picasso's mother and sister, fully aware of his remarkable talents, kept all of his childhood drawings, Richard kept twelve notebooks of Galois's classwork. These documents eventually made it into the library of the Academy of Sciences. Another mathematician whom Galois met around the same time was Jacques-François Sturm (1803–55). Sturm would later become one of the few to recognize immediately that Galois's ideas were diamonds in the rough.

In 1829 Galois published his first mathematical paper. This relatively minor paper dealt with mathematical objects known as *continued fractions.* The work had applications for quadratic equations, and it appeared in the journal *Annales de mathématiques pures et appliquées.* Incidentally, Abel died five days after the publication of Galois's first paper. For

Galois, this first foray into mathematical research soon turned into an explosion of new ideas. The seventeen-year-old was about to revolutionize algebra. While Abel had shown unambiguously that the general quintic cannot be solved by a formula that involves only the arithmetic operations and the extraction of roots, his premature death did leave open the much bigger question: How does one determine whether any *given* equation (quintic or of higher order) is solvable by a formula or not? Recall that many particular equations were still solvable. In principle, Abel's proof still allowed for the possibility that every specific equation had its own formulaic solution.

In order to answer the solvability question, Galois had not only to introduce the seminal concept of a group, but also to formulate an entirely new branch of algebra known today as Galois theory. As a starting point, Galois picked up the theory of equations where Lagrange had left off. He delved into relations among the putative solutions of an equation (such as the relation $x_1x_4 = 1$ between two of the four solutions x_1, x_2, x_3, x_4 of the equation of degree 4: $x^4 + x^3 + x^2 + x + 1 = 0$) and the permutations of these solutions that leave the relations unchanged (see notes for an example). Here, however, is where his genius truly took off. Galois managed to associate with each equation a sort of "genetic code" of that equation—the *Galois group* of the equation—and to demonstrate that the properties of the Galois group determine whether the equation is solvable by a formula or not. Symmetry became the key concept, and the Galois group was a direct measure of the symmetry properties of an equation. I shall describe the essence of Galois's brilliant proof in chapter 6. Richard was so impressed with Galois's ideas that he submitted that the young genius should be admitted to the École polytechnique without an entrance examination. To give Galois a chance at achieving this ambitious goal, he encouraged him to put his theory into the form of two memoirs, which Richard himself was prepared to take to the great Cauchy for presentation to the Academy of Sciences. The memoirs were indeed submitted on May 25 and June 1 of 1829, introduced briefly by Cauchy, and entrusted for judgment to Cauchy, Joseph Fourier (the secretary of the academy), and mathematical physicists Claude Navier and Denis Poisson.

More than six months after the submission, on January 18, 1830, Cauchy wrote the following apologetic letter to the academy:

> I was supposed to present today to the Academy first a report on the work of young Galois, and second a memoir on the analytic determination of primitive roots in which I show how one can reduce this determination to the solution of numerical equations of which all roots are positive integers. Am indisposed at home. I regret not to be able to attend today's session, and I would like you to schedule me for the following session for the two indicated subjects.

By the time the next session took place on January 25, however, Cauchy's egotistical tendencies apparently took over once again, and he ended up presenting only his own memoir and never mentioning Galois's work again. This was not the end of the misfortunes associated with these manuscripts. In June of 1829, the Academy of Sciences announced the establishment of a new Grand Prix for Mathematics. Tired of waiting for Cauchy's verdict, and having learned from *Ferrusac's Bulletin* of Abel's work on the theory of equations, Galois decided to resubmit the work, with some modifications, as an entry for the prize. (I find no direct evidence to support a speculation that Cauchy encouraged him to try for the prize, even though some indirect evidence described later points to Cauchy having been impressed with the work.) Galois's submission ("On the Conditions That an Equation Be Solvable by Radicals"—the four arithmetic operations and the extraction of roots) has since been judged to be one of the most inspiring masterworks in the history of mathematics. The work was entered in February 1830, shortly before the March 1 deadline. The prize committee consisted of mathematicians Legendre, Poisson, Lacroix, and Poinsot. For reasons that are not entirely clear, the academy's secretary, Fourier, took the manuscript home. He died on May 16, and the manuscript was never recovered among his papers. Consequently, entirely unbeknownst to Galois, his entry was never even considered for the prize. The prize was eventually awarded to Abel (posthumously, and justifiably, given the other entries) and Jacobi. You can imagine Galois's anger when he learned eventually that his own manuscript had been lost. The paranoid young man was now convinced that all the forces of mediocrity had united to deny him a well-deserved repute.

DISASTER STRIKES TWICE

If June of 1829 was a relatively happy month for Galois, with his important manuscript having been submitted to the academy, July was one of his worst. The coronation of Charles X in 1824 had resulted in a significant rise in the power of the church and the ultras. In Bourg-la-Reine, a new priest joined forces with other right-wing administrators in an attempt to demote the liberal Nicolas-Gabriel Galois from the mayorship. This young priest forged the mayor's signature on a few stupid couplets and despicable epigrams. Apparently unable to cope with the ugly scandal that had erupted, the delicate Nicolas-Gabriel committed suicide by gas asphyxiation. This tragedy occurred on July second, in Nicolas-Gabriel's Paris apartment on rue Saint Jean-de-Beauvais, only a stone's throw away from Évariste's school. The devastated youngster had to endure yet another emotional ordeal—a riot broke out during the funeral in protest against the malicious priest's attempt to participate in the service. Figure 58 shows the commemorative plaque for Mayor Nicolas-Gabriel Galois that still exists today on the wall of the Bourg-la-Reine city hall.

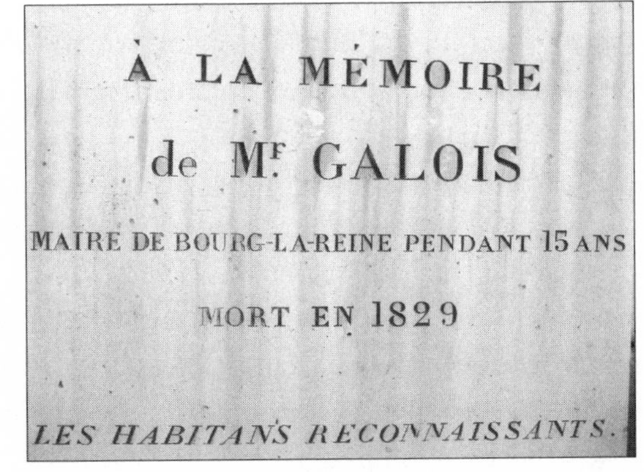

Figure 58

One can hardly think of a worse time for Évariste to have taken his second entrance examination to the École polytechnique. Yet, as fate would have it, the examination took place just one month after the funeral, on Monday, August 3, with Galois still in mourning. In the history of mathematics, this infamous examination has become almost synonymous with Galileo's questioning by the Inquisition. Compared to Galois, the two examiners, Charles Louis Dinet and Lefebure de Fourcy, were, in historian E. T. Bell's words, "not worthy to sharpen his pencils." Even though Dinet was himself a former student of the Polytechnique and was the teacher to have prepared none other than the great Cauchy for his own entrance examination, these two mathematicians are mostly remembered today for one thing only—for having failed one of the greatest mathematical geniuses of all time. Galois's name does not figure at all in Dinet's list of twenty-one candidates that he regarded as admissible.

We do not know with certainty what happened in this examination. Speculation has it that Galois's tendency to calculate mostly in his head and to commit only the final results to the blackboard left a bad impression in an oral examination in which he was supposed to show all of his deliberations. Dinet in particular had a reputation for posing relatively simple questions, but also of being utterly uncompromising when it came to the answers. Galois's patience, which was never exemplary, must have been stretched to the limit by the events surrounding his father's death. According to one version, when asked to outline the theory of arithmetical logarithms, Galois informed Dinet arrogantly, if correctly (but see notes), that there were no *arithmetical* logarithms. Legend has it that in his frustration with the examiners' inability to understand his unorthodox methods he threw the blackboard eraser at one of them; this story is not out of character, but is probably false—at least according to mathematician Joseph Bertrand (1822–1900). Clearly, the failure in the examination left Évariste deeply embittered and only enhanced his sense of persecution. Two decades later, Olry Terquem, the editor of *New Annals of Mathematics,* would say, "A candidate of superior intelligence is lost with an examiner of inferior intelligence." A biographical note that appeared in 1848 in the *Magasin pittoresque* also concluded, "For not possessing what is known as 'blackboard experience,' for not having exercised to solve out loud in front of a large audience those questions of details . . . Galois was declared inadmissible." Since two entrance attempts were the maximum allowed, Galois was now forced to enter the less prestigious École préparatoire (later called École normale). There was, however, still a "small" snag. In order to be admitted, Galois had to obtain a baccalaureate (the equivalent of a high-school diploma) in the arts and sciences and to pass an oral examination. His total disregard for anything that was not mathematics made the passing of these exams difficult, to say the least. Even the physics examiner, Jean Claude Peclet, wrote in astonishment, "He knows absolutely nothing . . . I have been told that he is good in mathematics. This greatly surprises me." Nevertheless, primarily based on his results in mathematics, Galois was

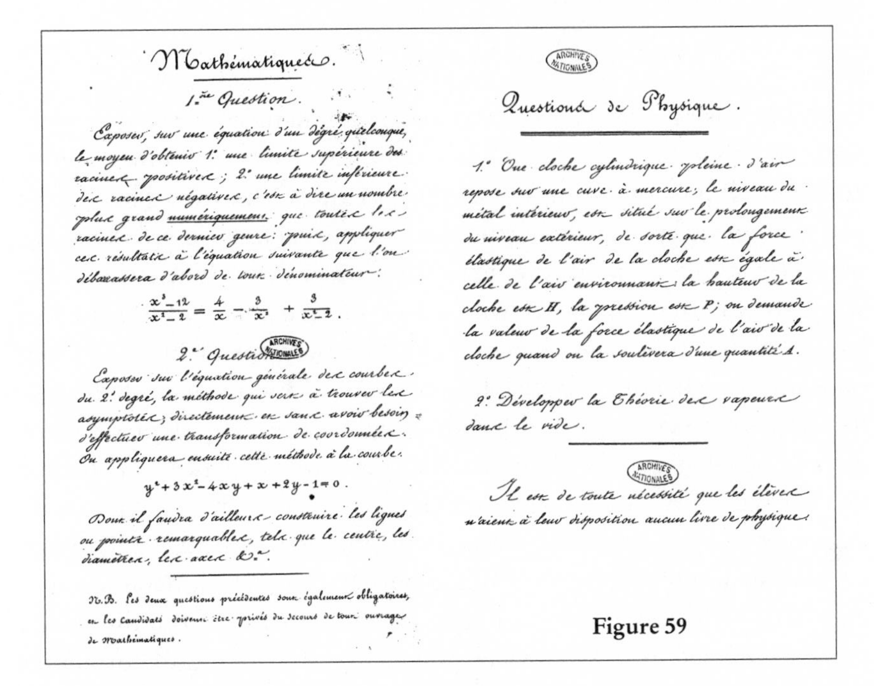

Mathématiques.

1.re Question.

Exposer, sur une équation d'un degré quelconque, le moyen d'obtenir 1.° une limite supérieure des racines positives; 2.° une limite inférieure des racines négatives, c'est à dire un nombre plus grand numériquement que toutes les racines de ce dernier genre: puis, appliquer ces résultats à l'équation suivante que l'on débarrassera d'abord de tout dénominateur:

$$\frac{x^3-12}{x^2-2}=\frac{4}{x}-\frac{3}{x^3}+\frac{3}{x^2-2}.$$

2.e Question.

Exposer sur l'équation générale des courbes du 2.d degré, la méthode qui sert à trouver les asymptotes, directement et sans avoir besoin d'effectuer une transformation de coordonnées. On appliquera ensuite cette méthode à la courbe.

$$y^2+3x^2-4xy+x+2y-1=0.$$

Dont il faudra d'ailleurs construire les lignes ou points remarquables, tels que le centre, les diamètres, les axes &c.a.

N.B. Les deux questions précédentes sont également obligatoires, et les Candidats doivent être privés du secours de tout ouvrage de Mathématiques.

ARCHIVES NATIONALES

Questions de Physique.

1.° Une cloche cylindrique pleine d'air repose sur une cuve à mercure; le niveau du métal intérieur, est situé sur le prolongement du niveau extérieur, de sorte que la force élastique de l'air de la cloche est égale à celle de l'air environnant: la hauteur de la cloche est H, la pression est P; on demande la valeur de la force élastique de l'air de la cloche quand on la soulèvera d'une quantité A.

2.° Développer la Théorie des vapeurs dans le vide.

ARCHIVES NATIONALES

Il est de toute nécessité que les élèves n'aient à leur disposition aucun livre de physique.

Figure 59

admitted at the beginning of 1830 in the section of sciences. Figure 59 shows the first pages of two of Galois's exams—mathematics (in the general competition of 1828) and physics (in his last general competition in 1829).

Not everything was dark in Galois's life. The year 1830 saw three of his articles—two on equations and one on the theory of numbers—published in the important *Ferrusac's Bulletin.* The first article was the precursor to Galois's revolutionary theory of equations. The appearance of his name in print right next to those of the leading mathematicians of the time must have given Galois some satisfaction. In the June issue in particular, Galois's two papers sandwiched a paper by Cauchy. During the same year, Galois also met Auguste Chevalier, who was to become his best friend. Auguste and his brother Michel introduced Galois to new socialistic ideas, inspired by a religious-egalitarian philosophy known as Saint-Simonianism (after the nobleman the comte de Saint-Simon). The socioeconomic concepts of this ideology were based primarily on the complete elimination of social inequalities. Given Galois's passionate disposition, his increasing involvement in tempestuous political activity spelled nothing but trouble.

LIBERTÉ, ÉGALITÉ, FRATERNITÉ

Ever since his coronation in 1824, Charles X had incited strong opposition. The opponents to the Bourbons and their ultras-dominated government fell into two camps—republicans and Orléanists. The former party, composed primarily of students and workers, expressed their revolution-inspired views in the newspaper *La Tribune.* The latter party wanted to replace Charles X with Louis-Philippe, duc d'Orléans, and had *Le National* as their chief voice. In the elections of July 1830, the opposition registered a landslide victory of 274 seats against the 143 for the government. Faced with abdication, Charles X attempted a coup d'état by issuing on July 26 an infamous series of ordinances. In the first he declared, "Freedom of the press is suspended . . . no newspaper or pamphlet . . . may appear in Paris, or in the departments." The other ordinances annulled the results of the elections and set the dates for new ones. The ordinances were accompanied by a warning from the police prefect, directed at public places permitting the reading of forbidden newspapers. This was more than the defiantly inclined Parisians were prepared to tolerate. On July 27, an article by the Orléanist Louis-Adolphe Thiers called in no uncertain terms for a rebellion of the people. Rioting in the streets began in the early hours of the afternoon. People could be seen carrying pieces of furniture on every street corner. Within three days, more than five thousand barricades were erected and heavy fighting erupted, accompanied by the chimes of all the church bells of Paris. The students of the École polytechnique were making history during those "*Trois Glorieuses*" ("Three Glorious Days"), as they took charge of the fighting in and around the Latin Quarter. The spirit and explosive energy of the *Trois Glorieuses* have been magnificently captured in the painting *Liberty Leading the People* (figure 60) by Eugène Delacroix (1798–1863). In the crowd, behind Liberty, one can distinguish the typical hat of a Polytechnique student.

As these fateful events were unfolding, to their unbearable frustration, Galois and his fellow students at the École normale were constrained to hearing the sounds of the revolution from behind barred windows and doors. The school's director, M. Guigniault, decided to use all means, including a threat to call in the troops, to prevent his students from participating in the rebellion. On the evening of the twenty-eighth, Galois could not take it anymore. In desperation, he tried several times

Figure 60

unsuccessfully to scale the outer wall. Bruised and defeated, he had to accept the fact that he had missed the revolution.

When the smoke cleared, there were almost four thousand people dead. As a compromise between the ultras and the republicans, the duc d'Orléans entered Paris on July 30 and was crowned on August 9, taking the supposedly conciliatory title Louis-Philippe I, king of the French. King Charles X left for exile, and Cauchy, always a Bourbon loyalist, left France as well, as a tutor to Charles's grandson. Guigniault, the ever-opportunistic director of the École normale, was quick to offer the services of his students to the new provisional government. Galois's contempt for his hypocritical director knew no bounds, and he was determined to use the first opportunity to expose his cunning duplicity.

That summer in Bourg-la-Reine, Galois's family discovered to their astonishment that the once fragile and reserved Évariste had turned into a passionate revolutionary who was prepared to sacrifice himself for his republican ideals. In the following fall, as he returned to school, he joined a militant wing of the republican party known as the Société des Amis du Peuple (Society of the Friends of the People). During the same period, he befriended other young republicans who were destined to become great political leaders: the biologist François-Vincent Raspail (1794–1878), the law student Louis Auguste Blanqui (1805–81), who later spent more than thirty-six years in prison, and the active republican Napoleon Lebon (1807–after 1856). The society had a reputation for not hesitating to use aggressive and even violent means to achieve its goals. After the arrest of its leader, Jean-Louis Hubert, the society became an underground secret association, with Raspail as its president.

At the École normale the strained relationship between the director and Galois was rapidly progressing to a showdown. Galois kept asking for things (such as a uniform similar to that of the École polytechnique; military training for students) that he must have known Guigniault would not even be prepared to consider. At the same time, Guigniault's declared policy that "good students should not be interested in politics" was clearly something Galois could not swallow. Finally, when Guigniault published a letter on December 2 attacking a liberal teacher of the Louis-le-Grand in one of the two student newspapers, the reply was swift and acrimonious. The newspaper *La gazette des écoles* (*The Schools' Gazette*) published the following letter from a "student at École normale":

> The letter that M. Guigniault had inserted in yesterday's *Lycée*, on the occasion of one of the articles in your newspaper, seemed to me to be quite indecent. I thought you would be interested in any attempt made to unmask this man.
>
> Here are the facts to which 46 students can testify.
>
> On the morning of 28 July, since many of the students of the École normale wished to join in the uprising, M. Guigniault told them, twice, that he could call the police to restore order. The police on 28 July!
>
> On the same day, M. Guigniault said to us, with his usual pedantism, "Many brave people have been killed on both sides. If I were a soldier, I would not know what decision to take. What should I sacrifice, freedom or legitimacy?"
>
> This is the man, who stuck a huge tricolor rosette on his hat the day after. These are our doctrinaire liberals!
>
> I should also like to inform you, sir, that the students of the École normale, inspired by noble patriotic spirits, very recently presented themselves to M. Guigniault, to inform him of their intention to address a petition to the Ministry of Education, asking for arms, and wishing to take part in military training, so as to be able to defend their territory if required.

Here is M. Guigniault's answer. It is as liberal as his answer of 28 July:

"The request addressed to me would make us look ridiculous; it is an imitation of what has been done in higher level institutions: it came from below. I should like to point out that, when the same request reached the Minister from these institutions of higher education, only two members of the Royal Board voted in favor, and they were precisely those on the Board not among the liberals. The Minister accepted, because he feared the students' turbulence, and their compassionate spirit, which appeared to threaten to ruin completely the University and École polytechnique." I believe that, from one point of view M. Guigniault is right to defend himself in this way, against being blamed for his prejudice against the new École normale. For him, nothing is as beautiful as the old École normale, which had everything.

We recently asked for a uniform, which was denied us; they did not wear them at the old school. In the old school, the course lasted three years. Although, when the new school was set up, the third year was acknowledged to be pointless, M. Guigniault brought it back.

Soon, following the rules of the old École normale, we will only be allowed out once a month, and will have to return by 5:00 p.m. It is wonderful to belong to the educational system that produced men like Cousin [referring to Victor Cousin, a philosopher and a conservative member of the Education Board] and Guigniault!

Everything he does shows his narrow outlook and ingrained conservatism.

Sir, I hope that these details will interest you, and that you will put them to the use you think fit, to the benefit of your estimable newspaper.

The newspaper editors added that they had deliberately removed the signature from the letter.

Galois neither confirmed nor denied being the author of this letter, even though he was widely suspected. To Guigniault, however, that was sufficient evidence to expel Galois, whom he regarded as a constant troublemaker. In his letter of explanation to the minister of education,

Guigniault claimed that he had "a full confession" from Galois and that in general, he had up to that point "tolerated his unconventional behavior, his laziness and his very difficult character."

The students at École normale showed little support for Galois. Those in the arts even posted a letter taking the director's side, in fear for their future careers and most probably prompted by Guigniault. Nevertheless, from a description published in *La gazette* we learn that at least one student showed some courage:

> We have just heard that the Director of the École normale asked each one [of his students] individually the following question: "Are you the author of the letter to *La gazette des écoles*?" The first four answered in the negative, while the fifth answered: "Sir, I do not think I can answer this question, because it would help betray one of my fellow students." M. Guigniault was extremely irritated by this proud, noble reply.

The bitter exchanges surrounding Galois's expulsion continued for three weeks. Letters on Galois's behalf interleaved by those backing Guigniault became a constant feature in the pages of the newspapers. Galois ended his last appeal to the students on December 30 by writing, "I am not asking anything for myself, but speak out for your own honor and according to your conscience."

On January 2, 1831, the *Gazette des écoles* published an article by Galois entitled "On the Teaching of the Sciences, the Professors, the Works, the Examiners." This was a remarkable manifesto calling for a complete reform in the instruction of the sciences. Most of Galois's complaints would sound relevant even today:

> Until when will the poor youngsters be obliged to listen or to repeat all day long? When will they be given some time to reflect on this accumulation of knowledge, to be able to coordinate [find a pattern in] this endless multitude of propositions, in these unrelated calculations? . . . Students are less interested in learning than in passing their exams.

Alluding probably to his own painful experiences with examiners, Galois lamented:

> Why don't the examiners pose questions to candidates other than in a twisted manner? It seems that they fear being understood by those

> they are interrogating; what is the origin of this deplorable habit of complicating the questions with artificial difficulties?

Unfortunately, in spite of his legitimate objections to the school system of his time, when circumstances would force Galois to open his own "school," it would not turn out to be a great success either.

A TURBULENT LIFE

Out of school and free to pursue his liberal dreams, Galois enlisted in the artillery of the national guard. While this organization was proud in having its own distinct uniform, it was more like a militia. Galois would continue to wear the same uniform even after the artillery had been disbanded and the national guard reorganized to comprise only the taxpaying populace, to which he did not belong. Not being a student, however, had its price—Galois now had no means of support. To make ends meet he decided to give lessons in mathematics, and a bookseller friend allowed him to use a room in his bookstore at 5, rue de la Sorbonne, for this purpose. Galois placed an ad in the *Gazette des écoles* announcing that he would hold an algebra course intended for those students who "feeling how incomplete is the study of algebra in the colleges, wish to go deeper into this science." This was not a good recipe for making money. A few dozen of Galois's republican friends first attended as a courtesy, but they quickly dropped out of the extremely advanced course. Galois's political activities did not help either, since they occupied more and more of his time. Galois's teaching ambitions were therefore reduced to low-level tutoring.

On the research front, a promising event happened at the beginning of 1831, only to turn later into yet another disappointment. Galois was asked to resubmit his memoir to the academy. The new version of "The Conditions for the Resolvability of Equations by Radicals" was introduced on January 17, and this time mathematicians Denis Poisson (1781–1840) and Sylvestre Lacroix (1765–1843) were charged with reviewing it. More than two months had passed, however, with no word from the academy. The frustrated Galois gave vent to his disgust by sending an inquiring letter to the president on March 31, 1831, adding sarcastically, "Sir, I would be grateful if you could relieve my concerns by inviting Mr. Lacroix and Mr. Poisson to announce whether they have also lost my memoir [as did Fourier], or whether they intend to report on it to the Academy." Even this provocative letter produced no response.

Meanwhile, political events were starting to have a great impact on Galois's life. The famous mathematician Sophie Germain (1776–1831), the first woman to have broken through the prevailing gender barrier and into the old boys' club, characterized his general attitude at the time as having a "habit of insult." She added a sad comment: "They say he will go completely mad, I fear this is true." In April, nineteen artillerymen of the national guard, who refused to disarm when their unit had been disbanded, were brought to trial. One of them was Pescheux d'Herbinville, to whom we shall return in relation to Galois's death. To the republicans' delight, all were acquitted on April 16 in a much-publicized trial known as the "trial of the nineteen." The Society of the Friends of the People organized a large banquet at the Aux Vendanges de Bourgogne restaurant to celebrate the event. Two hundred republican activists were in attendance on May 9, including the famous writer Alexandre Dumas (1802–70), the biologist-politician Raspail, Galois, and many others. In Dumas's words, "It would be difficult to find in all of Paris two hundred guests more hostile to the government than those." As the champagne started to flow at the end of the meal, many toasts were proposed: to the revolutions of 1789 and of 1793, to Robespierre, and many others. One of the more intellectually articulate toasts was proposed by Dumas, who declared: "I drink to art! May both quill and brush contribute as much as gun and sword to the social renewal to which we have dedicated our lives, and for which we are prepared to die." At one point, Galois, who was sitting at the extremity of one of the tables, jumped to his feet and proposed a toast. Holding in the same hand a glass of wine and an open jackknife, he was heard shouting: "To Louis-Philippe!" The event was later described in some detail in Dumas's memoirs:

> Suddenly, in the midst of a private conversation which I was carrying on with the person on my left, the name Louis-Philippe, followed by five or six whistles, caught my ear. I turned around. One of the most animated scenes was taking place fifteen or twenty seats from me.
>
> A young man, holding in the same hand a raised glass and an open dagger, was trying to make himself heard. He was Évariste Galois,

> since killed in a duel by Pescheux d'Herbinville, that charming young man who made silk-paper cartridges which he would tie up with pink ribbons.
>
> Évariste Galois was scarcely 23 or 24 at the time. He was one of the most ardent republicans. The noise was such that the very cause for that had become incomprehensible.
>
> All I could perceive was that there was a threat and that the name of Louis-Philippe had been pronounced; the intention was made clear by the open knife.
>
> This went way beyond the limits of my republican opinions. I yielded to the pressure from my neighbor on the left who, as one of the King's comedians, didn't care to be compromised, and we jumped from the windowsill into the garden.
>
> I went home somewhat worried. It was clear this episode would have its consequences. Indeed, two or three days later, Évariste Galois was arrested.

There are some irritating inaccuracies in Dumas's description (e.g., concerning Galois's age), and I shall return to the issue of the identity of the man who killed Galois later, but the basic facts are undoubtedly correct. The *Gazette des écoles,* which had supported Galois during his resentful exchanges with Guigniault, published its own version of the event in its May 12 issue: "Many toasts have been proposed; it appears that one firebrand, said to be a student, stood up from the table, pulled out of his pocket a dagger, and brandishing it in the air started to say: 'This is how I will be sworn in to Louis-Philippe.' " By brandishing the knife Galois was perceived as making a threat against the king's life. He was arrested the following day at his mother's home, held in preventive detention at the Sainte-Pélagie prison, and brought to trial on June 15, 1831.

The legal proceedings opened with a series of routine questions by the presiding judge, who basically wanted Galois to describe the events at the banquet. Then came the unexpected. The prisoner was asked, "Did you take out a knife . . . and utter 'To Louis-Philippe'?" To everyone's amazement, Galois answered, "I had a knife which I had used to cut meat in the meal. I waved it when saying, 'To Louis-Philippe, *if he betrays us.*' The last words were only heard by people in my immediate vicinity, because of all the whistling that had started . . . by people who understood my words as being a toast to Louis-Philippe's good health." Taken aback, the judge inquired whether Galois truly feared that there was a danger of the king abandoning his duties and betraying his nation. Évariste responded, "All the king's actions, while not showing bad faith yet, permit us to doubt his good faith." The exchange between the judge and Galois continued for a while. Witnesses were called, both for the prosecution and for the defense. The question of whether the meeting was private or public became a central issue. In the latter case, Galois's ambiguous toast could be taken as a provocation intended to incite violence against the king. Galois's own concluding remarks were cut short by the presiding judge, who wisely perceived that the hot-blooded youngster could bring about his own downfall with careless and inflammatory remarks. After deliberations that lasted only half an hour, Galois was acquitted. According to legend, as soon as the verdict had been read, Galois calmly collected his knife from the court's table of exhibits and silently left the courtroom. Given that the transcript shows that during the trial itself Galois claimed that he had lost the knife upon leaving the restaurant, this legend cannot be substantiated. One way or another, the temperamental nineteen-year-old found himself free on the streets again.

On June 15, the same day that Galois's trial began, the newspaper *Le globe* decided to make public the story of Galois's frustrating experience with the academy. An article written most probably by one of the Chevalier brothers, Galois's friends, started by describing Galois's genius and the fact that he had independently discovered the properties of elliptic functions (which had made Abel famous). The text then related Galois's unbelievable tribulations with the mathematical establishment. In particular, the article chronicled the misfortunes of Galois's memoir on the solvability of equations:

> Last year, before March 1st, M. Galois sent a memoir to the secretariat of the Institute of France on the solvability of algebraic equations. This memoir was his entry for the Grand Prix of Mathematics. It overcame certain difficulties that even Lagrange himself had not resolved. M. Cauchy had conferred the highest praise on the author on this subject. But what did that matter? The memoir was lost, the prize was awarded [to Abel and Jacobi], without the young savant being able to participate in the competition. In a reply to a letter to the Academy from young Galois, complaining about the negligent treatment

> of his work, all that M. Cuvier could write was: "The matter is very simple. The memoir was lost on the death of M. Fourier, who had been entrusted with the task of examining it." Now the memoir has been rewritten and presented again to the Institute. M. Poisson, who is to evaluate it, has not yet performed that duty, the result being that for more than five months its wretched author has been waiting for a kind word from the Academy.

Interestingly, assuming that Galois himself provided the Chevaliers with the contents of this article, we learn that Cauchy did express his appreciation of Galois's work, even if he failed to transmit the same enthusiasm to the academy. Perhaps in response to the public criticism of the academy's neglect, Poisson and Lacroix finally presented their verdict on Galois's work. Their report is dated July 4, 1831, and it was presented in the academy's session of July 11. This was a bomb—they did not approve Galois's propositions. In a lukewarm report that demonstrates clearly that Poisson and Lacroix either failed to comprehend or at the very least were prejudiced against Galois's innovative group-theoretical ideas, the rapporteurs write noncommittally:

> We have made all possible efforts to understand M. Galois's proof [of the conditions under which an equation is solvable by a formula]. His reasonings are neither sufficiently clear, nor sufficiently developed for us to be able to judge their exactness, and we are not in a position that enables us to give an opinion in this report. The author states that the proposition that makes the special topic of his memoir is a part of a general theory that could lead to many other applications. Frequently, it happens that different parts of a theory clarify one another, and are easier to grasp collectively, rather than when taken in isolation. One should therefore wait for the author to publish his work in its entirety to form a definitive opinion; but in the state at which the part submitted to the Academy currently is, we cannot recommend to you to give it your approval.

The academy adopted the conclusions of this negative report. Even when we accept the fact that clarity was never Galois's strongest suit, and that the explanations given left quite a bit to be desired, there is no escape from the conclusion that one of the most imaginative breakthroughs in the history of algebra had still to await its acceptance by a conservative audience. Basically, Galois's ideas fell victim to the fact that they were not what Poisson and Lacroix expected. The referees thought that they would find in the manuscript a simple criterion based on the coefficients that would tell them immediately whether or not any particular equation is solvable by a formula. Instead they found a whole new concept—group theory—and conditions based on the putative solutions of the equation. This was just too innovative to be accepted in 1831.

IMPRISONED

The academy's judgment delivered a huge blow to Galois. Nevertheless, convinced of the correctness of his propositions, Galois added under one of Poisson's critical annotations on the manuscript the words: "Let the reader judge" (figure 61 shows the page). Embittered scientifically,

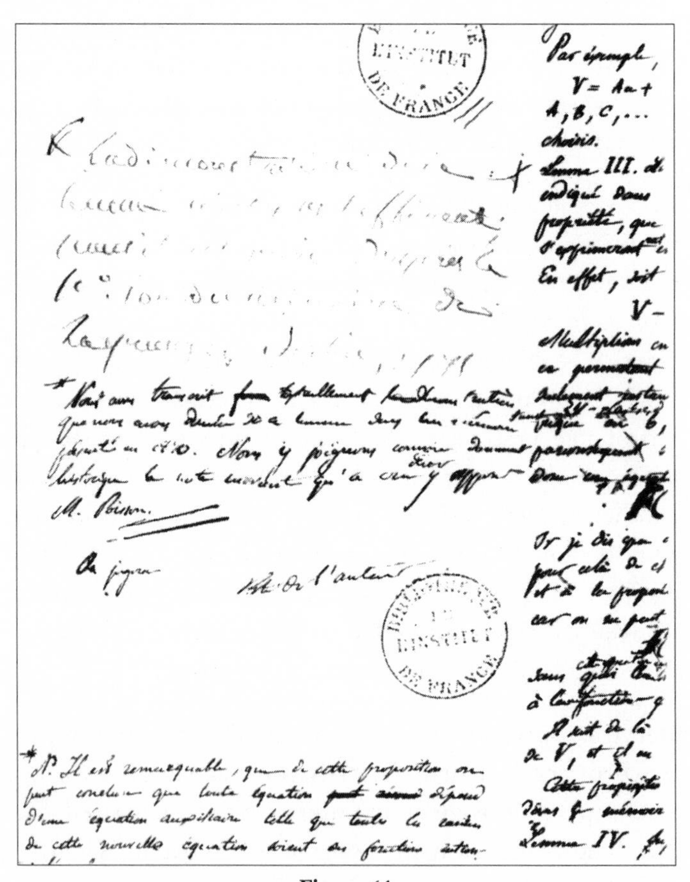

Figure 61

and violently disposed politically, his relationship with his mother also became unpleasantly strained. He therefore left his family's home and rented a room by himself at 16, rue des Bernardins (figure 62).

Troubles never come single file, but in battalions. Bastille Day (July 14) was approaching, and the republicans were making plans for a large demonstration. In particular, they wanted to conduct a provocative ceremony commemorating the planting some forty years earlier of a symbolic tree of freedom at the Place de la Bastille. The police took preventive measures and arrested many known activists during the night between July 13 and July 14. Galois managed to escape imprisonment either by not being on the police's "blacklist" or by not sleeping at his room. Around noon on July 14, however, a group of about six hundred people led by Galois and his friend Ernest Duchatelet, a student at the École des chartes, started to cross the Pont Neuf. Évariste was wearing his (by then illegal) national guard uniform and was armed to his teeth (carrying a few pistols, a loaded rifle, and a dagger). Being prepared for possible subversive gatherings, the police intervened swiftly. Galois and Duchatelet were arrested on the bridge, as were a few other republican leaders at other places. To make things worse, Duchatelet drew a pear symbolizing the king's head on his cell wall (figure 63 shows a caricature attributed to the painter Honoré Daumier in

Figure 62

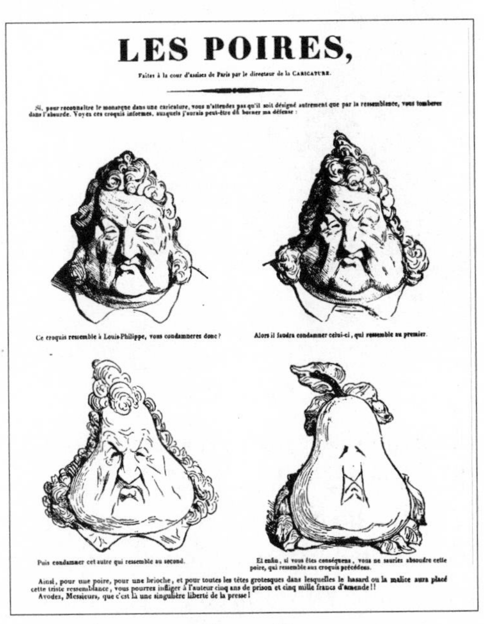

LES POIRES,

Faites à la cour d'assises de Paris par le directeur de la CARICATURE.

Si, pour reconnaître le monarque dans une caricature, vous n'attendez pas qu'il soit désigné autrement que par la ressemblance, vous tomberez dans l'absurde. Voyez ces croquis informes, auxquels j'aurais peut-être dû borner ma défense :

Ce croquis ressemble à Louis-Philippe, vous condamnerez donc?

Alors il faudra condamner celui-ci, qui ressemble au premier.

Puis condamner cet autre qui ressemble au second.

Et enfin, si vous êtes conséquens, vous ne sauriez absoudre cette poire, qui ressemble aux croquis précédens.

Ainsi, pour une poire, pour une brioche, et pour toutes les têtes grotesques dans lesquelles le hasard ou la malice aura placé cette triste ressemblance, vous pourrez infliger à l'auteur cinq ans de prison et cinq mille francs d'amende!!

Avouez, Messieurs, que c'est là une singulière liberté de la presse!

Figure 63

which Louis-Philippe metamorphoses into a pear). The head was drawn next to a guillotine and accompanied by a proclamation popular with the republicans at the time: "Philippe will carry his head to your altar, O Liberty." The trial of Galois and Duchatelet began on October 23, 1831. Since the charge of wearing an illegal uniform could hardly be denied, it was quite clear that Galois would be convicted (carrying weapons was quite common at the time). What came somewhat as a shock was the unreasonably harsh sentence of six months in prison. Duchatelet, who probably had a lesser reputation as a troublemaker, was sentenced to only three months of imprisonment. (I found no evidence whatsoever to support a speculation that Duchatelet's sentence was reduced in exchange for his agreement to collaborate with the police.) Galois appealed, but his sentence was confirmed on December 3. They were both sent to the Saint-Pélagie prison (figure 64) in the Fifth Arrondissement of Paris, not far from Jardin des plantes (the botanical gardens). Among other arrested republicans, the biologist Raspail, himself a prominent leader of the Friends of the People, was particularly provocative during his own trial in January of 1832. He went so far as to declare that the king, who betrayed his own people, "should be buried alive under the ruins of the Tuileries." Needless to say, this statement did not gain him much sympathy with the judges, and he was sentenced to fifteen months at Saint-Pélagie.

Figure 64

Saint-Pélagie was the type of prison you would expect for Paris of that period. A large wall surrounded the entire complex, and the buildings containing the cells enclosed three interior yards. Prisoners were housed according to the category of their crimes, with political prisoners occupying one of the side sections. Galois, who belonged to the lowest class in terms of his financial means, most probably found himself in one of the sixty-bed dormitories. Those who could afford it could pay their way even into private cells, with food brought in from local restaurants. Most of the information on Galois's miserable conditions in jail comes from the writings of three people who cared for the young man: his fellow inmate Raspail, whose *Letters on the Prisons of Paris* was published eight years later; the poet Gérard de Nerval (1808–55), who was arrested in February 1832 and even wrote a poem about the prison; and Galois's loving sister, Nathalie-Théodore, who visited her brother as often as she could and did her best to nourish both his body and his soul. Two dramatic incidents described in Raspail's memoirs are particularly noteworthy. On July 29, as the prisoners were on their third day of commemorating the "Three Glorious Days," a shot fired from the rue du Puits de l'Ermite in front of the prison injured one of the prisoners in Galois's cell. In an ensuing meeting of a delegation of prisoners with the prison's chief warden, Galois, who was a member of the delegation, apparently accused one of the prison guards of being the shooter and further insulted the warden. Consequently, he was thrown into the dungeon, provoking a violent reaction from the prisoners. Raspail quotes a prisoner talking to the warden: "This young Galois doesn't raise his voice, as you well know; he remains as cold as his mathematics when he talks to you." The other prisoners voiced their agreement: "Galois in the dungeon! Oh, the bastards! They have a grudge against our little scholar." Following this exclamation of support, the prisoners took control of the prison and order was restored only the following day. For fear of further riots, Galois was released from the dungeon.

Raspail also gives a rather vague but disturbing description of a suicide attempt by Galois. Apparently young Évariste, who had not been accustomed to heavy drinking, was often teased by his fellow inmates to drink himself into a stupor. "You are a water drinker, young man," they would mock him, "O Zanetto [the nickname given to Galois by the prisoners]! Leave alone the party of the republicans, and return to your mathematics." On one such occasion, the intoxicated young man revealed to Raspail the agony he had experienced since the death of his father: "I have lost my father and no one has ever replaced him." He then added a sentence that would turn out to be chillingly prophetic: "I will die in a duel on the occasion of some coquette of low class." As Raspail and a few other prisoners tried to lay him out on a bed, the blind drunk Galois shouted, "You despise me, you who are my friend! You are right, but I who committed such a crime must kill myself!" Only the rapid intervention of the prisoners prevented Galois from carrying out his deadly intention.

Nerval's description of his last few minutes in prison is equally moving: "It was five o'clock. One of the inmates led me to the gate and kissed me, he promised to come to see me as soon as he gets out of prison. He had still two or three months to serve. This was the unfortunate Galois, whom I did not see again, since he was killed in a duel the morning after he regained his freedom."

However, Galois's sister, Nathalie-Théodore, depicts the most heartbreaking picture of her brother's physical and mental state. After one distressing visit she writes in anguish in her diary: "To endure five more months without a breath of fresh air! This is a very bad perspective, and I fear that his health will suffer much. He is already so tired. He does not allow himself to be distracted by any thought, he has taken a somber character that makes him age before his time. His eyes are hollow as if he is fifty years old."

When not drunk, Galois spent most of his days in prison pacing ceaselessly around the yard, usually deep in thought. The evenings were devoted to noisy republican gatherings and patriotic ceremonies around the tricolor flag. Nevertheless, Galois found time to write a long preface (figure 65 shows the first page) to his outstanding mathematical memoirs. This was really a harsh indictment of the entire scientific establishment and its practices. The preface starts by mocking the hierarchy of scientists and the crippling constraints imposed by the need for support.

> Firstly, you will notice that the second page of this work is not encumbered by surnames, Christian names, titles, honors and the eulogy of some niggardly prince whose purse would have opened at the smoke of incense, threatening to close when the incense holder was empty. Neither will you see in letters three times as high as your head, homage respectfully paid to some high-ranking personality in science, or to some wise patron, a thing thought to be indispensable (I was going to say inevitable) for anyone wishing to write at twenty.

If one were to replace the word "prince" by "funding agency," Galois's points would remain as topical today as they were a hundred and seventy years ago. As one prominent scientist once told me, "In between

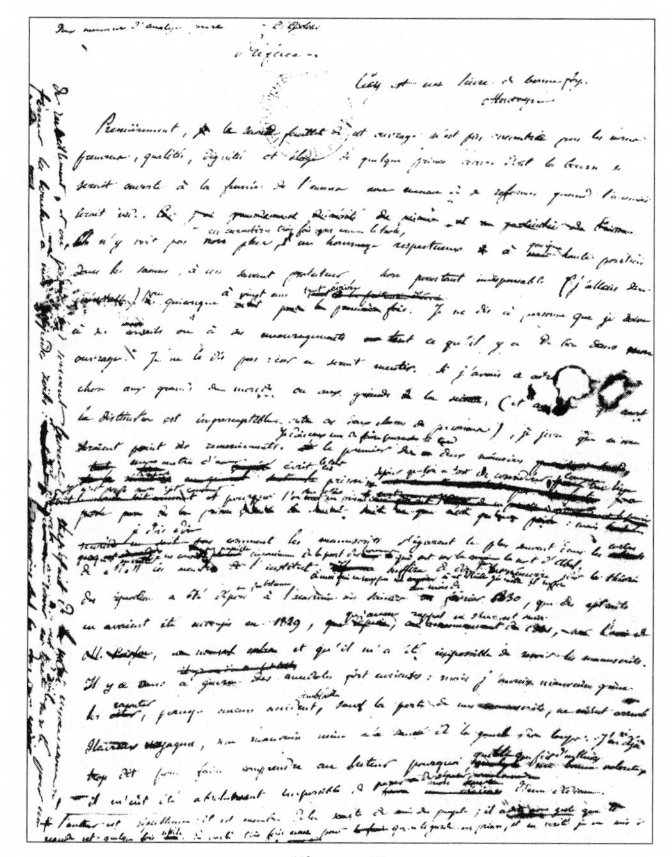

Figure 65

writing grant proposals describing what I *intend* to do, and writing reports about what I *have done,* there is no time left to actually *do* anything!"

Galois's preface ends in a hopeful, if scornful tone: "When competition—that is, selfishness—no longer rules in science, when people associate with one another for study and not in order to send sealed packages to the Academies, they will be eager to publish even small results, as long as these are new, while adding, 'I do not know the rest.' "

A ROMANTIC IN LOVE

In the spring of 1832 a devastating cholera epidemic swept through Europe. Paris was hit particularly hard. The contaminated water of the Seine River claimed a daily toll of approximately a hundred deaths.

Partly perhaps due to his fragile health, but more probably because this was a common practice with political prisoners, Galois was transferred on March 13 from Saint-Pélagie to a convalescent home at 84–86, rue de Lourcine (later number 94 at the current rue Broca), where he was placed on parole. At this home, known then as the Sieur Faultrier "house of health," something dramatic happened: Galois fell in love. Until then, possibly because of the dominating personality of his mother, Galois had had no relations with women. In fact, during one of the bacchanal sprees in prison he confided to Raspail, "I do not like women and it seems to me that I could love only a Tarpeia or a Graccha" (two legendary Roman women; Tarpeia betrayed her city to the Sabines, and Graccha is Cornelia Gracchus, the mother/educator of Tiberius and Gaius). The object of his flaming affection was young Stéphanie Potterin du Motel, who lived in the same building of the convalescent home. Her father, Jean-Paul Louis Auguste Potterin du Motel, was a former officer in the Napoleonic army, and her brother, who was sixteen at the time, later became a medical doctor. The Potterin du Motels maintained a close friendship with the convalescent home's owner.

Few love affairs in history have had more tragic consequences. Stéphanie may have initially shown some interest in the passionate and intelligent young man, but it did not take her long to coldly reject his advances. On the back of one of his already used papers, Galois made copies of two of Stéphanie's letters. These letters unfortunately contain gaps in which words and syllables are missing. Most likely Galois had torn up the originals in a rage. Later he desperately attempted to reconstruct the words of his loved one from the fragments, as hurtful as those words must have been.

The fate of one of the greatest geniuses ever to have lived was about to be sealed by the heart-piercing remarks of an "infamous coquette" who was less than seventeen years old at the time. The first letter, dated May 14, 1832, reads:

> Let's put an end to this, over that matter, please. I don't have enough spirit to continue a correspondence of this type, but I shall try to have enough [spirit] to converse with you, as I used to do before anything happened. So that's that, Monsieur, the . . . has [or: there are] . . . that must . . . you . . . than or: to me and not think anymore about things that could not exist and that never would have existed.

The letter leaves little doubt that the inexperienced and perhaps too hot-blooded Galois did or said something to offend Stéphanie or scare her away. The cold tone suggests that the young woman was probably not that enthusiastic to begin with. The second letter, most likely written a few days later, was even more devastating. Stéphanie was no longer interested even in a mere friendship.

> I followed your advice and I thought it over . . . what has happened . . . , come about between us, whichever name you want to call it. Furthermore, Monsieur, be assured that most probably, there would have never been anything more; you made wrong assumptions and your regrets are groundless. True friendship exists scarcely other than between persons of the same sex, especially . . . friends. . . . moan in the vacuum that . . . the absence of any feeling of this kind . . . my trust . . . but it has been badly hurt. . . . you have seen me sad, [you] have asked [me] the reason, I answered that my feelings had been hurt. I thought that you would take it like any person before whom a word is uttered from these . . . one is not. . . .
>
> The calmness of my thoughts leaves me the freedom to judge the people that I usually see without much reflection; this is the reason that I rarely regret having been mistaken about them or having been influenced in my view of them. I disagree with you about the fee[lings] . . . more than has . . . demand nor . . . [I] thank you sincerely for all those [feelings] where you were willing to take steps towards me.

Galois was devastated. The powerful effects of this affair on his mood and emotional attitude toward life in general can be judged from his letter on May 25 to his good friend Auguste Chevalier. At the time, Auguste, his brother Michel, and three dozen other Saint-Simonians had established a small community in Ménilmontant, east of Paris. Galois writes melancholically:

> My dear friend,
>
> There is pleasure in being sad, if one can hope for consolation. One is happy to suffer if one has friends. Your letter full of apostolic grace has given me a little calm. But how can I remove the trace of such violent emotions as those which I have experienced? How can I console myself when I have exhausted in one month the greatest

> source of happiness a man can have? When I have exhausted it without happiness, without hope, when I am certain I have drained it for life?

He continues with a Cassandrian description of his agonizing internal struggle: "I wish to doubt your cruel prophecy that I shall not do research anymore. But I must admit that there may be some truth in it; to be a scholar one must be only a scholar. My heart rebels against my head. I do not add as you do, 'It is a pity.' " He finishes with a glimmer of hope: "I shall see you on June 1. I hope we shall see each other often during the first fortnight in June. I shall leave around the fifteenth for Dauphine." But the shimmer of light at the end of the tunnel implied by the last paragraph is quickly extinguished by the last phrase of the postscript note: "How can a world which I detest soil me?"

Galois was never to see Auguste again.

We now come to the most intriguing part of the Galois story—his mysterious death. Let me note at the outset that from the purely mathematical point of view, or for the history of group theory and its application to symmetries, it is unimportant why Galois died or who killed him. However, any account of the life of this remarkable genius would be lacking without a discussion of these issues. In particular, there are striking similarities between the lives of the two main characters in the saga of the equation that couldn't be solved—Abel and Galois. They were both initially educated by a parent and inspired by a talented teacher. Both lost their father at a young age and had attempted to solve the same notoriously difficult problems. But this is not all. They were both victims of the same conservative mathematical establishment (Cauchy in particular), miserable (for different reasons) in their love lives, and both died tragically in the flower of their youth. Yet we know almost every minute detail of the circumstances surrounding Abel's death, while Galois's death is veiled in mystery, controversy, and speculation. This—how shall I put it?—lack of symmetry truly bothered me. Consequently, I have made a conscious decision to invest as much time and effort as it would take to investigate every aspect of Galois's life, and in particular his death. I have done my best to leave no stone unturned, have read every document I could put my hands on, and visited most of the relevant places. I can only hope that the results justify the effort.

A MYSTERIOUS DEATH

The known facts concerning Galois's activities between May 25 and that fateful morning of May 30, when he faced his opponent for a duel with pistols, are very few. On May 29, the eve of the duel, he wrote three letters. One was an apologetic address "to all republicans":

> I beg my patriotic friends not to reproach me for dying otherwise than for my country.
>
> I die the victim of an infamous coquette and her two dupes. It is in a miserable piece of gossip that my life is extinguished. Oh! why die for something so little, for something so contemptible!
>
> Heaven is my witness that only constrained and forced have I yielded to a provocation which I have tried to avert by every means. I repent having told a baneful truth to men who were so little able to listen to it calmly. Yet I have told the truth. I take with me to the grave a conscience clear of lies, untainted by patriotic blood.
>
> Adieu! What kept me alive was the public good. Forgive those who kill me, they are of good faith.

The last words, reminiscent of those of Christ on the cross ("forgive them; for they do not know what they are doing"), reflect traces of the religious education he had received from his mother. Otherwise, when taken at face value together with Stéphanie's letters, the picture that emerges from this note seems pretty clear. By words or action, Galois offended the young woman, and her two "dupes" provoked a duel. Galois bears no grudge against the two men who "are in good faith," and he only regrets having been entirely truthful. One feels an element of concession and surrender to authority in Galois's words: "only constrained and forced have I yielded to a provocation." I shall return to this important point later.

Next, Galois wrote a letter addressed to two republican friends, N.L. (almost certainly Napoléon Lebon) and V.D. (almost certainly Vincent Delaunay):

> My good friends,
>
> I have been provoked [to a duel] by two patriots. . . . It is impossible for me to refuse.

> I beg your forgiveness for not having informed either of you. But my adversaries have put me on my honor not to inform any patriot.
>
> Your task is simple: prove that I have fought against my will, that is, after having exhausted all possible means of compromise, and say whether I am capable of lying, even on such a trivial a subject as the one in question.
>
> Remember me, since fate did not give me a long enough life for my country to remember me.
>
> I die your friend.

This depressing letter, again taken at face value, adds one important piece of information: the opponents were "patriots," meaning active republicans. The feeling of Galois yielding to authority is further enhanced: "It is impossible for me to refuse . . . my adversaries have put me on my honor not to inform any patriot . . . I have fought against my will." Galois also vehemently stresses his truthfulness: "say whether I am capable of lying."

The third and most important letter from a scientific perspective contains Galois's mathematical legacy. The very long letter, addressed to his devoted friend Auguste Chevalier, presents a concise summary of the contents of the famous memoir rejected by Poisson and Lacroix, as well as other works:

> My dear friend,
>
> I have made some new discoveries in analysis. The first concerns the theory of equations, the others integral functions.
>
> In the theory of equations, I have investigated under which conditions the equations are solvable by radicals [by a formula]: this has given me the opportunity to make this theory more profound, and to describe all the transformations possible on an equation even when it is not solvable by radicals.
>
> All of this makes for three memoirs.

Galois then outlines what is known today as *Galois theory*, adding a few new theorems to the contents of the original manuscript submitted to the academy. Toward the end he notes, "You know, my dear Auguste, that these subjects are not the only ones that I have explored." Then, giving a brief description of a few more topics, he concludes regretfully, "I have no time and my ideas are not sufficiently developed in that terrain—which is immense."

Finally, just like Abel before him, he puts his faith in the judgment of the German mathematician Jacobi: "Make a public request of Jacobi or Gauss to give their opinion not as to the truth, but as to the importance of these theorems. After that, I hope some men will find it profitable to decipher this mess. I embrace you with effusion." Only one thing remained to be done—introduce some order into the manuscripts themselves. Galois went quickly through his mathematical papers and made some last-minute corrections and commentaries. One of those annotations (figure 66) contains what has become his most memorable and most sorrowful quote: *"Je n'ai pas le temps"*—"I have no time."

The duel took place in the early morning hours of May 30, 1832, near the pond of the Glacier at Gentilly (in the current Thirteenth

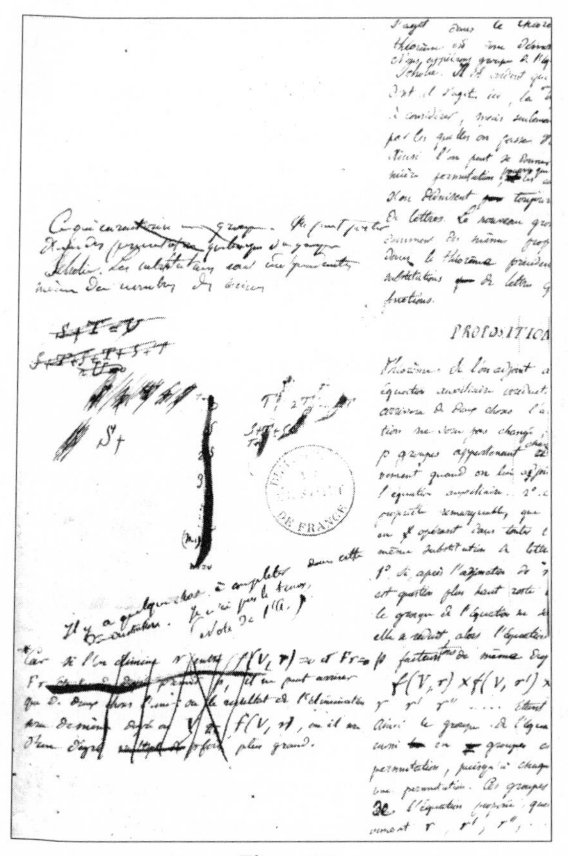

Figure 66

Figure 67

Arrondissement of Paris). The precise circumstances of the drama are not known. According to the autopsy report, Galois was shot in the stomach from the right side. The bullet pierced several parts of the intestines before lodging in his left buttock. What happened next is unclear. Did the witnesses leave the scene? Or was it one of them who took Galois to the hospital? The Cochin Hospital records show that Galois was brought in at 9:30 in the morning (figure 67 shows the entrance and one of the wings of the hospital at the end of the nineteenth century) and was assigned bed number 6 in the Saint-Denis ward. According to a much later testimony by Galois's cousin Gabriel Demante, it was a passing peasant who transported Galois to Cochin, but a note in the *Magasin pittoresque,* written by Pierre Paul Flaugergues, a former classmate of Évariste's, assigns this Samaritan role to a "former officer." Galois's younger brother Alfred, the only member of the family to have been notified, rushed to the hospital. The attending surgeon, Dr. Denis Guerbois, realized immediately that the end was near, as did the two brothers. Still fully conscious, Galois refused the services of a priest. From the tearful Alfred, Galois requested in consolation, "Don't cry, I need all my courage to die at twenty." Évariste Galois passed away at 10:00 a.m. on May 31, and the death certificate was signed on June 1. The death went almost unnoticed. The *Bulletin de Paris* of May 31 notes erroneously: "Death of Legallois."

The Lyon newspaper *Le Précurseur,* which had close ties with the Friends of the People, published in its issue of June 4 and 5 the following account (figure 68):

> Paris, 1 June—A deplorable duel yesterday has deprived the exact sciences of a young man who gave the highest expectations, but whose precocious fame is nevertheless because of his political activities. The young Évariste Galois, condemned for a year because of a toast proposed at the Vendanges de Bourgogne, fought with one of his old friends, a young man like himself, like him a member of the Society of the Friends of the People, and who was known to have figured equally in a political trial. It is said that love was the cause of the combat. The pistol was the weapon chosen by the two adversaries, they found it very hard, because of their old friendship, to have to aim at each other, and they left the decision to blind fate. At point blank range, each of them was armed with a pistol, and fired. Only one of those weapons was loaded. Galois was pierced through and through by the bullet of his adversary; he was transported to the Cochin Hospital, where he died in about two hours. He was 22 years old. L.D., his adversary, is a bit younger still.

As in many instances in which an event we are actually familiar with is being reported in the news media, this narrative is loaded with inaccu-

— Un duel déplorable a enlevé hier aux sciences exactes un jeune homme qui donnait les plus hautes espérances, et dont la célébrité précoce, ne rappelle cependant que des souvenirs politiques. Le jeune Évariste Gallois, condamné il y a un an pour des propos tenus au banquet des Vendanges de Bourgogne, s'est battu avec un de ses anciens amis, tout jeune homme comme lui, comme lui membre de la société des Amis du Peuple, et qui avait, pour dernier rapport avec lui, d'avoir figuré également dans un procès politique. On dit que l'amour a été la cause du combat. Le pistolet étant l'arme choisie par les deux adversaires, ils ont trouvé trop dur pour leur ancienne amitié d'avoir à viser l'un sur l'autre, et ils s'en sont remis à l'aveugle décision du sort. A bout portant, chacun d'eux a été armé d'un pistolet, et a fait feu. Une seule de ces armes était chargée. Gallois a été percé d'outre en outre par la balle de son adversaire; on l'a transporté à l'hôpital Cochin, où il est mort au bout de deux heures. Il était âgé de 22 ans. L. D., son adversaire, est un peu plus jeune encore.

Figure 68

racies. The duel took place on May 30, not the thirty-first; the shooting was not "at point blank range" according to the autopsy report, but from twenty-five paces; Galois did not die in two hours, but the following day; he was not "condemned for a year," but for six months; he was not twenty-two but only twenty years old. We therefore have to take the rest of the information in this article with a grain of salt. This is particularly true when we realize that this report appeared in Lyon, far from the capital. Nevertheless, if we were to take seriously the detailed description of the opponent, who would that fit? The answer is easy: Duchatelet. He was indeed a bit younger than Galois, had been arrested with him on Pont Neuf, and had been tried just before him. But Duchatelet's first name was Ernest, yet the article gave the initials "L.D."

There are a few more pieces of evidence that we should consider: First, Galois's cousin, Gabriel Demante, wrote to Galois's first biographer, Paul Dupuy, that during Galois's last meeting with Stéphanie, Évariste found himself in the presence of "a so-called uncle and a so-called fiancé," each of whom provoked the duel. Galois himself consistently talked of two men (both in his letter "to all republicans" and in his letter to his friends). Any attempt to uncover the truth should therefore identify the two opponents, not just one.

Second, recall that the writer Alexandre Dumas, when describing the events surrounding Galois's disastrous toast in his memoirs, named Pescheux d'Herbinville as Galois's killer. While one would normally not consider "D" to be the initial of d'Herbinville, the practices and style of spelling of the nineteenth century allowed for such liberties. For instance, Stéphanie's last name is sometimes spelled du Motel, at other times Dumotel. Even the family name on Évariste's maternal side changed from de Mante to Demante (appendix 8). Pescheux d'Herbinville was never in a trial with Galois but was in the "trial of the nineteen."

Lastly, the police chief Henri-Joseph Gisquet (1792–1866) wrote in his memoirs in 1840 that Galois "had been killed by a friend."

So what does all of this add up to?

CONSPIRACY THEORIES GALORE

Quite a few of Galois's biographers concluded that Galois had been killed by political enemies. Some of these allowed their imaginative plots to include even more intrigue and assumed that the "infamous coquette" was in fact a prostitute or a mysterious police agent who acted as a provocateur. This is not surprising. Alfred Galois himself remained convinced throughout his entire life that his brother was the victim of the king's secret police. But is there any convincing evidence for such conspiracy theories? Not really. Most of these fanciful descriptions were created *before* the unambiguous identification of the "infamous coquette" as Stéphanie du Motel. The "forensic" investigation that revealed Stéphanie's identity was carried out by an unlikely detective—a Uruguayan priest. Carlos Alberto Infantozzi of the University of Montevideo simply wouldn't give up. First, he used a magnifying glass and special lighting to uncover Stéphanie's name and signature from underneath Galois's erasures on a few of his manuscripts. Then he painstakingly sifted through the archives to discover her father's name, Jean Louis Auguste Potterin du Motel, and the family's address at the Faultrier convalescent home. There is little doubt that Stéphanie was neither a prostitute nor a police agent. She eventually married Oscar Théodore Barrieu, a language professor, on January 11, 1840. Stéphanie's father was not a medical doctor, as some biographers inferred from Infantozzi, but a former officer in the Napoleonic army and an inspector in the prison system. He had passed away by the time his daughter married. Stéphanie's brother, Eugene P. Potterin du Motel, did eventually become a doctor, but was only sixteen at the time of Stéphanie's "affair" with Galois. Researcher Jean-Paul Auffray, who performed what is probably the most extensive investigation of documents related to Galois, uncovered another interesting fact. Denis Louis Grégoire Faultrier, after whom the convalescent home was named, had himself been a former captain in the national guard. After the death of Stéphanie's father, this intimate friend of the Potterin du Motel family married Stéphanie's mother. As we shall soon see, this may provide a crucial piece in the puzzle.

So, you may wonder, why did Alfred Galois insist on his brother having been murdered by the police? One has to remember that Alfred, eighteen at the time, had an infinite admiration for his older sibling. To him, the entire concept of his genius, brave, but otherwise sickly and shortsighted brother being involved in a duel must have seemed so unfair that foul play had to be involved. Galois's first biographer,

Dupuy, whose extensive article was published in 1896, concluded back then that in all of Alfred's assertions (including a totally unfounded claim that Galois shot first into the air) "one feels a romantic invention." Physicist and author Tony Rothman, currently at Bryn Mawr College, reached a similar conclusion. After a thorough examination of many biographies in 1982 (and subsequent work), he concluded that "the tales of Bell, Hoyle and Infeld [all Galois biographers] are baroque, if not byzantine, inventions." I fully agree.

There is one other conspiracy theory, however, that needs to be considered seriously. In one of the most recent and most extensive Galois biographies, the Italian mathematician and historian of mathematics Laura Toti Rigatelli proposes that the famous duel was in fact not a real duel at all. Rather, Toti Rigatelli concluded that the depressed and disillusioned Galois decided to sacrifice himself for the republican cause. The republicans needed a corpse to stir up rebellion and he offered his—the duel was entirely staged. Toti Rigatelli's deduction was based on wide-ranging research, and in particular on an examination of the writings of the prefect of police Gisquet and of one of his undercover spies, Lucien de la Hodde.

While Toti Rigatelli's theory is certainly intriguing, I personally do not find it particularly convincing. For her story to hold, Toti Rigatelli is forced to claim that Galois fabricated his last three letters to "prevent anyone from suspecting the true circumstance of his death." This would be not only totally out of character for Galois, who always clung to the truth as he saw it, but even inconsistent with the conspiracy theory itself. Surely to instigate a revolution, a letter blaming the police for his death would have been much more effective. A closer examination of Toti Rigatelli's scenario reveals that what she regards as the strongest piece of evidence for Galois having sacrificed himself is, in her words, his "insistence on certain death" in his letters "to all republicans" and to Lebon and Delaunay. But what else could one expect from adieu letters written by a twenty-year-old crushed romantic the evening before a duel? Furthermore, as I shall soon argue, there are reasons to believe that at least one of Galois's opponents was much more experienced with the pistol than the young mathematician. Galois's expectation of certain death was therefore fully understandable. Who then killed Galois, and why was he killed?

THE DEATH OF A ROMANTIC

The accumulated evidence leaves very little doubt as to the reality of the duel. The clues also indicate that this was a classic case of *cherchez la femme.* Either by some careless words or by too impetuous a behavior, Galois somehow offended the young lady, who immediately informed two other men. When these "dupes" confronted Galois, he made the further mistake of referring to the whole affair as "a miserable piece of gossip," thus adding insult to injury. The consequences were disastrous. Quick to defend Stéphanie's honor, the two men challenged Galois to a duel. Who were these two men? From Galois's own letter we know that they were both republican "patriots." Galois's language also strongly suggests that at least one of his opponents had some position of authority, to which Galois felt compelled to yield. Both Stéphanie's father, Jean Louis Potterin du Motel, a former officer in Napoleon's Grand Army, and the convalescent home's owner, Denis Faultrier, a former captain in the national guard, fit the profile. Note, however, that the latter would also be consistent with another piece of evidence. Galois's cousin described one of the opponents as a "so-called uncle." Faultrier, the close family friend who later married Stéphanie's mother, fits this description like a glove. As to the identity of the second adversary, the situation is somewhat less clear. In his recent, well-researched Galois biography, Auffray suggests that the two men were in fact Stéphanie's father and Faultrier. This ignores both the cousin's testimony (as to a "so-called fiancé") and the description in *Le Précurseur,* which I find difficult to accept. While the article in *Le Précurseur* contains many inaccuracies, those are of the type expected from such reports. The combination of Gabriel Demante's description of a "so-called fiancé," together with the newspaper account, seems to add up to a presumed young lover. But who?

Ernest Armand Duchatelet, a young student at École des chartes and Galois's friend, fits the bill best. Recall that the prefect Gisquet also testified that Galois "had been killed by a friend." I have to admit that I was unable to find any documented evidence for Duchatelet having spent any time at the Faultrier convalescent home—he was released from prison months before Galois's transfer there. However, since political prisoners were customarily placed on parole in such "houses of health," Duchatelet might have been there prior to Galois's arrival. Furthermore,

Galois was allowed visitors at Faultrier's and indeed his friend Auguste Chevalier came to see him there. It would not be an incredible stretch to assume that Duchatelet came, too. Finally, the reluctance of the two friends (described in the newspaper) to aim at each other, and their decision to rather leave the determination of who would die to blind fate by loading only one pistol, is fully consistent with their characters (see also notes).

Could the opponent have been Pescheux d'Herbinville? Not very likely. He does not fit the description in the newspaper; had very little opportunity, if any, to meet Stéphanie (being from a very different social circle); and may even have been a homosexual (as is insinuated in Dumas's description of him). Then why on earth did Dumas name him? I do not know, but Dumas has been known to be wrong with such details on many occasions. His confusing one young republican with another would not be astonishing.

I humbly propose, therefore, that Galois's two opponents were Duchatelet and Faultrier. Is the almost two-hundred-year-old mystery of who killed Galois and why finally solved? Maybe. While I strongly believe that the Faultrier-Duchatelet duo is consistent with all the known facts, solid information is so seriously lacking that unless new evidence surfaces in the future, many uncertainties will remain.

Assuming that my conclusion as to the identity of the two adversaries is correct, the picture that emerges for the events on the day of the duel is the following: On the morning of May 30, 1832, Galois and Duchatelet faced each other at twenty-five paces, with Faultrier waiting for his turn. By a Russian-roulette-style procedure, Duchatelet happened to pick the loaded gun and shot Galois.

The autopsy report reveals two additional interesting pieces of information. First, while Galois was hit from the side, he did not stand fully sideways, in the way that would have minimized his chances of being hit. Did he not care to live? Given his state of mind this is not impossible. After all, from Galois's bleak point of view his life story could be summarized more or less as follows: two failed attempts to enter the École polytechnique; three memoirs rejected by the academy; two imprisonments; and a heart broken by unrequited love. In fact, shortly before he died, Galois drew himself as Riquet à la Houppe (figure 69), a fictional hunchback dwarf who was very clever and chivalrous but was

Figure 69

mocked by everyone around him. In the seventeenth-century tale, Riquet cured a young woman of her stupidity and eventually won her love, becoming the symbol of a Beauty and the Beast–type transformation. Sadly, Galois was less lucky in real life. Second, the autopsy report describes a large bruise on Galois's head that was probably caused when he fell. If knocked unconscious and presumed dead, this might explain a fact that puzzled many of the Galois biographers—most (if not all) of those present at the duel left the scene. The potential identification of Faultrier as one of the opponents solves another mystery that intrigued many researchers—why didn't one of the witnesses take Galois to the hospital? In the proposed scenario, Faultrier, the "former officer," might indeed have been the one to transport Galois to Cochin. A hint as to Galois's ever-present memories of his father may be provided by the following curiosity: When asked at the hospital for his address, Galois gave 6, rue Saint-Jean-de-Beauvais, the Paris address at which Nicolas-Gabriel had committed suicide.

POSTHUMOUS FAME

Galois's funeral took place on Saturday, June 2. It was attended by thousands of friends, members of the Friends of the People, and delegations of students from the schools of law and medicine. The leaders of the Friends of the People, Plagniol and Charles Pinel, delivered passionate eulogies. If the republicans had any plans to use the funeral to provoke a riot, these were rapidly dissipated by an unexpected turn of events. The police prefect Gisquet, who had arrested about thirty republicans as a preventive measure the previous evening, was keeping a close watch on the procession. He writes in his memoirs:

> On June 2, the republicans attended, in numbers of two to three thousands, the funeral procession of Legallois [misspelling Galois], with the intention of starting the barricades at the time of their return; but they learned of the death of General Lamarque [a famous general in Napoleon's army] and immediately realized the advantage they could take of such an event and the crowd that the funeral of the general would attract. Their plan was therefore modified: it was the coffin of a general of the Empire, of a patriot deputy, that would give the signal for the rebellion. The movement was therefore postponed till the 5th.

Fate thus robbed Galois even of the opportunity to incite a rebellion in death. The devastated Auguste Chevalier wrote a brief obituary that appeared in September 1832.

Fortunately, the gods were more generous with Galois's mathematical legacy. Two tenacious young men, Galois's brother Alfred and his friend Auguste Chevalier, took it upon themselves to ensure that Évariste's memory and his mathematical papers would be saved from oblivion (figure 70 shows a portrait of Galois, drawn from memory by Alfred in 1848). Painstakingly, they collected every piece of paper, catalogued all the manuscripts, and delivered their precious treasure to mathematician Joseph Liouville

Figure 70

(1809–82). The latter, overwhelmed with admiration, started his address to the Academy of Sciences in 1843 with, "I hope to interest the Academy in announcing that among the papers of Évariste Galois I have found a solution, as exact as it is profound, of this beautiful theorem: Given an irreducible equation of prime degree, to decide whether or not it is soluble by radicals." Liouville published the memoirs in his journal in 1846, announcing to the world, "I recognized the full exactness of the method by which Galois proves, in particular, this beautiful theorem [about the solvability of equations]." More recognition was soon to follow. Jacobi, in whom Galois had placed his confidence, proved true to the task. Having read Galois's papers in *Liouville's Journal,* he immediately contacted Alfred in an attempt to find out more on Galois's work on transcendental functions. By 1856, Galois theory was introduced into advanced courses of algebra in France and Germany.

The school that had expelled Galois also finally had a change of heart. On the occasion of its centenary celebrations, the École normale asked the famous Norwegian mathematician Sophus Lie (1842–99) to write an article that would summarize the impact of Galois theory on the history of mathematics. Lie concluded, "It is particularly characteristic of mathematics that two of the most profound discoveries that have ever been made (the theorem of Abel and the theory of algebraic equations of Galois) were the work of two geometers of whom one, Abel, was about twenty-two years old, and the other, Galois, had not reached twenty." When the great mathematician Émile Picard (1856–1941) evaluated in 1897 the mathematical achievements of the nineteenth century, he had this to say about Galois: "No one surpasses him in the originality and the profoundness of his conceptions."

The École normale came full circle when on June 13, 1909, its director, Jules Tannery, came to Bourg-la-Reine to deliver a special presentation on the occasion of the placement of a commemorative plaque on Galois's home (figure 71 shows a letter from Tannery to the Bourg-la-Reine mayor, and figure 72 shows the plaque on the original home). Hardly able to control his emotions, Tannery finished with a moving mea culpa:

> I owe the honor of giving a speech here to the position which I hold at the École normale. I thank you, Mr. Mayor, for allowing me to make an apology to the genius of Galois in the name of this school to

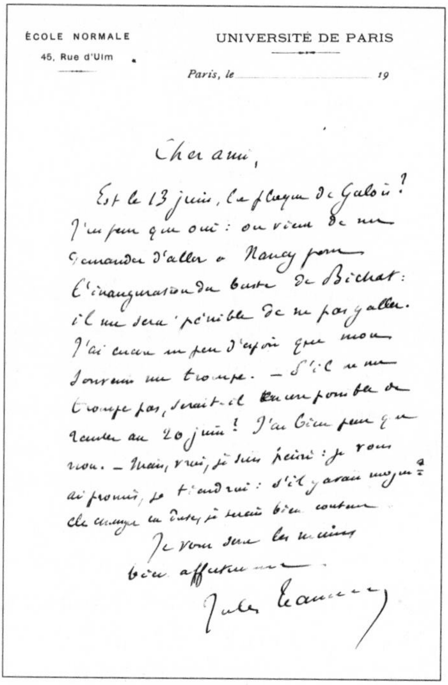

ÉCOLE NORMALE
45, Rue d'Ulm

UNIVERSITÉ DE PARIS

Paris, le 19

Cher ami,

Est-ce le 13 juin, la plaque de Galois? Je le pense que oui: on vient de me demander d'aller à Nancy pour l'inauguration du buste de Bichat: il me sera pénible de ne pas y aller. J'ai encore un peu d'espoir que mon souvenir me trompe. — S'il ne me trompe pas, serait-il bien impossible de remettre au 20 juin? J'ai bien peur que non. — Mais, vrai, je suis peiné: je vous ai promis, je tiendrai: s'il y avait moyen de changer la date, je serais bien content.

Je vous serre les mains bien affectueusement.

Jules Tannery

Figure 71

which he entered reluctantly, where he was misunderstood, which expelled him, but for which he was, after all, one of the brightest glories.

These sincere words echoed in my ears as I stood in the Bourg-la-Reine cemetery, where the memories of Nicolas-Gabriel and Évariste Galois are as inseparable today (figure 73) as the father and son were during Évariste's short life.

Figure 72

But how could a tool invented for finding whether certain equations can be solved, no matter how ingenious, evolve into a language describing all the symmetries of the world? After all, when we discuss symmetries, algebraic equations are not the first things that spring to mind. Galois himself was not sure where his theory was going to lead: "It would be possible to understand fully the general thesis which I put forth only when someone who has an application for it reads my work carefully." This is precisely where the unifying magic of group theory appears—that "grandeur of thought from such slight beginnings" that the British mathematician H. F. Baker was raving about. To fully appreciate the incredible encompassing power of the concept started by Galois, we shall now return to the realm of groups and symmetries.

Figure 73

– SIX –

Groups

Galois took algebra and turned it on its ear. If you want to know whether an equation is solvable or not, you simply try to solve it, right? Wrong, said Galois. All you need to do is to examine permutations of the putative solutions. How can permutations of solutions we don't even know tell us anything about solvability? The fact that permutations may provide at least some new information had long been known in the nonmathematical world. Anagrams—words or phrases formed by the letters of others in different order—do just that. Take the name GALOIS, for instance. Allowing for two-word anagrams leads us to such combinations as OIL GAS, GOAL IS, GO SAIL, and so on. How many different arrangements (disregarding meaning) of the letters in GALOIS can we construct? The answer is not difficult, but we could start with an even simpler case to uncover the general rule. The letters A and B allow for two arrangements: AB and BA. Three letters, A, B, C, can form six permutations: ABC, ACB, BAC, BCA, CAB, CBA. The pattern that emerges is simple. With A, B, C, there are three locations where the A can be placed (first, second, third). For each one of the three choices made for A, there are precisely two places left for the letter B (e.g., if A is second, B can be either first or third), and only one place remains for the C. The total number of arrangements is therefore $3 \times 2 \times 1 = 6$. The same logic applies to any number of objects. For the six letters in GALOIS, there are therefore $6 \times 5 \times 4 \times 3 \times 2 \times 1 = 720$ different arrangements, and for any number n of different objects there are $n \times (n - 1) \times (n - 2) \times (n - 3) \ldots \times 1$ permutations. To save space, the French mathematician Christian Kramp (1760–1826) introduced the notation $n!$ (*n factorial*) to denote this last product. The number of permutations of n different objects is therefore precisely $n!$

One of the earliest recorded studies of permutations occurs not in a math book, but rather in a book of Jewish mysticism that dates back to sometime between the third and sixth centuries. The Sefer yetzira (*Book of Creation*) is a short, enigmatic book that proposes to solve the mystery of creation by examining combinations of the letters of the Hebrew alphabet. The general premise of the book (which is attributed by kabbalistic legend to the Jewish forefather Abraham) is that different categories of letters form divine building blocks from which all things can be constructed. In this spirit, the book states, "Two letters build two words, three build six words, four build 24 words, five build 120, six build 720, seven build 5,040."

To see how uncovering the relations among different permutations and their properties can lead to new and deeper insights, examine the operation that permutes GALOIS into AGLISO. This operation is represented by (in the notation introduced in Chapter 2):

$$\begin{pmatrix} G & A & L & O & I & S \\ A & G & L & I & S & O \end{pmatrix}$$

where each letter in the upper row is replaced by the letter directly underneath it. Specifically, G is replaced by A, A by G, L stays the same, O by I, I by S, and S by O.

What happens if we apply the same operation twice? You can easily check that performing precisely the same substitutions a second time turns the AGLISO into GALSOI. Imagine now that starting with GALOIS, a computer that has gone haywire repeats the same operation, say, 1,327 times. Can we predict the final outcome? Of course, we could find the result the hard way, by applying the operation again and again, but this is extremely tedious and surely prone to many mistakes. Is there an easier way to find the answer? You may want to spend a few minutes thinking about this problem, since its deciphering reveals interesting properties of permutations that are in the spirit of Galois's proof. In any case, I will give the solution presently.

On the recreational mathematics side, permutations and their characteristics featured prominently in at least two very famous puzzles—the 14–15 puzzle and Rubik's Cube.

The 14–15 puzzle was introduced in the 1870s by America's greatest puzzlist, Samuel Loyd (1841–1911), and for a while it drove the entire world crazy. At the time, Loyd was already the foremost composer of

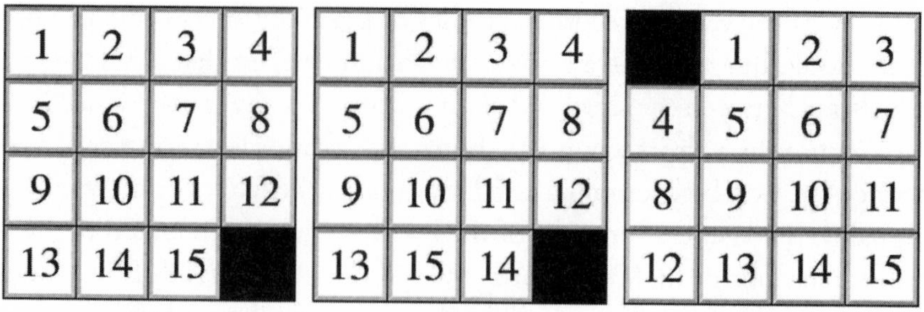

1	2	3	4
5	6	7	8
9	10	11	12
13	14	15	

(a)

1	2	3	4
5	6	7	8
9	10	11	12
13	15	14	

(b)

	1	2	3
4	5	6	7
8	9	10	11
12	13	14	15

(c)

Figure 74

chess problems in the United States, as well as a chess columnist in several magazines. Even before the celebrated 14–15 puzzle, however, he began publishing a wide variety of other types of mathematical puzzles.

The 14–15 puzzle consists of a grid of 4 × 4 tiles numbered 1–15 (figure 74a). The general goal was to slide the tiles up, down, or sideways and rearrange them in serial order from any initial configuration. The particular version of the 14–15 puzzle that caused all the commotion was one in which all the numbers were in regular order with the exception of the 14 and 15, which were reversed (as in figure 74b). Loyd offered a prize of one thousand dollars to the first person who could present a series of slides that would lead to the swapping of only the 14 and the 15. The puzzle created an unprecedented mania and fascinated people from all walks of life. Loyd's son, who later published a fascinating collection of his father's mind-boggling riddles (called *Cyclopedia of Puzzles*), wrote in his description of the universal infatuation that "farmers are known to have deserted their plows" to struggle with the stubborn puzzle. In reality, Loyd knew very well that he had risked absolutely nothing by offering the prize—he could prove that the puzzle couldn't be solved. To understand the crux of Loyd's proof, consider, for example, the following permutation:

$$\begin{pmatrix} 1 & 2 & 3 & 4 & 5 & 6 & 7 & 8 & 9 & 10 & 11 & 12 & 13 & 14 & 15 \\ 1 & 2 & 3 & 4 & 6 & 7 & 8 & 12 & 5 & 10 & 11 & 15 & 9 & 13 & 14 \end{pmatrix}$$

You can easily discover that this permutation is achievable from Loyd's 1–15 grid, if originally arranged in serial order (as in figure 74a). Even if you don't have Loyd's grid at hand, by mentally tracing the following sequence of moves (where every number represents the one to be slid to the blank space)—15, 14, 13, 9, 5, 6, 7, 8, 12, 15—you will find that it produces the desired permutation above. Let us count up how many pairs of numbers in this permutation are out of their natural order. For instance, in the natural order, 6 comes after 5, but in this permutation the order of 6 and 5 is reversed. We can take each digit in the second row in turn and count up the number of reversals:

1	contributes no reversals	0 reversals
2	contributes no reversals	0 reversals
3	contributes no reversals	0 reversals
4	contributes no reversals	0 reversals
6	is followed by 5	1 reversal
7	is followed by 5	1 reversal
8	is followed by 5	1 reversal
12	is followed by 5, 10, 11, 9	4 reversals
5	contributes no reversals	0 reversals
10	is followed by 9	1 reversal
11	is followed by 9	1 reversal
15	is followed by 9, 13, 14	3 reversals
9	contributes no reversals	0 reversals
13	contributes no reversals	0 reversals
14	contributes no reversals	0 reversals
	Total number of reversals	12

The total number, 12, is even, so this particular permutation is called an *even permutation.* Similarly, when the number of reversals is odd, we speak of an *odd permutation.* A little experimentation will convince you that, by design, the permutations that can be achieved with Loyd's toy are always even, as long as you start with the natural order and end up leaving the bottom right-hand corner empty. Since reversing just the one pair of number 14 and 15 results in an odd permutation (1 reversal), no matter how hard you try, you can never recover the natural order. Loyd was assured that he would never have to pay the offered prize.

If the 14–15 puzzle somehow caught your fancy and you happen to be in possession of Loyd's toy, you may want to try the following: From the initial configuration with the 14 and 15 reversed (Figure 74b), can you reach the natural order if the vacant square in the final configuration is at the upper left-hand corner (as in figure 74c)? The answer is presented in appendix 9.

About a century after the appearance of Loyd's puzzle, the Hungarian architect Ernö Rubik came up with an even more sophisticated, and hugely popular, device. Rubik's Cube (figure 75) consists of a 3 × 3 × 3 matrix of smaller cubes. The faces of the small cubes are painted in different colors, and the faces of the large cube are pivoted such that they can be rotated in different directions. The objective of the puzzle is to produce a configuration in which each face of the large cube is composed of a single color. Rubik invented the cube in 1974, and by 1980 it became an international sensation. For about three years the Rubik craze swept the world. From little children at school to CEOs in fancy offices, everybody was trying to solve the cube, and to do it in ever-shorter time. In honor of the inventor, on June 5, 1982, Budapest hosted the first world championship for the fastest cube solver. Nineteen national contests that had been held earlier produced champions who came to Budapest. The winner, Minh Thai of the United States, accomplished the task in an astonishing 22.95 seconds, even though the cubes used in the competition were new, and therefore slower in their rotations than their "broken in" counterparts. Still shorter times have been recorded since. At the time of this writing, Jess Bonde of Denmark has registered the shortest time achieved in an official championship—16.53 seconds! Even if one were to exclude the innumerable imitations of Rubik's Cube, a staggering number of more than 200 million cubes have been sold to date worldwide.

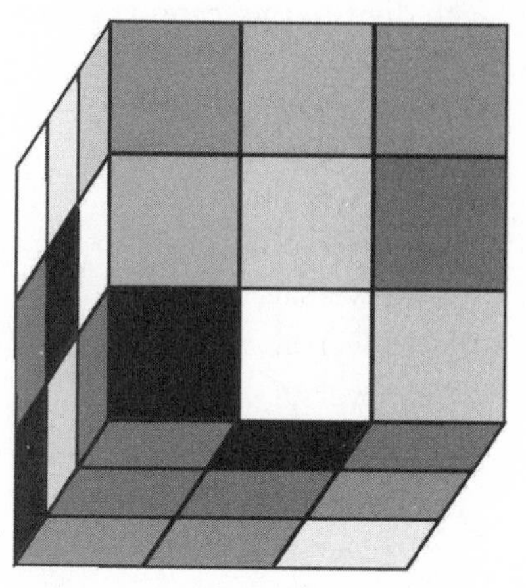

Figure 75

Since there are no fewer than 43,252,003,274,489,856,000 different patterns that the cube can exhibit, you can imagine that nobody has actually tried them all to solve it. Rather, each move of Rubik's Cube may be represented as a permutation of its vertices. Indeed, the solution to the cube's puzzle can be cast entirely in the language of group theory. Mathematician David Joyner of the U.S. Naval Academy has even schematized a complete course in group theory around Rubik's Cube and similar mathematical toys.

Returning now to the GALOIS-AGLISO puzzle presented at the

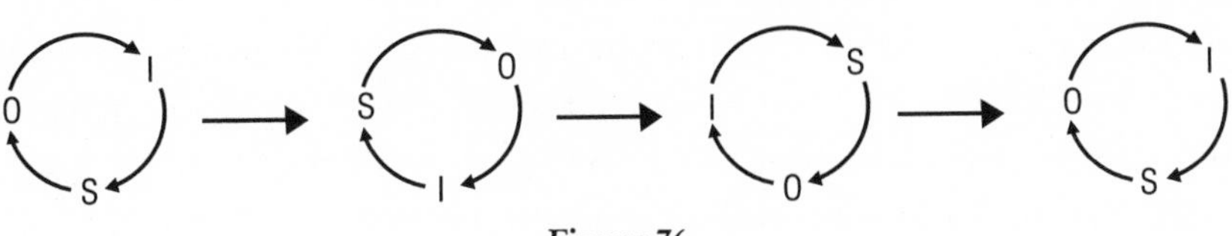

Figure 76

beginning of this section, how can we find what permutation would be obtained after 1,327 applications of the same transformation? First, note that the operation leaves the letter L unchanged, in the third position. Second, we discover that the letters O, I, S are permuted in such a way that the effect is to move them "round in a circle" (as in figure 76). This is similar to a basketball practice routine in which the players form a line and after shooting to the basket each player returns to the back of the line. Permutations of this type are called *cyclic permutations.* An important property of cyclic permutations is that they return to the original order after a fixed number of applications called the *period.* Figure 76 shows that the cyclic permutation of O, I, S is of *period 3*—the order OIS is recovered after three steps. The last point to notice about the GALOIS-AGLISO operation is that the letters G and A are transposed, returning to their original order after every two operations. If we put all of these pieces of information together, we discover an easy way to crack this problem. Since O, I, S return to their initial order every three steps and G, A every two steps (and L remains unchanged), we recover the original word GALOIS every $3 \times 2 = 6$ steps (you can check this by repeating the substitutions six times). The number 1,327 is equal to $6 \times 221 + 1$. This means that after the 1,326th ($= 6 \times 221$) step, the letters spell GALOIS, and then the one extra step simply changes that to AGLISO—the final word. There is an important lesson to be learned here: *The analysis of the properties of the permutation allowed us to predict with confidence the final outcome without actually having to perform the experiment.* This was the basic philosophy behind Galois's theory as well. He discovered an ingenious way to determine whether an equation is solvable from an examination of the symmetry properties of permutations of its solutions.

Just as two consecutive shuffles of a deck of cards produce nothing more than a different shuffle, performing one permutation followed by another results in yet a third permutation. Consequently, permutations obey the closure requirement of groups automatically. Recall that *clo-*

sure means that combining two group members by the group operation yields another group member. For instance, the set of all the positive numbers (integers, fractions, and irrational numbers) forms a group under the operation of ordinary multiplication. In particular, the requirement of closure is satisfied, because the product of any two positive numbers is also a positive number. The identification of permutations as crucial mathematical objects worthy of study thus set Galois on the road to formulating group theory.

GROUPS AND PERMUTATIONS

Permutations and groups are intimately related. In fact, the group concept was born out of the study of permutations. For Galois, this was only the first step in a series of ingenious inventions and ideas that paved the way to his brilliant proof.

Let me provide a brief reminder of the precise definition of a group introduced in chapter 2. A *group* consists of members that have to obey four rules with respect to the group operation. As an example, take the collection of all the possible deformations that can be performed on a piece of Play-Doh, with the operation being defined as "followed by." The rules are as follows. First, the combination of any two members by the group operation has to produce another member (this property is called *closure*). Obviously, a deformation of the Play-Doh followed by a second deformation simply generates another deformation. Second, the operation has to be *associative,* meaning that when three ordered members are combined, the result does not depend on which two are combined first. Successive transformations such as the Play-Doh deformations satisfy this rule automatically. Third, the group must contain a "status quo" or *identity element,* that when combined with any other member, it leaves that member unchanged. For the Play-Doh, the deformation "do nothing" plays this role. Finally, for every member of the group, there must be an "as you were" or *inverse element,* such that when a member is combined with its inverse the combination yields the identity. For every Play-Doh deformation, there exists a counterdeformation that restores the original shape.

Examine now the collection of all possible permutations of the three numbers 1, 2, 3:

$$\begin{pmatrix}123\\123\end{pmatrix}\quad\begin{pmatrix}123\\231\end{pmatrix}\quad\begin{pmatrix}123\\312\end{pmatrix}\quad\begin{pmatrix}123\\132\end{pmatrix}\quad\begin{pmatrix}123\\321\end{pmatrix}\quad\begin{pmatrix}123\\213\end{pmatrix}$$
$$I \qquad s_1 \qquad s_2 \qquad t_1 \qquad t_2 \qquad t_3$$

Here, in order to be able to refer to them, I have labeled each one of the different operations. The identity, which takes each number into itself, is denoted by *I*. Each of the operations t_1, t_2, and t_3 *transposes* or interchanges two of the numbers while leaving the third one intact. The two operations s_1 and s_2 are both *cyclic* permutations, moving the numbers round in a circle.

Observe now what happens when we apply two permutation operations successively. Recall that what is important is which number replaces which, and not the order in which they are written. Take, for instance, t_1 followed by s_1. The operation t_1 takes 1 into itself and then s_1 changes the 1 into 2. The net result is therefore the transformation $1 \rightarrow 2$. At the same time, t_1 replaces 2 by 3 and then s_1 replaces the 3 by 1, producing the net outcome: $2 \rightarrow 1$. Finally, 3 is transformed into 2 by the operation t_1 and then back into 3 by the operation s_1. We find that t_1 followed by s_1 gives the permutation:

$$\begin{pmatrix}123\\213\end{pmatrix}$$

which is precisely the operation t_3. In other words, if the symbol $\circ$ denotes the operation "followed by," we found that $s_1 \circ t_1 = t_3$ (recall that the operation applied first is always to the right).

The complete "multiplication table" for the six permutations takes the form:

$\circ$	I	s_1	s_2	t_1	t_2	t_3
I	I	s_1	s_2	t_1	t_2	t_3
s_1	s_1	s_2	I	t_3	t_1	t_2
s_2	s_2	I	s_1	t_2	t_3	t_1
t_1	t_1	t_2	t_3	I	s_1	s_2
t_2	t_2	t_3	t_1	s_2	I	s_1
t_3	t_3	t_1	t_2	s_1	s_2	I

where the entry in, say, row s_2 and column t_3 gives the outcome of $s_2 \circ t_3$, which is t_1. At first glance this table may look like a mess, but a closer inspection reveals an important truth: *The collection of all the permutations of three objects forms a group.* In fact, this statement is true for the permutations of any number of objects. The table demonstrates both closure (combining any two permutations of three objects gives another permutation of three objects) and the fact that every permutation has an inverse—one that "undoes" the effect of the first. In this case, you can check that s_1 and s_2 are each other's inverses—applying one after the other restores the original order ($s_1 \circ s_2 = I$; $s_2 \circ s_1 = I$). Similarly, each of the operations t_1, t_2, t_3 is its own inverse. That is, applying any one of them twice restores the status quo ($t_1 \circ t_1 = I$; $t_2 \circ t_2 = I$; $t_3 \circ t_3 = I$). The group of all the $n!$ permutations of n different objects is commonly denoted by S_n. The number of members in a group is called the *order* of the group. The order of the group of permutations of three objects, S_3, for instance, is 6, because there are precisely six such permutations.

Why should we care whether permutations form groups or not? Not only because historically these were the objects that gave birth to the group concept in the first place, but also because these particular groups are in some sense at center stage in group theory.

To demonstrate the special role of groups of permutations, inspect again the symmetries of the equilateral triangle. Recall that there were six such symmetries leaving the triangle unchanged, corresponding to the identity, rotation through 120 degrees, rotation through 240 degrees, and reflection about three axes (see figure 9, page 15). In chapter 2 we discovered that the set of symmetries of any object forms a group. Since the group of symmetries of the triangle has precisely the same number of members as the group of permutations of three objects—both are of order 6—it makes sense to wonder whether these two groups are somehow related. But what does a counter-clockwise rotation of the triangle by 120 degrees actually do (figure 77)? It simply takes vertex A and moves it from position 1 to position 2. At the same time, it moves vertex B from position 2 to position 3 and vertex C from position 3 to position 1. In other words, we can look at this rotation

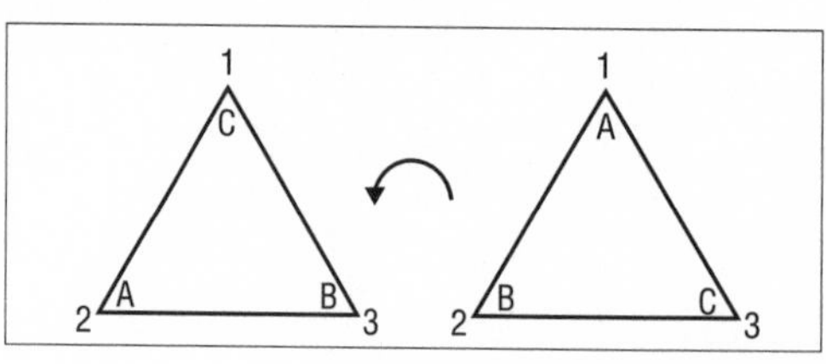

Figure 77

as nothing but a permutation of the positions 1, 2, 3 with respect to the vertices of the rotating triangle:

$$\begin{pmatrix} 123 \\ 231 \end{pmatrix}$$

Similarly, each one of the remaining five symmetries of the triangle corresponds to one of the other permutations—*the structure of the two groups is identical!* This establishes an unexpected and intimate link between symmetries and permutations, through the theory of groups. This realization forms the basis for an important theorem proven in 1878 by the English mathematician Arthur Cayley (1821–95). In simple language, the theorem states a very remarkable fact: *Every group is cast in the same mold as a group of permutations.* That is, in spite of the immense latitude that the definition of groups allows, there is always a group of permutations that for all practical purposes is identical to any group. In the mathematical jargon, two groups that have the same structure or the same "multiplication table," such as the group of permutations of three objects and the group of symmetries of the equilateral triangle, are called *isomorphic.* To give another example, recall from chapter 2 that the group of symmetries of the human figure contains two members—the identity and reflection about a vertical plane (the latter representing bilateral symmetry). The "multiplication table" for this group under the operation "followed by" (where I and r denote the identity and reflection, respectively) takes the form (because applying the reflection twice restores the original figure):

$\circ$	I	r
I	I	r
r	r	I

Examine now the simple group composed of the two numbers 1 and −1 with the operation of ordinary multiplication. The multiplication table (this time literally) for this group is:

×	1	−1
1	1	−1
−1	−1	1

If you inspect the two tables you discover immediately that they have precisely the same structure once you make the correspondence $I \leftrightarrow 1$, $r \leftrightarrow -1$. The group of symmetries of the human body is *isomorphic* to this two-member multiplication group.

In addition to the fundamental concept of groups of permutations, there was another clever mathematical tool that Galois needed to enable him to embark on the proof that the general quintic (and any higher degree equation) cannot be solved by a formula. This was the idea of a *subgroup*. Like some splinters of political parties or organizations that occasionally become parties themselves, certain subsets of the members of a group may by themselves satisfy all the four requirements of being a group (closure, associativity, identity, inverse). In that case the subset is said to form a subgroup. For instance, the two permutations I and t_3 on page 165 form a subgroup of the group of permutations of three objects S_3, because $I \circ t_3 = t_3$, and $t_3 \circ t_3 = I$ (see table on page 165), implying closure and meaning that both t_3 and I are their own inverses. If we divide the order (number of members) of the parent group (6 in the case of S_3) by the order of the subgroup (2 for the above subgroup) we obtain the *composition factor.* In the above example the composition factor is $6 \div 2$ or 3. The fact that this turned out to be a whole number is not an accident. An important theorem due to Lagrange ensures that this will always be the case: The order of a finite subgroup always evenly divides the order of its finite parent group. You will never find that a group of order 12 has a subgroup of order 5, 7, or 8; it may have subgroups of order 2, 3, 4, or 6.

Galois was now in possession of all the tools he felt he needed for the proof, but a hugely imaginative leap was still required to put all of these elements together to create a coherent picture. Mathematical history was about to be made.

GALOIS'S BRILLIANT PROOF

In a famous Sidney Harris cartoon, two scientists are seen next to a blackboard covered with equations. One points to the phrase "THEN A MIRACLE OCCURS," which is written between two complex equations, and the caption reads: "I think you should be more explicit here in step two." Galois's insight was nothing short of a miracle. In the history of science, even great discoveries can usually be traced to something that was "in the air" at the time. These are ideas whose time has come. Most physicists would agree, for instance, that had Einstein not suggested his theory of special relativity, from which the famous equation $E = mc^2$ emerged, someone else would have sooner or later come up with the same idea. One notable exception, where there was almost nothing "in the air"—where the same grand vision would have probably materialized only much, much later—is Einstein's theory of general relativity. This is the notion that the force of gravity merely reflects the geometry of space and time. Massive bodies warp spacetime around them just as a heavy bowling ball causes a trampoline to sag. In their motion around the Sun, planets follow curved orbits not because of some unfathomable attraction, but because of this warping. This idea represented such a revolution in the perception of the very fabric of the universe that the famous American physicist Richard Feynman (1918–88) once said, "I still can't see how he thought of it." Even today, ninety years after the first paper on general relativity, Einstein's intuition is still astonishing (I shall return to general relativity in Chapter 7).

Many mathematicians are similarly awed when they think about Galois. Joseph Rotman of the University of Illinois told me, "Galois's invention of groups was a stroke of genius. After all, the great mathematician Abel, who worked on the problem of solvability by radicals at the same time, did not come up with group theory. Indeed, only Cauchy, on his return to France in the 1840s, seemed to appreciate Galois's achievements, and Cauchy's intense group theoretical studies led to the use of group theory in other fields of mathematics." Algebraist Peter Neumann of Oxford University added, "Galois had an extraordinary insight towards the understanding of groups in their own right, but equally extraordinary was his understanding of how they could be used in the theory of equations—ultimately creating what we now call Galois theory (which, after all, is the modern theory of equations)."

So, how did Galois prove his inventive propositions? Even just the essence of Galois's proof is somewhat technical, but it provides such a unique window into his unsurpassed creativity that it is definitely worth the effort required to penetrate it. Following the logical steps of the proof is like having walked through the labyrinth of Mozart's mind while he composed one of his symphonies.

The proof contains three crucial ingredients, all marked by originality and imagination. Galois started by showing that every equation has its own "symmetry profile"—a group of permutations (now called the Galois group) that represents the symmetry properties of the equation. The importance of this step cannot be overemphasized. Before Galois, equations were always classified only by their degree: quadratic, cubic, quintic, and so on. Galois discovered that symmetry was a more important characteristic. Classifying equations by their degree is analogous to grouping the wooden building blocks in a toy box according to their sizes. Galois's classification by symmetry properties is equivalent to the realization that the shape of the blocks—round, square, or triangular—is more fundamental. Specifically, the Galois group of an equation is the largest group of permutations of the putative solutions that leaves the values of certain combinations of these solutions unchanged. For instance, take the group of permutations of two objects. This group is composed of two members—the identity and the operation that interchanges the two objects. Now examine the quadratic equation. We can denote its two putative solutions by x_1 and x_2. Clearly, the combination that is the sum of the two solutions, $x_1 + x_2$, remains unchanged under the operation of both members of the group of permutations of two objects. The identity leaves x_1 and x_2 intact, and exchanging x_1 and x_2 simply transforms $x_1 + x_2$ into $x_2 + x_1$, which has the same value. For equations of degree *n*, we know from Gauss's fundamental theorem of algebra that they have *n* solutions. The maximum number of possible permutations of *n* solutions is $n!$, and the group containing all of these permutations is the group we previously called S_n. Galois was able to prove that for any degree *n*, one can always find equations for which the Galois group is actually the full S_n. In other words, he showed that at any degree, there are equations that possess the *maximum symmetry possible*. There are quintic equations, for instance, for which the Galois group is S_5.

The second ingredient in Galois's proof was yet another innovation. Having already introduced the concept of a subgroup, Galois now gave that concept an additional twist by defining a *normal subgroup*. Take for example the group of six permutations of three objects, S_3. You can easily check that a subset composed of the three operations I, s_1, s_2 (see page 165) forms a subgroup of S_3. Closure is guaranteed by the fact (see multiplication table on page 165) that $s_1 \circ s_1 = s_2$, $s_2 \circ s_2 = s_1$, and s_1 and s_2 are each other's inverses ($s_1 \circ s_2 = I$). Let us denote this subgroup of three members by T. Now suppose that we take any member of T, such as s_1, and we "multiply" it from the left by a member of the parent group S_3, say t_1, and from the right by the inverse of that same member (which happens to be also t_1, because t_1 is its own inverse). That is, we construct the sequence of operations $t_1 \circ s_1 \circ t_1$. Using the "multiplication table" on page 165 we discover that $s_1 \circ t_1 = t_3$, and that $t_1 \circ t_3 = s_2$. In other words, $t_1 \circ s_1 \circ t_1 = s_2$, and s_2 is in itself a member of the subgroup T. If any member of a subgroup satisfies this property (that multiplying it from the left by a member of the parent group and from the right by the inverse gives a member of the subgroup), then the subgroup is called a *normal subgroup*. You can easily verify that T is indeed a normal subgroup of S_3. In fact, T is the *maximal* (of the highest order) normal subgroup of S_3. In general, if a group has normal subgroups at all (other than the group itself), one of these would be the largest. In turn, this maximal subgroup may have as offspring normal subgroups of its own. One of those would again be of the highest order. In this way, an entire genealogy of maximal normal subgroups can be traced. We can use the family tree of these subgroups to create a sequence of *composition factors* (order of the parent group divided by that of the maximal normal subgroup). In the case of S_3 and T the composition factor is $6 \div 3 = 2$. The only normal subgroup that T has is the simplest, in fact trivial, group—the one composed of the identity I alone. This group is of order 1. Therefore, the composition factor between T and its normal subgroup is $3 \div 1 = 3$. The hierarchy of generations of groups S_3, T, and the one composed of I alone therefore give us the sequence of composition factors 2, 3.

Nowhere did Galois's genius shine brighter than in the third step of his proof. Here he put to use all of those creations of his imagination. The question that even the great Abel had left open—What does it take for an equation to be solvable by a formula?—was about to be answered. Galois showed that to enjoy the luxury of a formulaic solution, equations must have a Galois group of a very particular type. Specifically, Galois called a group *solvable* if every single one of the composition factors generated by its descendant maximal normal subgroups was a prime number (divisible only by 1 and itself). He then was able to fully justify the use of the name "solvable" by proving that *the condition for an equation to be solvable by a formula is that its Galois group should be solvable.* Essentially, Galois showed that when the Galois group of an equation is solvable, the process of the solution of the equation can be broken up into simpler steps, each involving only the solution of equations of lower degree.

How is the theorem used in practice? In the case of the general cubic, for instance, the equation is most symmetric when its Galois group is S_3 (the group of all the permutations of the three solutions). S_3, however, is definitely solvable—as we have just seen, both composition factors 2 and 3 are prime numbers. Consequently, the general cubic *is* solvable by a formula, as indeed dal Ferro, Tartaglia, and Cardano had shown. For the general quintic, on the other hand, Galois started in a similar way, by first demonstrating that there are equations for which the Galois group is the group of permutations of 5 solutions, S_5. Here, however, came the punch line. Galois proved that S_5 as a group is not solvable (one of the composition factors turns out to be 60, which is not a prime number). The quintic, therefore, has the wrong sort of Galois group. This completed the proof that the general quintic equation (and similarly, any general equation of a higher degree) is not solvable by a formula. One of the most intriguing problems in the history of mathematics was finally put to rest once and for all. To accomplish this Herculean task, however, Galois had not only to come up with brilliant ideas, but also to invent an entirely new branch of mathematics and to identify symmetry as the source of the most essential properties of equations.

The definitive word on the insolvability of the quintic by a formula may sound at first as a disappointing result, but to what treasures this "disappointment" has led. The biblical story of King Saul comes to mind. When the donkeys of Kish, Saul's father, had strayed, Kish told his son, "Take one of the boys with you; go and look for the donkeys." This search for the lost donkeys led Saul to the prophet Samuel, who anointed the young man to be the first king of Israel. Galois's search for a solution to the quintic produced the "supreme art of mathematical abstraction"—group theory.

THE DATING GAME

Even though not invented with that grand purpose in mind, group theory turned out to be the "official" language of all symmetries. The prominent role that permutations play in group theory may appear at first glance to be somewhat surprising. After all, while we are all fully aware of symmetries, permutations do not strike us as being as conspicuous in our everyday lives. Permutations do show up, however, even if stealthily, and sometimes in the most unexpected places.

Consider the all-important problem of finding a partner for marriage. Moving on from one chance encounter to the next, everyone is searching for a true soul mate. But how does the searcher know when she or he is the one? Can it be (as in the movies) that when you see this one particular person you immediately know that there is no one else for you in the entire world? Or, to use the words of one of the characters in the movie *Serendipity,* when should you stop looking for Mr./Ms. Right and be happy with "Mr./Ms. Good-Enough-for-Right-Now"? To transform this life-changing problem into one that is more tractable, it helps to make a few simplifying assumptions. Suppose that the average female or male meets during the appropriate period in life four people who could be considered potential spouses (I shall discuss situations involving a different number of candidates later). Assume further that had the partner seeker been able to examine all four candidates, he or she would have been able to rank them from the worst (denoted by 1) to the most suitable (denoted by 4), with no two ranking precisely the same. Chance usually does not allow for the luxury of seeing all potential partners at once. Furthermore, social etiquette coupled with common decency normally prohibit one from going back to a previously rejected candidate. Rather, life's flow carries men and women through a series of meetings occurring in random order. Consequently, for four potential partners, each one of the following 4! = 24 permutations of the order of the meetings has the same probability:

1234	2134	3124	4123
1243	2143	3142	4132
1324	2314	3214	4213
1342	2341	3241	4231
1423	2413	3412	4312
1432	2431	3421	4321

The sequence 3142, for instance, means meeting the second-best candidate first, the worst candidate second, the best candidate third, and the second-to-worst candidate last. Waiting for Mr./Ms. Right to show up as the last candidate in this case would certainly not have produced the most desirable result. Indeed, too protracted a process may result in diminishing returns. So, what are the poor young (and not-so-young) people supposed to do? Or, more specifically, how can the spouse hunters maximize their chances for finding the best partner?

The first thing to realize is that a general strategy for addressing precisely such (clearly simplified) problems does exist. If the number of potential partners is 4, the idea is to choose a number, call it k, between 1 and 4. Then, after having met and scrutinized $k - 1$ potential partners, to choose the first one that is better than all of those previously examined (or, if none is, to choose the last one). For instance, if $k = 2$, the idea would be to look carefully at the first candidate ($k - 1 = 1$) and then choose the first potential partner that is better than the one already tested (recall that the assumption is that one cannot return to a previous potential partner). The rationale behind this strategy is obvious—on one hand it takes full advantage of the information that has already been gathered, and on the other, of the fact that the future is unknown. The general strategy does not tell you, however, which value to choose for k. To decide that, we have to find out which value of k gives the highest probability for choosing the best candidate (number 4). For $k = 1$ ($k - 1 = 0$), for instance, the first candidate ends up being chosen. For this selection to be the best, the seeker relies on the six permutations in the order of the encounters in which 4 appears first: 4321, 4312, 4231, 4213, 4132, 4123. Clearly, the probability of hitting one of these six permutations out of the existing twenty-four possibilities is one out of four. This is easy to understand—the partner seeker has not yet met any of the candidates, and there is one chance out of four to find the best one in the first rendezvous. The same is true for $k = 4$. In this case ($k - 1 = 3$), one is betting on the chance of the fourth and last candidate being better than any of the previous three. This corresponds to the six permutations 3214, 3124, 2314, 2134, 1324, 1234, in the order of the meetings, and the chances of hitting those are again one out of four. For $k = 3$ ($k - 1 = 2$), the spouse searcher meets two of the potential partners and then chooses the first one that comes subsequently and is better than both. The permutations that will result in the best choice (number 4) are in this case: 3241, 3214, 3142, 3124, 2341, 2314, 2143, 1342, 1324, 1243. For instance, if the order of encounters was 3241, the seeker meets first candidates 3 and 2 and then, since number 4 is better than either of those two, number 4 would be the one chosen. When the order is 3214, the third candidate (number 1) is not better than the first two, and therefore the search continues and leads to number 4. The list above (for $k = 3$) shows that in this case there are ten permutations that produce the best choice. The chances of success are therefore $10 \div 24$ or about 42 percent. Finally, for $k = 2$ ($k - 1 = 1$), the choice is for the first candidate that is better than the first one seen. You can check that the permutations resulting in "netting" number 4 in this case are: 3421, 3412, 3241, 3214, 3142, 3124, 2431, 2413, 2143, 1432, 1423. For instance, when the order is 3412, the second candidate is already better than the first, so this candidate would be selected. On the other hand, when the order is 3214, the spouse seeker rejects the second and third candidates, since they are not better than the first, and has to await the last potential partner to find one that is better. Since $k = 2$ yields the desired result in eleven of twenty-four occasions, or a probability of success of about 46 percent, this is the best strategy to adopt. A similar calculation shows that $k = 3$ gives the highest chances if the number of potential partners is 5, 6, 7, or 8. If the number of potential spouses is 9 or 10, you would maximize your chances with $k = 4$.

Life is, of course, much more complex than this oversimplified model, especially when it comes to affairs of the heart. The choice of a partner is too serious a matter to be reduced to a mere examination of permutations. Nonetheless, it remains true that permutations may pop up where least expected. The general strategy outlined above, by the way, could be applied to many other (especially less critical) circumstances, from choosing a used car to picking a family dentist. If the number of potential options is very large (say, larger than thirty), one can prove mathematically that the "37 percent rule" produces the best chances of success. That is, examine 37 percent of the potential cars, restaurants, or family physicians and then choose the first one that is better than anything you've seen before. (In case the more mathematically inclined readers wonder where such a strange number as 37 percent came from, it is approximately equal to $1 \div e$, where e is the base of the natural logarithms.)

SHAKEN, NOT STIRRED

Finding the love of your life through mathematics is not the only process where permutations are thrust into the limelight. Lotteries commonly provide such situations, and nowhere more dramatically so than in the 1970 Vietnam era draft lottery.

On November 26, 1969, President Richard Nixon signed an executive order that instructed the Selective Service to establish a random selection sequence for induction. The executive order stipulated that the lottery would be based on birth dates, but it provided no specific instructions on the precise method of drawing the dates.

This was not the first time in history when a draft was supposed to be based on some sort of lottery. The biblical story of the judge Gideon is particularly intriguing. God first told Gideon, "The troops with you are too many to give the Midianites into your hand. Israel would only take the credit away from me, saying, 'My own hand has delivered me.' Now therefore proclaim this in the hearing of the troops, 'Whoever is fearful and trembling, let him return home.' " Thus Gideon sifted them out; twenty-two thousand returned, and ten thousand remained.

In the second "lottery" stage, God imposed on Gideon a second selection criterion. Gideon was asked to take his troops down to the water to drink. He was then told to choose only those three hundred who lapped the water, "putting their hands to their mouths," and to release all the rest "who knelt down to drink water" directly, "as a dog laps." Clearly, Gideon's "lottery" was far from a random selection—all the possible permutations of the pool of candidates were not treated equally. Many interpretations have been suggested for the peculiar choice of the second criterion. The simplest is that the entire scheme was used merely to select a small number of people, in order to amplify the impression of the miraculous victory. More elaborate explanations relate the kneeling to practices used in the worship of other gods, or the use of the hands (rather than drinking directly from the stream) to a demonstration of being considerate and not greedy.

Oddly enough, even though conducted thousands of years later, the 1970 draft lottery also suffered from problems with randomization.

The procedure itself was simple enough. The officials tucked scraps of paper with the 366 dates of the year (including February 29) into capsules. One by one, these capsules were drawn from a bowl on December 1, 1969. Every man born between 1944 and 1948 was assigned a draft number corresponding to the order in which his birth date was picked. For instance, the first date drawn was September 14, and all men with that birthday were assigned no. 1. Men born on June 8 (the last capsule drawn) were assigned no. 366. Clearly, every drawing represents a permutation of the 366 dates. The smaller his draft number, the higher the chances that a man would actually be drafted. The table below shows the average lottery numbers by month obtained in the 1970 lottery.

MONTH	AVERAGE NUMBER	MONTH	AVERAGE NUMBER
January	201	July	182
February	203	August	174
March	226	September	157
April	204	October	183
May	208	November	149
June	196	December	122

One does not have to be a trained statistician to detect a clear trend in these numbers. While the average numbers for the months January–May stay fairly constant, there is a marked and almost steady decline in the numbers corresponding to the months June–December. November and December, in particular, had considerably lower average numbers than those of January–May. The consequences were disturbing—men born late in the year had a significantly higher likelihood of being drafted into a difficult war.

Under true randomization, each one of the possible orderings of the dates has an equal probability of one in 366! (the number of possible permutations), and you would expect the average number for the different months to be roughly the same, around 183 or 184. Instead, the data show that each of the first six months had an average number above this, while each of the last six months had an average number below it. Statisticians were able to demonstrate that the probability for a pattern such as the one exhibited in the table to occur in a truly random selection process was less than 1 in 50,000. How did this happen?

The description of the procedure that set the lottery up provides some important clues:

> The men counted out 31 capsules and inserted in them slips of paper with the January dates. The January capsules were then placed in a large, square wooden box and pushed to one side with a cardboard divider, leaving part of the box empty. The 29 February capsules

> were then poured into the empty portion of the box, counted again and then scraped with the divider into the January capsules. Thus, according to Captain Pascoe [chief of public information for the Selective Service System], the January and February capsules were thoroughly mixed. The same process was followed with each subsequent month, counting the capsules into the empty side of the box and then pushing them with the divider into the capsules of the previous month. Thus, the January capsules were mixed with other capsules 11 times, the February capsules 10 times and so on, with the November capsules intermingled with others only twice and the December ones only once. . . . In public view, the capsules were poured from the black box into the two-foot-deep bowl. Once in the bowl, the capsules were not stirred. . . . The persons who drew the capsules . . . generally picked ones from the top, although once in a while they would reach their hand to the middle or to the bottom of the bowl.

This detailed account leaves little doubt that it was the insufficient mixing, which resulted from placing the capsules month by month, that was the culprit in the ensuing nonrandom risk. Following some public criticism, the procedure was corrected in the 1971 draft lottery. To be sure, perfect randomization may be more difficult to achieve in practice than one might suspect. Take, for instance, what is considered to be the fairest thing people have come up with for deciding randomly between two difficult choices—tossing a coin. The probability of getting heads or tails is equal, right? Not quite. A recent study by statisticians Persi Diaconis and Susan Holmes of Stanford University and Richard Montgomery of the University of California Santa Cruz shows that due to imperfect tossing (which even results sometimes in the coin not flipping at all) a coin is more likely to land on the same face it started out on. The bias is not huge—a coin will land the same way it started about 51 percent of the time—yet it shows that even such simple things cannot be taken for granted. No one is better equipped to examine whether something is random or not than Diaconis. He is the statistician who demonstrated that it takes the average card player no fewer than seven shuffles to create a random order in a deck of cards. He is also known for exposing and debunking various "psychic" phenomena. Diaconis's extensive experiments with coins show that you should never make any important decision based on a penny spinning on the table on its edge. Because of the extra weight on the head side, such pennies land as tails much more frequently than as heads.

As important as they are, however, permutations by themselves are far from being the whole story when it comes to group theory—groups lead into much vaster expanses of abstraction. In particular, if two apparently distinct problems are characterized by groups that are isomorphic to each other (have the same structure), this is a heavy hint that the two problems may be more closely connected than you might have suspected.

THE SUPREME ART OF ABSTRACTION

In a book entitled *The Natural History of Commerce* which appeared in 1870, John Yeats writes, "No amount of abstract reasoning would have led us to discover the properties and uses of iron." He was probably right. Yet abstraction is precisely what has given mathematical structures their portability. They can be carried over from one discipline to the next and from one conceptual environment to another.

Cayley's theorem—that every group, irrespective of its members or the operation between them, is essentially a carbon copy of (is isomorphic to) a group of permutations—set the stage for the understanding of groups as abstract entities. Cayley's own work, and subsequent innovative developments by the mathematicians Camille Jordan, Felix Klein, Walter von Dyck, and others, showed that one could start with essentially any group and then literally strip it of most of its specifics until nothing was left but the bare essentials. That undraped skeleton is sufficient for capturing the structure and all the important properties of groups. An analogy that inevitably springs to mind is that with the twentieth-century art school of minimalism. There too, the goal of artists such as Carl Andre, Donald Judd, Robert Morris, and others was to focus attention on the most fundamental and to reduce visual form to its utmost simplicity. Essentially by design, the appreciation of minimal art, and indeed of mathematics, has always been primarily intellectual, and therefore learned, rather than intuitive.

Starting with groups of permutations (the only groups known at the time), Cayley took a giant leap and formulated his first intuitions about the abstract group concept as early as 1854. As in Galois's case, however, his original ideas were so far ahead of their time that they attracted no attention. As historian and math-education critic Morris Kline once put it, "Premature abstraction falls on deaf ears, whether they belong to mathematicians or to students." Intellectually, therefore, Cayley (figure 78) could be regarded as one of Galois's most direct successors. Cayley's life, on the other hand, stands out in stark contrast to that of the ill-fated French romantic. Arthur Cayley's teachers at King's College in London immediately recognized his unusual mathematical abilities. As he continued his education in Cambridge, the head examiner of the university ranked Cayley as being "above the first." The young man lived up to his teachers' expectations; even before the age of twenty-five, Cayley already had two dozen mathematical papers to his name. His overall prolificacy is matched only by Cauchy and Euler. Unlike the violent ups and downs (mostly downs!) in Galois's life, Cayley's life ran smoothly and successfully. After his insightful, if relatively unnoticed, papers of 1854, Cayley turned his attention to other important mathematical topics, but he returned to groups with a bang in 1878. In a series of four seminal papers he managed to move group theory toward the very center of mathematical investigation. Indeed, following Cayley's work, it took only four more years for abstract, axiomatic definitions of groups to emerge.

Figure 78

Mathematics scholar James R. Newman writes in his monumental compilation *The World of Mathematics* that "the Theory of Groups is a branch of mathematics in which one does something to something and then compares the result with the result obtained from doing the same thing to something else, or something else to the same thing." This baffling statement could hardly be taken as an acceptable definition in a dictionary, yet it captures the level of abstraction that has become group theory's hallmark. Let me use a few nonmathematical examples to explain the concept.

The same joke can be rephrased differently for distinct contexts and circumstances. A physicist wanting to express disparaging contempt for someone's intellectual abilities (not that they ever do that . . .) might say, "He is so dense, light bends around him." One who grew up in the Internet generation might use, for the same purpose, the image "I don't think his URL allows outside access." A tax consultant could say, "If brains were taxed, he'd get a refund," and a chemist might choose the words "He's got an IQ of about room temperature." Similarly, the riddle with the powers of seven that appeared, centuries apart, in the Ahmes Papyrus, in Fibonacci's book, and in the *Mother Goose* nursery rhymes (chapter 3) was essentially the same, even though the wording was different. Finally, one could argue that many fairy tales, such as "Snow White" and "Cinderella," are really the same story in different packaging: an evil stepmother tortures a princess-to-be until a handsome prince rescues the damsel in distress. Groups allow for a similar abstraction. An identical group structure can describe what appear to be rather disparate concepts. I will demonstrate this unifying power of groups with a few relatively simple examples.

Start with four operations that can be performed on any pair of jeans. X will denote the operation "turn the trousers back to front" (while you're not wearing them!). Y stands for "turn the trousers inside out." Z represents "turn the trousers back to front and inside out," and I will denote the identity, meaning do nothing. The composition of two operations (denoted by "$\circ$") is simply achieved through "followed by." You can easily check that these operations form a group. In particular, each of the operations is its own inverse: $X \circ X = I$ (turning back to front twice restores the original position); $Y \circ Y = I$ (turning inside out twice brings the trousers back to as they were), and the combination of any two of the operations gives the third. For instance, $Z \circ Y = X$, since "turning inside out" followed by "turning back to front and inside out" results simply in "turning back to front." The "multiplication table" for this group therefore takes the form (recall that the entry corresponding to row X and column Y is $X \circ Y$, meaning Y followed by X):

$\circ$	I	X	Y	Z
I	I	X	Y	Z
X	X	I	Z	Y
Y	Y	Z	I	X
Z	Z	Y	X	I

Consider next an interesting operation commonly denoted by the symbol Δ that can combine (in a certain fashion) any two sets of objects. For instance, if set A is composed of cats that have at least some black patches in their fur and set B of cats that have at least some white fur, then $A \Delta B$ yields the set of cats that have either black or white patches in their fur, but not both. Graphically, if A and B are symbolically represented by the areas of the circles in figure 79, then $A \Delta B$ corresponds to the shaded area—Δ joins the sets together excluding the overlap. Take now the following four simple sets. Set X has only one object in it: a chicken. Set Y also contains just one object: a cow. Set Z is composed of two objects: a cow and a chicken. Set I is the empty set, which has no object whatsoever in it (this set's function is similar to the role of zero in ordinary addition). Now use the operation Δ to combine any two of these sets. For instance, $X \Delta Z = Y$, because the set of objects that are in X or Z but not in both is the set containing a cow. Similarly, $Y \circ I = Y$, because there is a cow in Y which is clearly not in the empty set I. The set I therefore plays the role of the identity. Each of X, Y, Z is its own inverse, because the set of objects that are not in both X and X is clearly empty: $X \circ X = I$. You can easily check that the sets X, Y, Z, I combined by the operation Δ form a group, the table of which is:

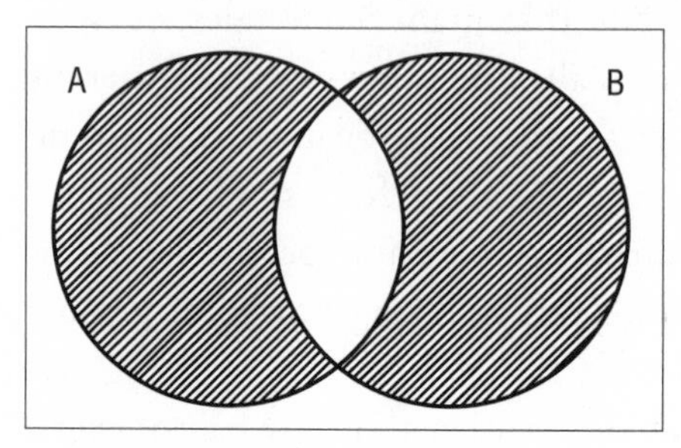

Figure 79

Δ	I	X	Y	Z
I	I	X	Y	Z
X	X	I	Z	Y
Y	Y	Z	I	X
Z	Z	Y	X	I

But, this is precisely the same table as the one we have just obtained for the trousers transformations! Even though both the group members in the two cases and the group operation were entirely different, the two groups have an identical structure—they are isomorphic to each other. Could this be merely a consequence of the fact that the two groups we have chosen are somewhat peculiar? To convince ourselves that this is not the case, let us consider a very ordinary group of rotations. To visualize the transformations we are about to perform more easily, you may want to use some rectangular box with different patterns on its faces, such as a matchbox or a thick book. Examine the following four operations (figure 80):

X—a half turn about the axis labeled *x*.
Y—a half turn about the axis labeled *y*.
Z—a half turn about the axis labeled *z*.
I—the identity, that leaves the box "as is."

Some experimentation will show that if you perform, for instance, *X* followed by *Y*, you obtain the same result as if you have performed operation *Z*. At the same time, if any of *X*, *Y*, *Z* is performed twice, the initial configuration (the identity) is restored. The table of this group is again identical to the previous two tables—this geometrical group is also isomorphic to the jeans group and the chicken-cow group.

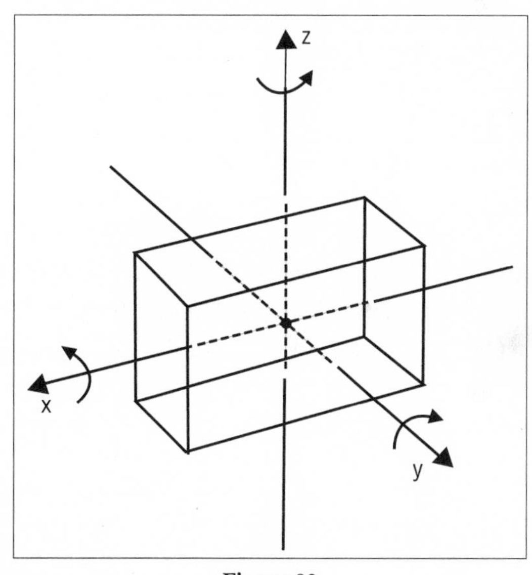

Figure 80

Perhaps nowhere was the application of group theory more surprising than in the field of anthropology. An extremely complex kinship-marriage system that was discovered among the Kariera, a tribe of Australian Aboriginals, left anthropologists baffled. Each Kariera belongs to one of four classes or clans: Banaka, Karimera, Burung, and Palyeri. Marriage and the association of descendants to classes were found to obey the following strict rules:

1. A Banaka can only marry a Burung.
2. A Karimera can only marry a Palyeri.
3. The children of a male Banaka and a female Burung are Palyeri.

4. The children of a male Burung and a female Banaka are Karimera.
5. The children of a male Karimera and a female Palyeri are Burung.
6. The children of a male Palyeri and a female Karimera are Banaka.

Perplexed by this unusual system, the famous French anthropologist Claude Lévi-Strauss (born 1908) described the rules to compatriot mathematician André Weil (1906–98) in the 1940s, hoping that the latter would identify some guiding pattern. Weil was the perfect person to turn to. In addition to his outstanding mathematical skills, he was obsessed with languages and linguistics. His passion for Sanskrit and knowledge of ancient texts, such as the religious epic *Mahabharata,* even gained him his first appointment at the Aligarh Muslim University in India. After some contemplation, Weil was indeed able to translate the entire Kariera scheme into the language of group theory. In order to reproduce Weil's explanation, I shall denote the four classes as follows:

Banaka—A
Karimera—B
Burung—C
Palyeri—D

The conjugal rules (1) and (2) above—that is, that an A can only marry a C (and vice versa), and a B can only marry a D—may be represented by the following "family" correspondence, denoted by "f":

$$f = \begin{pmatrix} \text{ABCD} \\ \text{CDAB} \end{pmatrix}$$

Notice that if this permutation is performed twice, it restores the original order—$f \circ f = I$ (where I is the identity; f takes A into C and C into A, so applying f twice takes A into itself and the same for all the other letters). According to the rules of descendants 3–6, the class of the children can be determined either by their paternal ancestors (e.g., the child of a male Banaka is always a Palyeri) or by their maternal ones (e.g., the child of a female Banaka is always a Karimera). Using the symbols for the classes and p and m for the paternal and maternal rules respectively, this can be expressed by the two permutations:

$$p = \begin{pmatrix} \text{ABCD} \\ \text{DCBA} \end{pmatrix} \qquad m = \begin{pmatrix} \text{ABCD} \\ \text{BADC} \end{pmatrix}$$

Note again that $p \circ p = I$ and $m \circ m = I$. Also, each two of the permutations f, p, m operating in succession produce the third (e.g., $f \circ p = m$). We are now in the position to construct the complete "multiplication table" for the four permutations I, f, p, and m:

$\circ$	I	f	p	m
I	I	f	p	m
f	f	I	m	p
p	p	m	I	f
m	m	p	f	I

We discover that not only do the Kariera marriage-kinship rules form a group, but closer inspection of the multiplication table will convince you that this group is also isomorphic to the jeans/chicken-cow/rotations group! In fact, we could think of this table as describing any abstract group in which each of three members X, Y, Z is its own inverse, and the combination of any two gives the third.

Incidentally, you may wonder whether the complex Kariera kinship rules could somehow be translated into a western-civilization equivalent. They can. Imagine two families, Smith and Jones. Members of both families live in New York and in Los Angeles. Four classes could then be defined: Smith family members living in New York; Smith family members living in Los Angeles; Jones family members in New York; Jones family members in Los Angeles. The rules could be formulated as follows: A Smith can only marry a Jones (and vice versa), and a New Yorker can only marry a Los Angeleno (and vice versa). The children live at the mother's residence but adopt the father's family name. These (obviously contrived) kinship rules produce precisely the same structure as that of the Kariera.

Clearly, no one suspects that the Kariera knew group theory. The group-theoretical description of their marriage rules may not even have been entirely necessary for anthropological research. Nevertheless, analyzing the rules in this fashion can reveal underlying structures that are otherwise difficult to recognize or may be missed altogether. Stripping groups from different disciplines to their bones is analogous in many ways to the analysis of structures of different languages. The recognition of the interconnections among, say, all Indo-European languages has been achieved through a similar process. Claude Lévi-Strauss's extensive analyses in the area of social anthropology, as expressed in his *Elementary Structures of Kinship,* are therefore generally acknowledged as the driving force behind modern structuralism—the search for underlying units and the rules that govern how they are put together.

Structuralism derived its organizing principles from, and was inspired by, the linguistic work of the Swiss linguist Ferdinand de Saussure (1857–1913). Saussure abandoned the traditional approach to languages, which was based primarily on historical and philological studies, in favor of a structural analysis. A structuralist examining an airplane built out of Legos would not care much if the model can actually fly. Instead, much like a group theorist, the structuralist would recognize that there are different kinds of building blocks, and that these basic units are connected together according to very specific rules. In language, the elements could be the phonemes that make all the speech sounds (of which there are thirty-one in English), and the rules would be the grammar according to which words can be ordered. The fact is that with a rather limited set of grammar rules and a finite set of phonemes or terms, humans have been able to produce impressive works such as Shakespeare's plays, Dante's *Divine Comedy,* and the *Encyclopaedia Britannica.* Even toddlers are able to utter entire phrases nobody has ever expressed before. The astonishing pace at which children are capable of learning languages, and the similarities in both the learning process itself and the characteristic errors made by children all across the globe, have motivated the idea of a universal grammar. Just as group-theoretical principles underlie all symmetries, the theory of universal grammar postulates that all the languages have underlying principles of grammar that are innate to all humans. In some sense, universal grammar is not really a grammar, but an initial state of the language faculty that all humans possess. Note that this does not mean that all languages have the same grammar, only that there are common and invariant basic rules. Insights of this type, partly derived from structuralism, have been applied both to linguistic theory and to cognitive psychology by the influential MIT researcher Noam Chomsky. In Italy, novelist and philosopher Umberto Eco is also known for his detailed structuralist analyses in the area of the meaning of signs (semiotics) in social and literary contexts.

Given the philosophical parallels between group theory and linguistics, it should come as no surprise that just around the same time that

Saussure was revolutionizing linguistics, the Norwegian mathematician Axel Thue (1863–1922) was introducing the concept of a formal language—a set of words (or strings of characters composed of some alphabet) that can be described by some formal grammar (a set of precisely defined rules). A very simple example of a formal language could be a set of strings composed of the letters *g* and *l*. The "grammar" could be defined, for instance, by the rules:

1. Start with *g*.
2. Every time you encounter the letter *g* in a word, replace it with *gl*.
3. Every time you encounter the letter *l*, replace it with *lg*.

You can verify that this language would include words such as *g*, *gl*, *gllg*, *gllglggl*, and so on. Formal languages play an important role in computer science and in complexity theory (concerned with the intrinsic complexity of computational tasks). If Thue's definitions of formal linguistics look reminiscent of the elements and definitions of group theory, this is not an accident. The two topics are intimately related, especially through an important problem known as the *word problem:* Decide whether, by using replacements allowed by the grammar, any two given words may be transformed into one another.

To what conclusion do all of these examples lead us? Groups can reach the same level of abstraction one normally associates only with ordinary numbers. Whether we speak of seven samurai, seven good years, seven days of the week, seven brides for seven brothers, or seven politicians (actually, I am not sure who wants to speak about them), these are all manifestations of the same abstract entity—the number seven. Similarly, the four groups we have just encountered (the trouser transformations, the Kariera, and so on) are all specific realizations of one and the same abstract group. Serendipitously, by constituting a group of permutations, the Kariera rules offer yet another manifestation of Cayley's theorem—there indeed exists a group of permutations that is identical in structure to the other three groups.

Mathematicians usually refer to groups that are isomorphic to each other as if they are only one group. The particular group realized by the jeans and the Kariera rules is known as the *Klein four-group,* after the German mathematician Felix Christian Klein (1849–1925). Klein was responsible for a major breakthrough in the application of group theory—the recognition that geometry, symmetry, and group theory are unavoidably linked. In fact, not merely linked. Klein showed that in many respects, geometry *is* group theory. This surprising statement represented such a dramatic break with the traditional view of geometry that it deserves a more detailed exposition.

WHAT IS GEOMETRY?

Around 300 BC the Greek mathematician Euclid of Alexandria published what was to become the best-selling mathematical textbook of all time—*The Elements.* In this thirteen-volume opus, Euclid laid the foundations of the Euclidean geometry we learn in school, and until the nineteenth century this was the only geometry known. Euclid attempted to build an entire theory of geometry on a well-defined logical base. Accordingly, he started with only five postulates or *axioms,* assumed to hold true, and sought to prove all the other propositions on the basis of those postulates by logical deductions. Axioms are like the rules of the game, the "truth" of which is not to be disputed. If you want to change the axioms you would be playing a different game. For instance, the first axiom states, "Between any two points a straight line may be drawn." Euclidean geometry describes propositions that are deduced to be true if this and the other axioms hold. The second, third, and fourth axioms were equally concise, but the fifth one was different, more complicated in its formulation, and consequently it had a more convoluted history. Even Euclid himself was probably not entirely happy with his fifth axiom, since he tried to avoid it for as long as he could—the proofs for the first twenty-eight propositions in *The Elements* do not make use of the fifth axiom. The version of the fifth axiom, known as the *parallel axiom,* most often cited today is named after the Scottish mathematician John Playfair (1748–1819), even though it appeared first in the commentaries of the Greek mathematician Proclus in the fifth century. It states: "Given a line and a point not on the line, it is possible to draw exactly one line parallel to the line through that point." Over the centuries, a number of discontented geometers attempted unsuccessfully to prove the fifth axiom from the first four, in an effort to formulate a more economical geometry. These were not complete failures, however, since they did provide new insights. In particular, these attempts led to an understanding that many other alternative formulations of the fifth axiom are possible, all equivalent to one another. Eventually, this winding path opened the way for the development of new, non-Euclidean geometries.

The first to have made significant progress in the direction of non-Euclidean geometries, albeit without realizing it himself, was the Jesuit Giovanni Girolamo Saccheri (1667–1733). In a work that was quite remarkable for its time, *Euclides ab omni naevo vindicatus* (*Euclid Freed from All Flaws*), Saccheri examined an intriguing "what if?" question—what if the sum of the angles of a triangle is not equal to 180 degrees (as we learn in Euclidean geometry), but is greater or smaller? Could one still construct a logical geometry that is self-consistent? About a century later, Legendre picked up where Saccheri had left off and showed in his famous geometry book (the one studied by Galois) that the statement that the sum of the angles is equal to 180 degrees is completely equivalent to Euclid's fifth postulate (that is, one can assume either of the two to hold true and prove the other). Both Saccheri and Legendre, however, failed to grasp the full implications of these alternative possibilities, and they ended up getting bogged down by incorrect contradictions. Nonetheless, these works and complementary research by the Alsatian mathematician Johann Heinrich Lambert (1728–77) helped to focus attention on the "parallel postulate," which by 1767 was dubbed "the scandal of elementary geometry" by the French mathematician Jean d'Alembert. Four mathematicians from three countries—Gauss, Bolyai, Lobachevsky, and Riemann—were eventually responsible for formulating correctly the first non-Euclidean geometries. In these new geometries the fifth postulate is in fact replaced with one of its negations: "Through a point not on a line, there is either more than one line parallel to the given line or none." Or equivalently, the sum of the angles in a triangle is either less than 180 degrees or greater than 180 degrees.

Visualizing how such geometries may be realized is not difficult. Examine the three surfaces in figure 81. Euclidean geometry is the geometry of flat space, of the type that is encountered on a desktop. In this geometry, parallels (assumed to be infinite) never meet, and the angles of any triangle always sum up to 180 degrees. On a surface shaped like a curved saddle, on the other hand, the sum of a triangle's angles is always less than 180 degrees. On the surface of a sphere, such as the surface of the Earth, the sum of a triangle's angles exceeds 180 degrees (in the particular case shown in the figure the sum is actually 270 degrees). The saddle-shaped geometry is known today as *hyperbolic geometry.* János

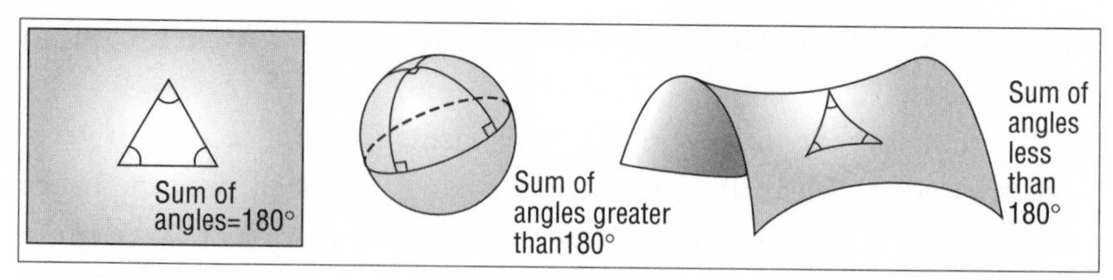

Figure 81

Bolyai (1802–60), a young Hungarian mathematician, worked out many of the features of this geometry by 1824. In a letter to his father, the mathematician Farkas Bolyai, János could not restrain his exhilaration with the discovery: "I have created a strange new world, out of nothing." By 1831 the exuberant János completed a detailed description of his new geometry. Since the father was about to publish a massive treatise (the *Tentamen*) on the foundations of geometry, algebra, and analysis, János prepared his manuscript in the format of an appendix to his father's book. A letter from Gauss, to whom the work had been sent for review, quickly dampened Bolyai's enthusiasm. Gauss first expressed his admiration for the ideas in the paper, but he was also quick to point out that "the entire content of the work . . . coincides almost exactly with my own meditations which have occupied my thoughts for the past thirty or thirty-five years." While there is no doubt that Gauss had indeed anticipated on his own most if not all of Bolyai's results, he had never published them (apparently for fear that the radically new geometry would be regarded as philosophical heresy). Bolyai's realization that he was not the originator of the idea was devastating to him. He became deeply embittered, and his subsequent mathematical work lacks the imaginative quality of hyperbolic geometry.

Unbeknownst to either Bolyai or Gauss, the Russian mathematician Nikolai Ivanovich Lobachevsky (1792–1856) published in 1829 an entire treatise heralding hyperbolic geometry as an alternative to Euclidean geometry. Since the work appeared in the obscure *Kazan Messenger*, however, it went almost entirely unnoticed until a French version was published in *Crelle's Journal* in 1837. In 1868, the Italian Eugenio Beltrami: (1835–1900) finally put the Bolyai-Lobachevsky geometry on the same footing as that of the Euclidean geometry.

Gauss's brilliant student Georg Friedrich Bernhard Riemann (1826–66) first discussed *elliptic geometry*, such as one encounters in its simplest form on the surface of a sphere, in a classic lecture delivered on

June 10, 1854. Riemann's paper makes the top ten list of many mathematicians. One of the key differences between elliptic geometry and Euclidean geometry is that on the surface of a sphere the shortest distance between two points is not a straight line. Rather, it is a segment of a great circle, whose center coincides with the center of the sphere (as is the case for the equator or meridians on a globe). Flights from Los Angeles to London take advantage of this fact and do not follow what would appear to be a straight line on the map, but rather a great circle that bears northward from Los Angeles (figure 82). You can easily check that any two great circles meet in two diametrically opposite points (e.g., two meridians, which are parallel at the equator, meet at the two poles). Consequently, there are no parallel lines at all in this geometry. Riemann took the abstract non-Euclidean concepts much further and introduced curved spaces in three and even higher dimensions. In some of these, the nature of the geometry could change from place to place, being elliptic in some regions and hyperbolic in others. A crucial difference between Riemann's work and the research of all of his predecessors (including the great Gauss) was a change in perspective. When Gauss analyzed a curved two-dimensional surface, he looked at it as someone would study the surface of a globe—from an external, three-dimensional point of view. Riemann, on the other hand, examined the surface of the same globe from the perspective of a painted dot on that surface.

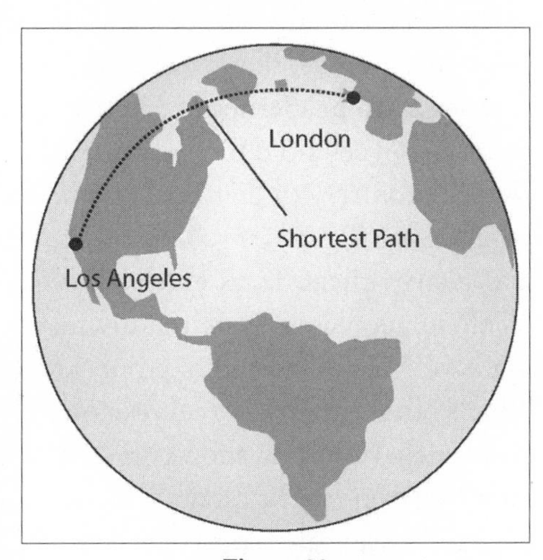

Figure 82

The non-Euclidean geometries may appear at first blush as nothing but the futile if ingenious inventions of too-imaginative mathematical minds. As we shall see in the next chapter, however, the solutions to Einstein's equations describing the structure of space and time turned out unexpectedly to require precisely the classes of geometries described above. Riemann's perspective, that which stems from being a part of the curved space under consideration, is at the foundation of modern cosmology—the study of the universe as a whole. If you think about it for a moment, this is an absolutely astounding fact. Saccheri's seemingly innocent "what if?" question about Euclid's fifth postulate has led to no less than the consideration of geometries that provided the tools Einstein needed to explain the cosmic fabric. Just as Galois's theory of groups has become the language of symmetries and non-Euclidean geometry the language of cosmologists, this type of "anticipation" by mathematicians of the needs of physicists of later generations has repeated itself many times throughout the history of science.

The generalization and abstraction of geometry was a welcome development, but by the 1870s the proliferation of geometries appeared to be getting totally out of hand. In addition to all the non-Euclidean geometries above, there was a motley collection of *projective geometry* (dealing with properties of geometric figures under projection, as when an image on celluloid film is projected onto a movie-theater screen); *conformal geometry* (which deals with angle-preserving compactifications of spaces); *differential geometry* (the study of geometry using calculus); and many others. If, as Plato believed, "God is a geometer," which of all these geometries gets the divine approval? This was the point at which the twenty-three-year-old Felix Klein (figure 83 shows him at a later age) came to the rescue with his group-theoretical approach and order started to crystallize out of chaos.

Figure 83

In a seminal lecture entitled "Comparative Review of Recent Researches in Geometry" delivered in 1872 at the University of Erlangen, Klein boldly reversed the roles of symmetry and geometries. In his words, "There are transformations of space that do not alter at all the geometrical properties of figures. By their nature, these properties are, in effect, independent of the position occupied in space by the figure being considered, of its absolute size, and of its orientation." Before Klein, mathematicians had thought primarily in terms of geometrical objects, such as circles, triangles, or solids. Instead, Klein suggested in his so-called Erlangen program that the geometry itself is characterized and defined not by the objects, but rather by the group of transformations that leaves it invariant. Take for instance the group of rigid motions—the motions that preserve distances and angles, and consequently shapes. Since such motions are the bread and butter of Euclidean geometry, the latter can be defined as the geometry that remains invariant under all the transformations that are in the group of rigid motions. A circle of a given radius remains the same circle no matter how you turn it. Two triangles that overlap precisely (and are the subject of so many theorems in Euclidean geometry and a constant source of headache to high school students) stay congruent even if you translate, rotate, or reflect them. Klein's radical idea, however, allowed for a much wider variety of geometries to exist. Other transformations, which might twist or stretch the objects, could define new geometries. In other words, the unifying basic concept that is the backbone of every geometry is the *symmetry group.* Even though each one of the many geometries may be based on a different group of transformations, the fundamental blueprint for all geometries is the same. In projective geometry, for instance, distances are clearly not invariant. The model captured on film for the original *King Kong* movie was only eighteen inches tall, very different from its fifty-foot screen image. Projective geometry is therefore characterized by a different group of symmetry transformations from that of Euclidean geometry (concepts such as "hexagonal" or "elliptical" are preserved in projection). According to Klein, what mathematicians have to do to define a geometry is to provide a group of transformations and to identify the ensemble of entities that remain unchanged under those transformations. These ideas have been later expanded upon and endowed with much greater depth by two mathematical giants: the Norwegian group theorist Sophus Lie (1842–99) and the towering figure of late-nineteenth-century mathematics—the Frenchman Henri Poincaré (1854–1912).

With Klein's innovative Erlangen program, Cayley's abstraction of groups, Lie's tendency toward structural thinking, and Poincaré's all-embracing mathematics, it was starting to become clear that symmetry and group theory provide the underpinning for much of mathematics. In fact, to Poincaré, "all mathematics was a matter of groups." Areas that previously appeared to be totally unrelated, such as the theory of algebraic equations, a multitude of geometries, and even number theory (through seminal works by Euler and Gauss), became suddenly unified by one basic structure. Even though Klein was regarded by some of the (rather arrogant) Berlin mathematicians of his time as "a charlatan without real merit," he still had another elegant result in his repertoire. With one group-theoretical masterstroke he combined algebra with geometry and connected it all back to—believe it or not—Galois's work on the quintic. This was not a one-man show, however. The Prussian Leopold Kronecker and the Frenchman Charles Hermite paved the road to these deep interconnections.

THE RETURN OF THE QUINTIC

Leopold Kronecker (1823–91) represented one of those truly rare combinations of gifted mathematician and successful businessman. His unusual ability to recognize and immediately befriend people who were on the rise in either the financial or mathematical world also proved very useful for the promotion of his own career. Some of Kronecker's main mathematical contributions came in the theory of elliptic functions (the topic on which Abel wrote his famous 125-page paper) and the theory of *algebraic numbers* (numbers that are the solutions of certain algebraic equations).

In 1845, Kronecker's uncle on his mother's side passed away. The uncle had been a prosperous banker and a farming-enterprise executive. The management of his affairs was thrust upon the shoulders of the young mathematician, who had just passed the oral examination for his doctoral thesis on August 14 of that year. Kronecker took the responsibility with great energy and uncompromising thoroughness. While the demanding job forced him to spend the following eight years as a businessman, he did not neglect his mathematics. Where others in his position might have taken up an easier subject to occupy their leisure time, Kronecker devoted his efforts to gaining what was probably the deepest understanding of Galois's theory among all of the late-1840s mathematicians (recall that Galois's memoirs were published by Liouville in 1846). The result was a crystal-clear memoir on the solvability of equations that was published in 1853. In his description of the work, historian of mathematics E. T. Bell is lavish with praise: "Kronecker took the refined gold of his predecessors, toiled over it like an inspired jeweler, added gems of his own, and made from the previous raw material a flawless work of art with the unmistakable impress of his artistic individuality upon it." Having returned fully to mathematics, Kronecker spent the next five years mounting a direct attack on the quintic equation. As you recall, both Abel and Galois proved that the general quintic couldn't be solved *by a formula* that involves simple operations on the coefficients, not that it could not be solved at all. Yet the actual method of solution remained elusive. As is often the case with scientific discoveries whose time has come, just as Kronecker was attempting to finally crack the quintic, a French mathematician was also busy doing precisely the same thing.

Charles Hermite (1822–1901) was the sixth child of Ferdinand Hermite and Madeleine Lallemand, who had five boys and two girls. During Charles's childhood, as the family's drapery business prospered, the family moved from Dieuze to the larger city of Nancy. After attending a school in Nancy, followed by the Lycée Henri VI in Paris, Hermite entered the Louis-le-Grand, some eleven years after Galois had left that institution. His mathematics teacher there was—you guessed it—the same Louis Richard who had been Galois's mentor. The gifted teacher was again quick to recognize in Hermite "a young Lagrange." If you have ever doubted that history has a habit of repeating itself, consider this. While at Louis-le-Grand, Hermite published two mathematical papers. One of them was entitled "Considerations on the Algebraic Solution of the Equation of the Fifth Degree." This was an interesting paper that demonstrated that Lagrange's method of solution (chapter 3) could not work. The title and contents of the paper also showed, however, that at least at age twenty, Hermite was still totally unaware of either the works of Abel or of Galois (not that anyone else in the mathematical world was aware of Galois's work at the time). To continue the parallelism between Hermite's school experiences and those of Galois one step further, Hermite also tried for the École polytechnique. Unlike the wretched Évariste, he passed the entrance examination, but was ranked only sixty-eighth. Then, to add insult to injury, after only one year at the Polytechnique, Hermite was forced out because of a physical handicap, a disabling deformity in his right foot.

Hermite returned to the quintic equation in the late 1850s, and his paper on the subject appeared in 1858—the same year that Kronecker also published a paper with an identical title: "On the Solution of the General Equation of the Fifth Degree." Hermite's result was spectacular. Using a special kind of elliptic function, he was able, for the first time, to solve the general quintic. The centuries of repeated attacks had finally paid off.

Kronecker even went one step beyond. First, he obtained practically the same solution as Hermite, but using a different approach, closer in spirit to Galois's ideas. Second, in a subsequent paper published in 1861, he dug into the underlying reasons for the success of the method he had employed. In other words, Abel and Galois proved that the general quintic couldn't be solved by a formula; Kronecker tried to understand why it could be solved by elliptic functions. Another achievement of Kronecker was the publication (in 1879) of a simpler, shorter, and better-organized version of Abel's proof. He also corrected a minor error in the original rather lengthy proof (which fortunately had no effect on the outcome). This set the stage for Felix Klein's decisive assault.

The philosophy behind Klein's investigation was really quite simple. Earlier in this chapter we used the familiar properties of the group of symmetries of the equilateral triangle and those of the group of permutations of three elements to show that the two groups are really one and the same (isomorphic). Klein turned this logic on its head. He first showed that two seemingly disparate groups are isomorphic and then exploited this fact to unveil the reasons for this unexpected connection. Klein's findings were published in 1884 in a massive tract with the outlandish title *Lectures on the Icosahedron and the Solution of Equations of the Fifth Degree.* How are the two topics in the title related? Klein started with a simple examination of the solid known as the icosahedron (figure 84). Plato regarded this beautiful solid as one of the basic constituents of the cosmos (the others being the tetrahedron, the cube, the octahedron, and the dodecahedron, all collectively known as the *Platonic solids*). The icosahedron has twelve vertices, twenty faces (each an

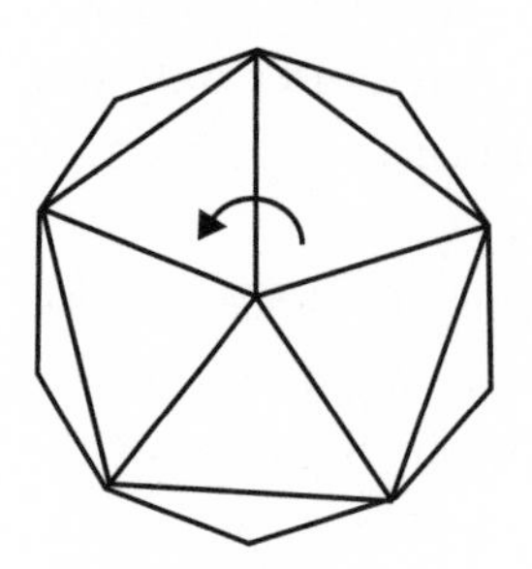
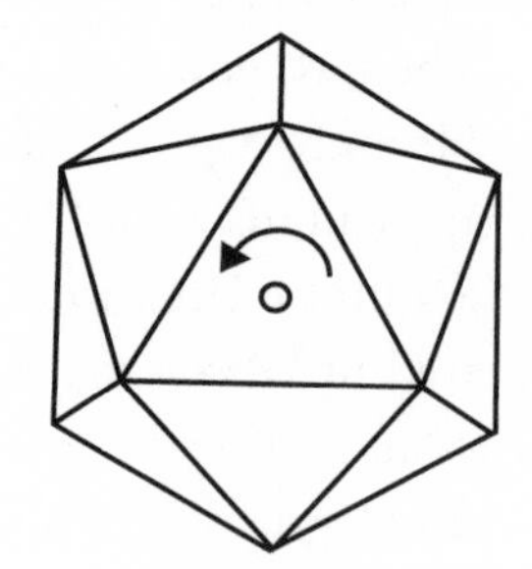
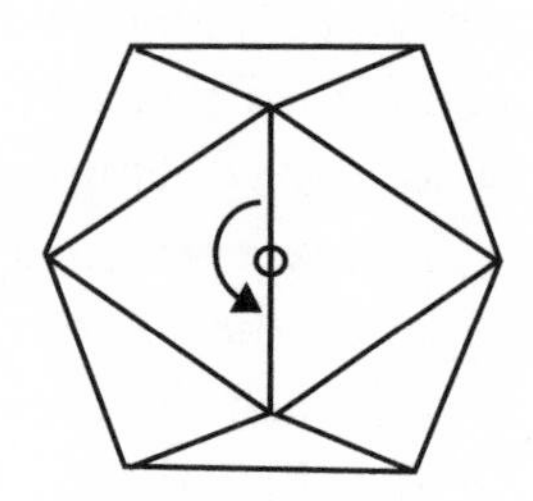

Figure 84

equilateral triangle), and thirty edges (the lines where two faces meet). Klein first showed that there are precisely sixty rotations that leave the icosahedron unchanged. These are (figure 84): four rotations through multiples of 72 degrees about lines joining opposite vertices (for a total of twenty-four); two rotations through 120 degrees about lines joining the centers of opposite faces (for a total of twenty); half turns about lines joining the midpoints of opposite edges (for a total of fifteen); the identity, which leaves it "as is." Klein then showed that these rotations form a group. Next, he examined a particular group of permutations of the five putative solutions to the quintic equation. More specifically, he examined only the even permutations (which contain an even number of transpositions). Since there is a total of 5! = 120 permutations of five elements, there are precisely 60 even permutations (and 60 odd ones). Then came the checkmate. Klein proved that *the icosahedral group and the permutations group are isomorphic.* But, recall that Galois's proof on the solvability of equations relied entirely on the classification of equations according to their symmetry properties under permutations of the solutions. The unexpected link between permutations and icosahedral rotations allowed Klein to weave a magnificent tapestry in which the quintic equation, rotation groups, and elliptic functions were all interwoven. Just as the completion of a jigsaw puzzle reveals the full picture, the fundamental interconnections discovered by Klein provided the definitive answer as to why the quintic could be solved by elliptic functions.

The unifying power of group theory was so overwhelming that by the end of the nineteenth century it was becoming clear that its reach would overflow the boundaries of pure mathematics. Physicists, in particular, were starting to take notice. First, through Einstein's theory of general relativity, geometry was recognized as a key property of the universe at large. Then symmetry was identified as the foundation from which all the laws of nature ultimately spring. These two simple truths virtually guaranteed that the search for an all-encompassing theory of the cosmos would turn largely into a search for underlying groups.

– SEVEN –

Symmetry Rules

Nature has been kind to us. Being governed by universal laws, rather than by mere parochial bylaws, she has given us an opportunity to decipher her grand design. Unlike in the real estate business—where everything is location, location, location—neither our location in space nor our orientation with respect to the Earth, Sun, or the fixed stars makes any difference for the laws of nature we deduce. If not for this symmetry of the laws of nature under translations and rotations, scientific experiments would have to be repeated in every new laboratory across the globe, and any hope of ever understanding the remote parts of the universe would be forever lost. This is a powerful concept. When Newton first proposed that the dynamics of celestial bodies could be described by mathematical formulae, and that on top of that, these formulae expressed universal laws, this provoked understandable reactions throughout Europe. The explanation of falling apples would hardly have been sufficient to cause much of a sensation. The motions of the planets, on the other hand, had always been regarded as the unmistakable work of God's guiding hand. The eighteenth-century poet Alexander Pope probably expressed the feelings of many when he wrote:

> Nature and Nature's laws lay hid in night:
> God said, Let Newton be! and all was light.

Newton, himself a piously devout man, had no intention of bringing the omnipresence of God into question. In his scientific masterpiece *Principia* (figure 85 shows the front page) he wrote, "This most beautiful system of the sun, planets, and comets, could only proceed from the counsel and dominion of an intelligent and powerful Being. And if the fixed stars are the centres of their like systems, these, being formed by the like wise counsel, must be all subject to the dominion of One." Nevertheless, the notion of the universe as some sort of machine did make it even into some contemporary artworks, such as the impressive painting *A Philosopher Lecturing on the Orrery,* by Joseph Wright of Derby (figure 86). This was part of the transformation from the Greek organismic universe that treated the cosmos as a biological organism to the mechanistic universe.

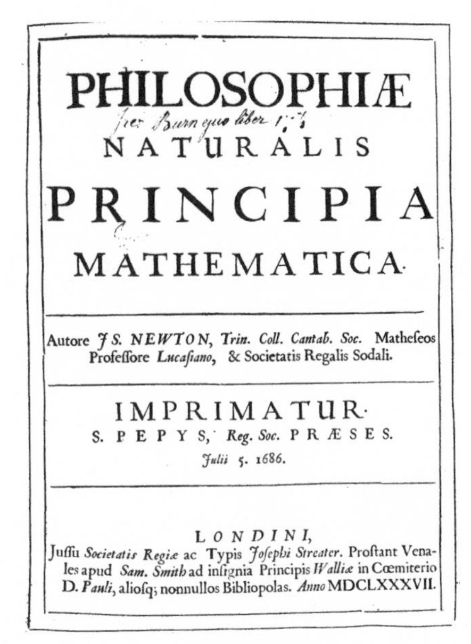

PHILOSOPHIÆ
NATURALIS
PRINCIPIA
MATHEMATICA.

Autore *J S. NEWTON, Trin. Coll. Cantab. Soc.* Mathefeos Profeffore *Lucafiano,* & Societatis Regalis Sodali.

IMPRIMATUR.
S. PEPYS, *Reg. Soc.* PRÆSES.
Julii 5. 1686.

LONDINI,
Juffu *Societatis Regiæ* ac Typis *Jofephi Streater.* Proftant Venales apud *Sam. Smith* ad infignia Principis *Walliæ* in Cœmiterio D. *Pauli,* aliofq; nonnullos Bibliopolas. *Anno* MDCLXXXVII.

Figure 85

The world around us appears as transient as the clouds. The histories of humankind, of the Earth, of the solar system, of the entire Milky Way galaxy, and even of the universe as a whole are marked by relentless, sometimes violent changes, albeit on different time scales. Fortunately, the laws of nature are less ephemeral. When astronomers observe a galaxy that is a billion light-years away, the light entering the aperture of their telescope at that

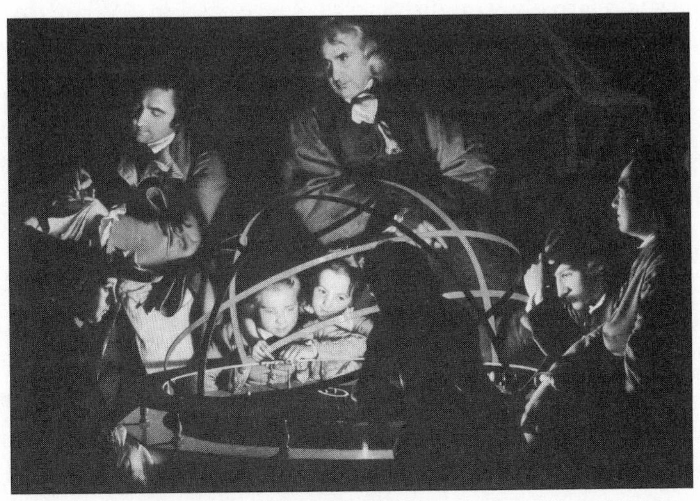

Figure 86

moment has been on its way for a billion years. In other words, telescopes are true time machines—they give glimpses of the universe's distant past. As far as we can tell, Mother Nature does not allow any amendments to her constitution—the laws of nature have not changed in any noticeable way, at least since the time the universe was no more than a second old. Laws with a more fleeting existence would have made it very difficult for physicists (if those existed at all) to unravel the cosmic history.

SPACETIME

The symmetry of the laws of nature extends well beyond mere translations and rotations. The laws don't care, for instance, how fast or in which direction we move. You must have encountered the simplest manifestation of this fact in a train station. You sometimes can hardly tell whether it is your train or the one on the adjacent track that is moving. Two observers moving at constant velocities (i.e., with neither the speed nor the direction of motion changing) will find nature to obey precisely the same laws, irrespective of whether one is shooting for the sky in a futuristic rocket at 99 percent of the speed of light while the other is sitting lazily on the back of a giant turtle. Galileo and Newton had already recognized this important symmetry between observers moving at constant velocities, but Einstein gave it an enormous emphasis and a totally unanticipated twist in his theory of special relativity. One part of this symmetry is relatively straightforward. The question, "When does New York stop at this train?" may be phrased surrealistically but is in fact perfectly legitimate even in Newtonian physics. A person on a train could definitely regard that train as standing still while everything else is moving. Einstein, however, formulated this symmetry so as to agree with the unexpected experimental result that the speed of light always comes out to be the same, irrespective of how the source of light or the observer is moving. In other words, to the symmetry dictating that the laws of physics (including the laws of electromagnetism and light) should appear the same to all uniformly moving observers, he added another one: *The speed of light is precisely the same for all observers.*

The constancy of an absolute speed of light was an implicit feature of Maxwell's equations (the theory of electromagnetism), but at first blush it appears extremely counterintuitive. In fact, it seriously strains our common sense of how things behave. When someone flings an apple forward while driving a convertible car (luckily not many drivers do that), the apple's speed with respect to the ground is the sum of the speed of the car and the speed at which the apple is being thrown. In the same way, we might expect that if that convertible were coming directly toward us, the speed that we would measure for the light emitted by its headlights would be the sum of the speed of light (about 670 million miles per hour or 300 million meters per second) and the speed of the car. Einstein tells us, however, and numerous experiments confirm, that this is *not* the case. Rather, even if the car were moving at the incredible speed of 99.99 percent of the speed of light, the speed that we would record for the light from its headlights would remain unchanged, at 670 million miles per hour. Furthermore, the same would be true if we were to measure the speed of the light emitted by the car's taillights as the car is receding away at speeds close to the speed of light. Before we delve into the implications of this crucial finding, let us examine for a moment what might have happened had the speeds of sources been added to (or subtracted from) the speed of light. Figure 87 shows crossing runways in an airport. The airplane traveling south has just landed at high speed. As it is about to enter the intersection, the pilot notices a baggage cart coming into the intersection from the west. The pilot swerves rapidly to avoid a collision. Suppose now that an observer is watching the entire incident from the south leg of the crossroads. To make the point clearer, assume that the landing airplane is moving very close to the speed of light. If the speed of light were not constant, the observer would see the light reflected from the airplane moving toward him at nearly twice the speed of light (the sum of the speeds of the airplane and of light). The light reflected from the slow cart, on the other hand, would approach the observer at the speed of light (since it is reflected perpendicularly to the direction of motion). Consequently, light from

Figure 87

the plane would reach the observer significantly earlier than the light from the cart. The observer would see the plane swerve violently with no apparent reason whatsoever. The constancy of the speed of light to all observers eliminates such paradoxes in which effects precede their causes.

To ensure the symmetry of the laws of physics for uniformly moving observers, as well as the invariance of the speed of light, the theory of special relativity had to pay a price. Einstein discovered that space and time cannot be treated as separate entities. Rather, they are inseparably tethered together by symmetry. Einstein's original paper on special relativity had the unassuming title "On the Electrodynamics of Moving Bodies" (figure 88 shows the front page), yet, as the following example will show, it literally changed our perception of reality.

Imagine that over a period of a few years you are videotaping an apple resting on a table as it ages and disintegrates. What this (none too exciting) film is really capturing is the "motion" of the apple through time, as opposed to its motion through space. Time, according to special relativity, is a fourth dimension that has to be added to the familiar three dimensions of space. When the apple is propelled at some speed, it necessarily travels through all four dimensions, since as the apple cruises through space, time is progressing too. Will the moving apple age at the same rate as the stationary apple? The surprising answer of special relativity is that it will not. The faster the apple journeys through space, the slower its "clock" will tick, as seen by an observer at rest. As the apple's speed approaches the speed of light, its time (for an observer at rest) will slow down to a crawl. This might sound utterly unbelievable had it not been unambiguously confirmed by a multitude of experiments. For instance, an elementary particle called a muon is constantly being produced in the Earth's upper atmosphere by the bombardment of high-energy particles known as cosmic rays. The fact that these muons can travel through tens of miles of the atmosphere is due entirely to the relativistic slowing down of the muons' internal "clocks." At rest, muons live only about two-millionths of a second before decaying into lighter particles. For such short lives, even had they whizzed through space at the speed of light, the travel time through the atmosphere would be more than ten times longer than the muon lifetime (in the absence of relativistic effects). Researchers who timed and counted such muons between the summit and foot of Mount Washington in New Hampshire

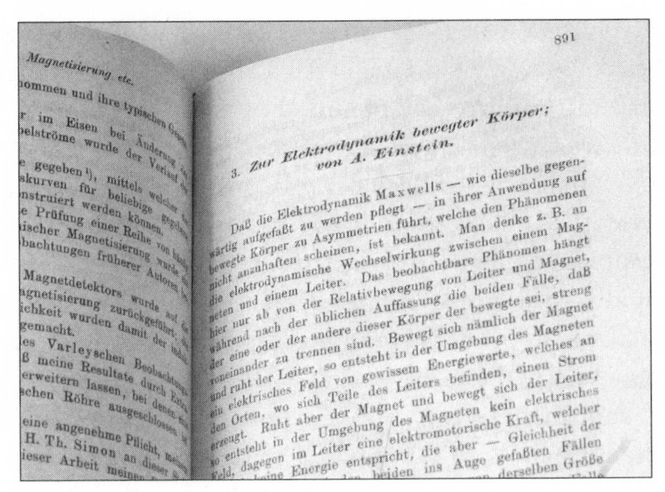
891

3. Zur Elektrodynamik bewegter Körper;
von A. Einstein.

Daß die Elektrodynamik Maxwells — wie dieselbe gegenwärtig aufgefaßt zu werden pflegt — in ihrer Anwendung auf bewegte Körper zu Asymmetrien führt, welche den Phänomenen nicht anzuhaften scheinen, ist bekannt. Man denke z. B. an die elektrodynamische Wechselwirkung zwischen einem Magneten und einem Leiter. Das beobachtbare Phänomen hängt hier nur ab von der Relativbewegung von Leiter und Magnet, während nach der üblichen Auffassung die beiden Fälle, daß der eine oder der andere dieser Körper der bewegte sei, streng voneinander zu trennen sind. Bewegt sich nämlich der Magnet und ruht der Leiter, so entsteht in der Umgebung des Magneten ein elektrisches Feld von gewissem Energiewerte, welches an den Orten, wo sich Teile des Leiters befinden, einen Strom erzeugt. Ruht aber der Magnet und bewegt sich der Leiter, so entsteht in der Umgebung des Magneten kein elektrisches Feld, dagegen im Leiter eine elektromotorische Kraft, welcher an sich keine Energie entspricht, die aber — Gleichheit der ... beiden ins Auge gefaßten Fällen ... derselben Größe

Figure 88

in 1941 confirmed that the traveling muons lived longer, just as predicted by special relativity. Experiments in 1975, in which muons were accelerated to 99.94 percent of the speed of light, showed that such fast-lane muons lived twenty-nine times longer than their counterparts at rest, again in full agreement with the expectations from special relativity.

OK, you may think, but muons are bizarre elementary particles and not normal clocks. Would the watches on our wrists or our heartbeats also slow down if we were to move at speeds approaching the speed of light? Well, an experiment in 1971 used actual clocks. Physicists Joseph Carl Hafele and Richard Keating flew around the globe in opposite directions on commercial Pan Am flights. They carried with them four atomic clocks that were synchronized at the beginning of the trip with a stationary clock in Washington, D.C. At the end of the trip, the clocks that traveled eastward (and therefore faster than the Earth's spin) showed, as expected, an elapsed time shorter by 59 billionths of a second, while those that traveled west (effectively moving slower than the clock in D.C.) recorded times that were longer by 273 billionths of a second.

One of the key predictions of special relativity is that the velocities of a body through the space and time dimensions always combine to give precisely the speed of light. A muon at rest, for instance, has its entire "velocity" pointing in the time direction, as it "travels" only through the time dimension. For muons in motion, the larger the component of their velocity through space, the slower they "age," with their time coming effectively to a halt (for observers at rest) as the muons' speed approaches the speed of light. Light itself always travels through three-dimensional space at precisely the speed of light. Special relativity tells us that nowhere can light travel at any other speed, nor is it ever possible to catch up with light—light can never be at rest. In this sense, perceiving light is a bit like the perception of motion in a movie. Each frame in the film captures a slightly different scene, and when these frames are flashed rapidly and successively before our eyes we can see the motion. When the film is stopped, the motion disappears. We can see light only when it is moving at the speed of light.

Oddly enough, in spite of his incredible intuition and deep insights in physics, Einstein's attitude toward pure mathematics was at first rather lukewarm. As a student in Zurich, his less-than-perfect attendance in the math classes of mathematician Hermann Minkowski (1864–1909) gained him the title "lazy dog." Through an ironic twist of history, once Einstein published his theory of special relativity, it was none other than Minkowski himself who used symmetry to put the theory on a firm mathematical basis. Minkowski showed that space and time may be "rotated" as a four-dimensional entity, just as a sphere can be rotated in three-dimensional space. More important, in the same way that a sphere is symmetric (i.e., it does not change) under rotation through any angle about any axis, Einstein's special relativity equations are symmetric ("covariant" in the physics lingo) under these spacetime rotations. This remarkable symmetry of the equations has become known as *Lorentz covariance,* after the Dutch physicist Hendrik Antoon Lorentz (1853–1928), who first described these transformations in 1904. You will probably not be too surprised to hear that the collection of all the symmetry transformations of the Minkowski spacetime forms a group, similar to the group of ordinary rotations and translations in three dimensions. This group is known as the *Poincaré group,* after the outstanding French mathematician who refined the mathematical basis of special relativity.

Suspicious at first ("ever since the mathematicians have invaded the relativity theory, I myself no longer understand it"), Einstein slowly began to grasp the incredible power of symmetry. If the laws of nature are to remain unchanged for moving observers, not only do the equations describing these laws need to obey Lorentz covariance, *the laws themselves may actually be deduced from the requirement of symmetry.*

This profound realization has literally reversed the logical process that Einstein (and many of the physicists who followed him) employed to formulate the laws of nature. Instead of starting with a huge collection of experimental and observational facts about nature, formulating a theory, and then checking whether the theory obeys some symmetry principles, Einstein realized that the symmetry requirements may come first and dictate the laws nature has to obey. Let me demonstrate this type of input-output reversal using a few simple analogies.

Suppose you have never seen a snowflake before, but you are asked to guess its general shape. Clearly, you cannot even start without at least some information. Even a picture of one ray of the snowflake (figure 89) is not very helpful—you cannot guess the form of an elephant from its tail. Now, however, you are given some additional facts—you are told that the general shape is symmetric under rotations through 60 degrees about its center. This instruction immediately limits the possibilities to six-cornered, twelve-cornered, eighteen-cornered, and so on snowflakes. Since nature usually opts for the simplest, most economical solution, six-cornered snowflakes (as in figure 90) would be an excellent guess. Symmetry imposes such rigid constraints that the theory is being guided, almost inevitably, to the truth.

As a somewhat more intricate example, imagine that biologists in a distant solar system investigate the "DNA" structure of all life forms on

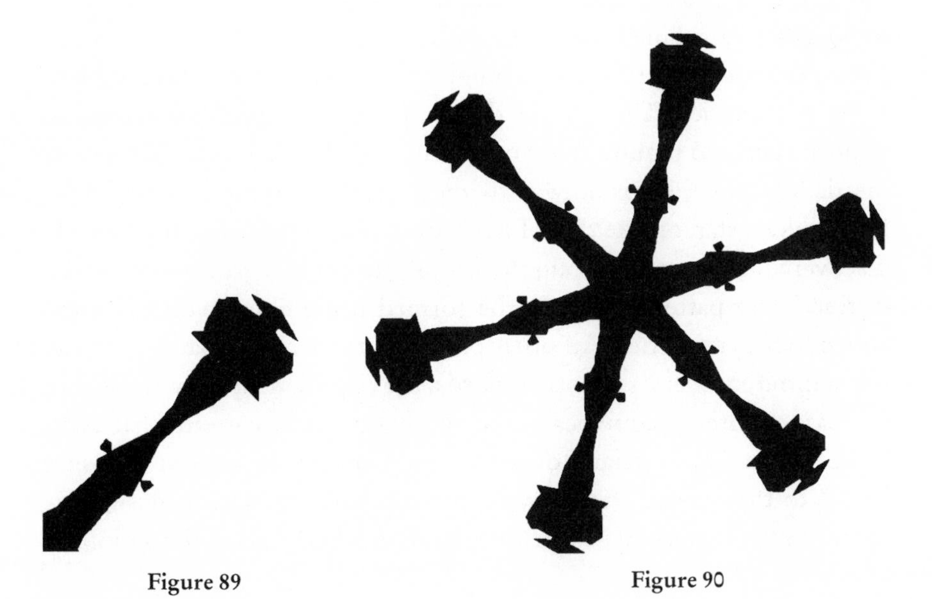

Figure 89 Figure 90

their planet. After years of work they discover that life is always based on very long strands of "DNA" that come in seven different configurations, such as the ones in figure 91. A careful inspection of the different strip "designs" reveals that each one of them can be obtained by a symmetry operation or a combination of symmetry operations on the basic symbol *b*. For instance, the first strand involves only translation symmetry—a motif is simply shifted repeatedly. The second strand represents a *glide reflection*, which, as you recall (chapter 1), involves mirror images that are translated in relation to one another. The fourth strand is obtained by translation and reflection about a horizontal mirror line. The sixth "DNA" pattern can be obtained in several different ways—for instance, through successive translations of four symbols, or through successive glide reflections of a pair of reflected symbols. In an attempt to formulate their findings in the language of a law, the extraterrestrial biologists might conclude that all DNA strands are arranged in patterns that are symmetric under combinations of translations, rotations, reflections, and glide reflections. Suppose, however, that the biologists had a hunch to begin with (maybe after discovering a few strands), that DNA strands have to obey some symmetries. They could then approach the problem from the opposite end and require at the outset that DNA strands would be symmetric. Clearly, there is no way to guess the basic motif—it could look like *b*, like a star, or like the AFLAC duck. However, once the motif is discovered, one can use group theory to prove that there are only seven distinct strip patterns that can be formed using combinations of the above four symmetries. All other patterns are merely variations on the seven different themes. In other words, the requirement of symmetry in this case dictates unequivocally the number of frieze patterns that exist. Princeton mathematician John Horton Conway has given amusing names to the seven different types of strip patterns. The names correspond to the pattern of footprints obtained when each of the actions is

```
  (i) b b b b b b b b b b b ...
 (ii) b p b p b p b p b p b ...
(iii) b d b d b d b d b d b ...
 (iv) b b b b b b b b b b b
      p p p p p p p p p p p ...
  (v) b q b q b q b q b q b ...
 (vi) b q p d b q p d b q p ...
(vii) b d b d b d b d b d b
      p q p q p q p q p q p ...
```

Figure 91

repeated: hop, step, jump, sidle, spinning hop, spinning sidle, spinning jump.

The symmetries of the laws of physics under translations, rotations, and uniform motion (including the invariance of the speed of light) are absolutely essential to our understanding of space and time, but they do not in themselves impose the existence of new forces or new particles. As we shall soon see, however, the attempts to understand gravity, and to unify all of the basic forces of nature, have elevated the significance of symmetry principles to a yet higher level—symmetry has become the *source* of forces.

A WEIGHTY SYMMETRY

The brilliance of special relativity expanded the horizons of the symmetry of the laws of physics to all uniformly moving observers. But, you may wonder, what about accelerating observers? By and large, most motions we observe around us are not uniform—they start from rest, come to a rest, or involve deflections, curves, or rotations. If the laws of electromagnetism, say, were to break down, or even just change significantly in a rocket accelerating from its launch pad, we would not be able to send astronauts into space. Einstein was not prepared to accept this as an option. Indeed, why should the laws depend on how the observer is moving? Moreover, accelerated motion is so ubiquitous—from the motion of planets around the Sun to sprinters on a track—that any theory that does not discuss acceleration is hopelessly incomplete. Another obvious deficiency of special relativity was the fact that the theory totally ignored gravity. Yet gravity is everywhere, and unlike electromagnetism from the forces of which one can shield, there is no way to evade gravity's grip. One of Einstein's main goals became, therefore, to extend symmetry's reach even further. In particular, he felt that the laws of nature have to look precisely the same not only to observers moving with constant velocities, but to all observers, whether in a laboratory that is accelerating along a straight line, rotating on a merry-go-round, or moving in whichever way. Just as the fictional biologists in the previous section could have started with the symmetry principle and then deduce from it the seven possible strip patterns, Einstein also wanted to put symmetry first. Inspired by the Lorentz covariance of special relativity (the fact that the equations do not change under rotations of space-time), he now required general covariance, implying symmetry of the laws of nature—whichever those may be—under any change in the space and time coordinates. This was not a trivial requirement. After all, about a million whiplash injuries per year in the United States alone demonstrate that people do feel sudden accelerations. Every time we take a sharp turn with the car we feel our body being pushed sideways by the centrifugal force, and airplanes hitting air pockets make our stomachs physically leap into our throats. On the face of it, there appears to be an unmistakable distinction between uniform and accelerating motion. When you ride a train or an elevator that moves at a constant speed, you don't feel the motion. Your point of view—that you are at rest while everything around you is moving—is as valid as that of the people waving good-bye on the platform or those waiting patiently in the hotel lobby. When an astronaut's cheeks get pulled down forcefully at launch, on the other hand, she definitely feels the acceleration. So how can the laws of physics be the same even in accelerating frames of reference? What about these additional forces? The culminating solution to this puzzle was Einstein's crowning achievement, one that took him years to conceive. Let us try to follow his train of thought as he sought to establish symmetry as the source of the laws of physics.

Imagine life in an accelerating boxcar (figure 92). If the boxcar is constantly accelerating to the right, we know from everyday experience that everything will be pushed backward (to the left in the figure). The lamp hung from the ceiling, for instance, would be tilted from the vertical direction. Every object dropped to the floor would fall at an angle, and every person sitting in a chair facing forward would feel pressure both from the seat underneath and from the back of the chair. This is very easy to understand. If a man in the boxcar drops his keys, the horizontal speed of the keys remains unchanged (apart from small changes due to the air's resistance) and equal to the speed the keys had at the instant they were dropped. At the same time, the boxcar itself continuously accelerates to higher and higher speeds. The keys are therefore left behind, resulting in

Figure 92

a path that is tilted backward. Here, however, comes an important realization. The experiences of the person in the accelerating boxcar are no different from those one would have if gravity itself were stronger and were tilted, instead of pointing straight down. Put differently, the gravitational force produces precisely the same phenomena as those observed in accelerated motion.

Consider another situation. When you stand on a bathroom scale inside an elevator that is accelerating upward, the scale will register a higher weight (because your feet exert a greater pressure on the scale)—as if gravity became stronger. An elevator accelerating downward would feel like a weaker gravity. In the extreme case that the elevator's cable snaps, you and the scale would be free-falling in unison, and the scale would register zero weight. (This is not a recommended weight-loss procedure, however—think of what the scale would record when the elevator does eventually hit the bottom of the shaft!) Astronauts float "weightless" inside the space station not because they are outside the reach of the Earth's gravity, but because both the station and the astronauts undergo the same acceleration toward the Earth's center—they are both free-falling.

While pondering various thought experiments of this type, Einstein was eventually led in 1907 to an electrifying conclusion: *The force of gravity and the force resulting from acceleration are in fact the same.* This powerful unification was dubbed the *equivalence principle*—acceleration and gravity are two facets of the same force; they are equivalent. Inside a free-falling elevator it is impossible to tell whether you are weightless because the elevator is accelerating downward or because gravity has been miraculously "switched off." Einstein described that moment of epiphany he had in 1907 in a lecture delivered in Kyoto in 1922: "I was sitting in the patent office in Bern [Switzerland] when all of a sudden a thought occurred to me: If a person falls freely, he won't feel his own weight. I was startled. This simple thought made a deep impression on me. It impelled me toward a theory of gravitation." Medical laboratories take advantage of the equivalence principle all the time. They use centrifuges to whirl fluids rapidly to separate substances of different densities. The centrifuges act as artificial-gravity machines. The acceleration of the rotational motion is equivalent to an increased gravitational force.

A statement of a pervasive symmetry accompanied the equivalence principle—the laws of physics, as expressed by Einstein's equations of general relativity, are precisely the same in *all* systems, including accelerating ones. That is, the laws are symmetric under any change in the spacetime coordinates. So why are there apparent differences between what is observed, say, on a merry-go-round and in a laboratory at rest? Those, general relativity tells us, are only differences in the *environment*, not in the laws themselves. In the same way that up and down appear to be different on Earth (in spite of the symmetry of the laws under rotations) because of the Earth's gravity, observers on the merry-go-round feel the centrifugal force that is equivalent to gravity. In other words, the symmetry among all frames of reference, including accelerating ones, *necessitates* the existence of gravity. As the examples of the accelerating boxcar and the elevator have shown us, the laws of physics in an accelerating frame are indistinguishable from those in a frame that experiences gravity.

Armed with the powerful insights afforded by the equivalence principle, Einstein felt that he was finally ready to tackle the two most intriguing questions that Newton's theory of gravitation had left totally unanswered. First and foremost was the million-dollar "how" question: How does gravity do its trick? Or alternatively: How can the Sun, which is at a distance of almost a hundred million miles from Earth, exert an inescapable gravitational pull that holds the Earth in its orbit?

Newton was fully aware of the fact that he had no answer:

> Hitherto we have explained the phenomena of the heavens and of our sea by the power of gravity, *but have not yet assigned the cause of this power* [emphasis added]. This is certain, that it must proceed from a cause that penetrates to the very centres of the Sun and planets, without suffering the least diminution of its force . . . and propagates its virtue on all sides to immense distances, decreasing always as the inverse square of the distances. . . . But hitherto I have not been able to discover the cause of those properties of gravity from phenomena, and I frame no hypotheses.

Second, there was the disturbing conflict between special relativity and Newton's notion of gravity. While the former states definitively that no mass, energy, or information of any sort can propagate faster than light, Newton envisaged gravity as exerting its force instantaneously across the vast expanses of space. Such a "speedy" gravity could have opened the door to some truly bizarre and undesired phenomena. For instance, if the Sun were to suddenly disappear, all the planets in the solar system would immediately start to move along nearly straight lines, since the force holding them in elliptical orbits would vanish. However, the Sun would actually disappear from view to people on Earth only about eight minutes later, since it takes light that long to traverse the Sun-Earth distance. If inhabitants of Neptune existed, they would start their journey into cold space a full four hours before they would see the Sun disappearing. Such cause-and-effect topsy-turviness would turn our perception of reality into an incomprehensible nightmare. Being a firm believer in the correctness of both special relativity and the equivalence principle, Einstein realized that the time had come for a complete overhaul of Newton's theory of gravitation.

The first hints of the possibility of a warped spacetime may have dawned upon Einstein from another intriguing thought experiment. This had originally been proposed by physicist Paul Ehrenfest (1880–1933) and later became known as *Ehrenfest's paradox.* One of the known results of special relativity is that the length of moving bodies, as measured by observers at rest, contracts along their direction of motion. The contraction is larger the higher the speed. This is no illusion—a moving rod can be momentarily confined in a space in which it would not fit when at rest. Consider then what happens to a flat object, such as a compact disc, when it is spinning very rapidly. Since the circumference rotates faster than the interior, it would contract more. This would distort and warp the shape of the disk. Once the idea of acceleration as a source of warps was introduced, Einstein would not let go of it. He concluded that acceleration would warp the very fabric of spacetime. And, according to the equivalence principle, if acceleration can cause space to be curved, so can gravity. This became the essence of general relativity—*gravity warps and bends spacetime* in the same way that circus trapeze artists cause the safety net on which they land to sag. Just as heavier objects would cause a more pronounced distortion in a trampoline, the higher the mass of a body, the more curved spacetime becomes in its vicinity. The path of a Jeep negotiating sand dunes in the Sahara is determined by the shape of the undulating terrain. Similarly, the paths of the planets around the Sun are a consequence of the curvature the Sun produces in spacetime. The planets are simply seeking the most direct route, and the shapes of their orbits reveal the curved geometry of spacetime. Within the framework of a warped spacetime, gravity's influence is definitely not instantaneous. Einstein calculated that disturbances in the shape of spacetime propagate like ripples in a pond, *precisely at the speed of light.* If the Sun were to miraculously disappear, the vanishing of its gravitational influence would reach Earth in eight minutes—simultaneously with its visual disappearance. This gratifying result eliminated the last nagging problem of Newtonian physics.

The fact that Einstein turned curved spacetime into the cornerstone of his new theory of the cosmos created a need for mathematical tools to describe such spaces. The math classes he had missed at school came back to haunt him. Fortunately, the former math skeptic had someone to turn to—Marcel Grossman (1878–1936), Einstein's old classmate, and an accomplished mathematician. In an uncharacteristically helpless tone Einstein repented: "I have become imbued with great respect for mathematics, the more subtle parts of which I had previously regarded as sheer luxury!" The ever-reliable Grossman did not fail to deliver. He pointed Einstein both to the non-Euclidean geometry of Riemann and to mathematical methods developed by the mathematicians Elwin Christoffel, Gregorio Ricci-Curbastro, and Tullio Levi-Civita. Recall that Riemann had in fact "anticipated" precisely the machinery that Einstein needed—a geometry of curved spaces in any number of dimensions. The introduction of calculus into geometry through the branch known as *differential geometry,* and the development of *tensor calculus* further allowed for precise calculations to be carried out (tensors are "boxes of numbers" that can represent spaces in any number of dimensions). After a few dead ends in the years 1912–15, Einstein decided to follow his main guiding light—the symmetry of all frames implied by the principle of general covariance. His intuition bore fruit, and at the end of 1915, general relativity, an all-embracing theory of spacetime and gravity, was born (figure 93 shows the front page of the paper). In a note to theoretical physicist Arnold Sommerfeld, Einstein could not hide his exuberance: "Be sure you take a good look at them [the equations of general relativity]; they are the most valuable discovery of my life."

Einstein was the first to acknowledge his debt to mathematics. In an address to the Prussian Academy of Sciences in 1921 he declared, "We may in fact regard [geometry] as the most ancient branch of physics. . . . Without it I would have been unable to formulate the theory of relativity." In a lecture in 1933 he added, "The creative principle [of science] resides in mathematics."

Almost from the day of its first appearance, the underlying symmetry and logical simplicity of general relativity won it many admirers among the greatest physicists of the time. Ernest Rutherford (who discovered the atomic nucleus) and Max Born (a quantum mechanics pioneer) later compared the theory to a work of art.

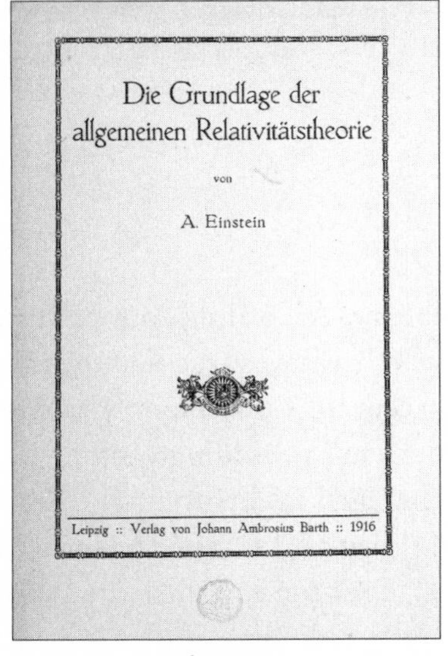
Die Grundlage der allgemeinen Relativitätstheorie

von

A. Einstein

Leipzig :: Verlag von Johann Ambrosius Barth :: 1916

Figure 93

One of the key predictions of general relativity was the bending of light rays under the influence of gravity. In particular, the Sun was predicted to bend starlight from distant stars positioned directly behind it. For the light from the Sun not to totally overwhelm the light from the stars, the observations had to be carried out during a total solar eclipse, when the Moon blocks out the Sun's light. The idea at the basis of the experiment was simple: By comparing a photograph taken during a solar eclipse to a photograph of the same patch of the sky taken when the starlight is undeflected, one could attempt to measure the apparent slight shifts in stellar positions caused by the bending of light.

The observations, by two British teams, took place during the solar eclipse of May 29, 1919, but Einstein did not receive the final confirmation of the results until September 22. The two teams, one led by the famous British astrophysicist Arthur Eddington (1882–1944), found an average deflection of 1.79 seconds of arc, which agreed perfectly (within the expected experimental errors) with the prediction of general relativity. The joyously enthusiastic Einstein was quick to inform his mother. The confirmation of general relativity was formally announced in a joint meeting of the Royal Society and the Royal Astronomical Society in London on November 6, 1919, and was rightfully pronounced to be "one of the greatest achievements in the history of human thought." The next day, the entire world woke up to the news of a "Revolution in Science" (figure 94 shows the article in the London *Times* of November 7, 1919), and Einstein was instantly propelled to the unexpected status of a media star. Not that everybody fully understood all the implications of the new

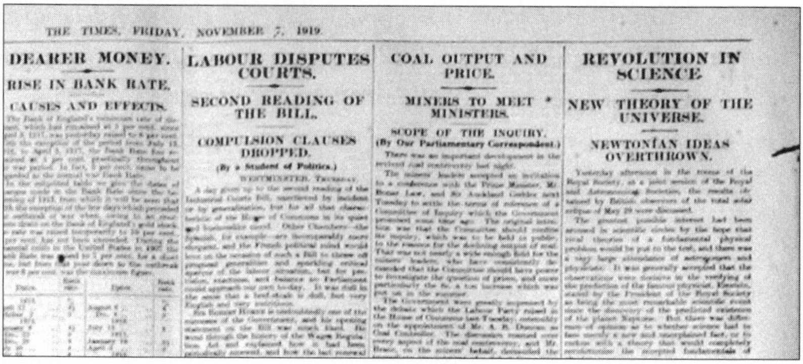
THE TIMES, FRIDAY, NOVEMBER 7, 1919.

DEARER MONEY.
RISE IN BANK RATE.
CAUSES AND EFFECTS.

LABOUR DISPUTES COURTS.
SECOND READING OF THE BILL.
COMPULSION CLAUSES DROPPED.
(By a Student of Politics.)

COAL OUTPUT AND PRICE.
MINERS TO MEET MINISTERS.
SCOPE OF THE INQUIRY.
(By Our Parliamentary Correspondent.)

REVOLUTION IN SCIENCE.
NEW THEORY OF THE UNIVERSE.
NEWTONIAN IDEAS OVERTHROWN.

Figure 94

theory. According to a popular anecdote, a reporter asked Eddington whether it was true that the theory of relativity was so complicated that, except for Einstein, only two other people in the world could really understand it. Eddington sat silently for a few minutes. The reporter encouraged Eddington not to be too modest, to which Eddington replied, "Not at all, I was just trying to figure out who the other person was."

Even today, I am in total awe of the following wondrous chain of ideas and interconnections. Guided throughout by principles of symmetry, Einstein first showed that acceleration and gravity are really two sides of the same coin. He then expanded the concept to demonstrate that gravity merely reflects the geometry of spacetime. The instruments he used to develop the theory were Riemann's non-Euclidean geometries—precisely the same geometries used by Felix Klein to show that geometry is in fact a manifestation of group theory (because every geometry is defined by its symmetries—the objects it leaves unchanged). Isn't this amazing?

Recall that Galois was rather uncertain about the potential applications of his group-theoretical ideas. The combined power of the imaginations of mathematicians such as Klein, Lie, Riemann, Minkowski, Poincaré, and Hilbert "joined forces" with the unsurpassed physical intuition of Einstein to turn symmetry and group theory into the most basic descriptors of spacetime and gravity.

INTO THE QUANTUM WORLD

As important as symmetries are for the laws describing spacetime and gravity, their importance is magnified even further in the realm of the subatomic particles. Unlike in classical physics, where the word *particle* usually conjures up an image of something like a tiny billiard ball, in quantum theory—the theoretical framework used in particle physics—particles can behave as waves. The state of any system and its time evolution are described by an entity called the *wavefunction.* The wavefunction of an electron is a probability wave, used, for instance, to determine the probability of finding the electron at a certain position with a particular spin direction. Since all the electrons in the universe are identical, the only way to distinguish one from the other is by their energy, momentum (product of mass and velocity in classical physics), and spin. These basic quantities are defined in quantum mechanics by the response of the wavefunction to various symmetry transformations in space and time. The energy, for instance, reflects the change in the wavefunction that results from shifting the time coordinate (equivalent to the resetting of the clocks). Let me explain this concept briefly. Imagine that two photographers take a picture of the circular waves that propagate from the point of impact of a pebble thrown into a pond. The flashes on both cameras are set to go off precisely at 8:00 a.m. However, one of the two clocks that control the flashes happens to be off by one second. This means that although the two cameras will record the same wave, they will record it at slightly different phases. Where one photograph shows a crest in the wave, the other may show a trough, and vice versa. Quantum mechanics defines the energy of a system, such as the electron, through the change in the phase of its wavefunction (measured in cycles of the wave) caused by resetting the clocks by one second. Similarly, the momentum of the electron characterizes the change in the phase of the wavefunction under a slight translation in space. While these definitions do relate basic physical properties to symmetry transformations, they probably sound surprisingly abstract. Anyone who had some high-school physics may remember that quantities such as energy and momentum are normally associated with a rather different concept—*conservation laws.* Conservation laws reflect the fact that some quantities can neither be created nor destroyed—they have the same values whether we measure them today, tomorrow, or a million years from now. The conservation of energy is the physical equivalent of the phrase "there is no free lunch." If we could get energy out of nothing we wouldn't be paying more at gas pumps every time oil production dwindles. Conservation of momentum is familiar to anyone who has watched billiard balls collide. You will never see the two balls rolling backward (toward the player)—the total momentum of the recoiling balls has to combine to equal the momentum of the cue ball. Conservation laws are the physicist's breakfast, lunch, and dinner. Experimental particle physicists, for instance, use huge accelerators to smash particles into one another. These accelerators are gigantic structures (the one in Geneva, Switzerland, uses a circular tunnel that is seventeen miles long), in which subatomic particles are accelerated to extremely high energies. The goal is to probe the fundamental forces on shorter and shorter distances, and to produce heavier particles that are predicted theoretically to exist. The experimenters take advantage of the fact that the total energy and momentum of the collision products have to be precisely equal to those of the incoming particle and the target (because of the conservation laws) to determine even the properties of particles that cannot be detected directly by the experimental apparatus.

On the face of it, therefore, we seem to have two unrelated definitions. On one hand, basic quantities such as energy and momentum are defined through the response of the wavefunction to symmetry transformations. On the other, the same quantities are associated with conservation laws. What is the precise relation between symmetries of the laws of physics and conservation laws? The unexpected answer was given by the German mathematician Emmy Noether (1882–1935), and it is usually referred to as *Noether's theorem.* Before I explain this result, however, I want to describe very briefly the life of this extraordinary woman, in order to shed some light on the type of difficulties a woman experienced in a male-dominated mathematical world.

Emmy Noether was born in Erlangen, Germany, where her father was a professor of mathematics. Emmy's original intention was to become a language teacher of French and English, but at age eighteen she decided instead to study mathematics. This turned out to be easier said than done. Although women had been allowed to enroll at universities in France since 1861, this was still not permitted officially in the conservative Germany of 1900. The academic senate at the University of Erlangen declared in 1898 that the admission of female students would "overthrow all academic order." Nevertheless, Emmy was at least given special permission to attend some courses. After successfully passing exams in Nürnberg, Göttingen, and Erlangen and benefiting from the slow but gradual changes in the gender bias, she was eventually awarded a doctorate in mathematics in 1907. This was not, however, the end of her battles with the German academic establishment. Even though Noether was invited in 1915 by David Hilbert and Felix Klein to join the faculty at Göttingen, these two renowned mathematicians had to fight the university authorities for four more years before she was formally allowed to teach. During the period of the letter exchanges and verbal skirmishes with the administration, Hilbert tricked the bureaucrats by permitting Emmy to lecture in courses advertised officially under his own name.

Noether proved the theorem bearing her name in 1915, shortly after her arrival at Göttingen. She started by examining continuous symmetries. These are symmetries under transformations that can be varied continuously, such as rotations (where the rotation angle can be changed continuously). The symmetry of a sphere, for instance, holds under arbitrarily small rotations, unlike the discrete symmetry of a snowflake, which is symmetric only under rotations by multiples of 60 degrees. The result that Noether obtained was stunning. She showed that *to every continuous symmetry of the laws of physics there corresponds a conservation law and vice versa.* In particular, the familiar symmetry of the laws under translations corresponds to conservation of momentum, the symmetry with respect to the passing of time (the fact that the laws do not change with time) gives us conservation of energy, and the symmetry under rotations produces conservation of angular momentum. Angular momentum is a quantity characterizing the amount of rotation an object or a system possesses (for a pointlike object it is the product of the distance from the axis of rotation and the momentum). A common manifestation of conservation of angular momentum can be seen in figure skating—when the ice skater pulls her hands inward she spins much faster.

Noether's theorem fused together symmetries and conservation laws—these two giant pillars of physics are actually nothing but different facets of the same fundamental property.

With the rise of the Nazis to power, Noether, whose parents were both Jewish, was forced to leave Germany, and she moved to Bryn Mawr College in the United States. She continued to lecture at Bryn Mawr and Princeton until her sudden death, following surgery, in 1935. In his memorial address, mathematical physicist Hermann Weyl alluded to the struggles Emmy Noether had to endure because of her gender: "If we in Göttingen often chaffingly referred to her as 'der Noether' (with the masculine article) it was also done with a respectful recognition of her power as a creative thinker who seemed to have broken through the barrier of sex."

Most of the symmetries we have encountered so far had to do with a change in our viewpoint in space and time. Many of the symmetries underlying elementary particles and the basic forces of nature are of a different type—we change our perspective on the identity of particles. This may sound alarming; an electron is always an electron, right? Not really, when it comes to the fuzziness of the quantum realm.

Recall that the only thing that is certain in quantum mechanics is that everything is uncertain. Only probabilities can truly be determined. An electron can be in a state in which it is spinning neither definitely in one sense nor in the other. Rather, the state is a mixture of spinning clockwise with spinning counterclockwise. More surprisingly, electrons can be in states that mix them with another elementary particle called a neutrino. The neutrino is a particle of almost zero mass and no electric charge. Just as the Moon can be full, dark, and anything in between, particles can carry the label "electron," "neutrino," or be a mixture of both, until we perform a specific measurement (such as that of the electric charge) that can distinguish the two. The realization of this ability of particles to metamorphose between different states took physicists an important step toward the unification of all the forces of nature.

Newton was the first to introduce the concept of unification. His theory of gravity unified the force that keeps our feet on the ground with the force that holds planets in their orbits. Before Newton no one suspected that one force is responsible for both. Michael Faraday and James Clerk Maxwell introduced the second major unification—they proved that electric and magnetic forces are in fact the same force in different guises. Varying the electric field generates a magnetic field and vice versa. In addition to the gravitational and electromagnetic forces, we currently distinguish in nature two nuclear forces. One, the *strong nuclear force,* is what holds protons and neutrons tightly bound together in the atomic nucleus. Without it, protons would fly apart due to their mutual electromagnetic repulsion, so no element other than hydrogen (which has only one proton) would have ever formed. The *weak nuclear force* is responsible for the radioactive decay of uranium, and it transforms a neutron into a proton, while creating in the process an electron and an antineutrino (the "antiparticle" of the neutrino). These radioactive decays were first discovered experimentally in 1896, but their association with the weak nuclear force was clarified only in the 1930s.

In the late 1960s, physicists Steven Weinberg, Abdus Salam, and Sheldon Glashow conquered the next unification frontier. In a phenomenal piece of scientific work they showed that the electromagnetic and weak nuclear forces are nothing but different aspects of the same force, subsequently dubbed the *electroweak force.* The predictions of the new theory were dramatic. The electromagnetic force is produced when electrically charged particles exchange between them bundles of energy called *photons.* The photon is therefore the messenger of electromagnetism. The electroweak theory predicted the existence of close siblings to the photon, which play the messenger role for the weak force. These never-before-seen particles were prefigured to be about ninety times more massive than the proton and to come in both an electrically charged (called W) and a neutral (called Z) variety. Experiments performed at the European consortium for nuclear research in Geneva (known as CERN for *Conseil Européen pour la Recherche Nucléaire*) discovered the W and Z particles in 1983 and 1984 respectively. (Incidentally, Dan Brown's bestselling thriller *Angels and Demons* has brought the research at CERN to the attention of millions of readers.)

The W and Z are eighty-six and ninety-seven times more massive than the proton (respectively), just as the theory predicted. This was undoubtedly one of symmetry's greatest success stories. Glashow, Weinberg, and Salam managed to unmask the electromagnetic and weak forces by recognizing that underneath the differences in the strengths of these two forces (the electromagnetic force is about a hundred thousand times stronger within the nucleus) and the different masses of the messenger particles lay a remarkable symmetry. The forces of nature take the same form if electrons are interchanged with neutrinos or with any mixture of the two. The same is true when photons are interchanged with the W and Z force-messengers. The symmetry persists even if the mixtures vary from place to place or from time to time. The invariance of the laws under such transformations performed locally in space and time has become known as *gauge symmetry.* In the professional jargon, a *gauge transformation* represents a freedom in formulating the theory that has no directly observable effects—in other words, a transformation to which the physical interpretation is insensitive. Just as the symmetry of the laws of nature under any change of the spacetime coordinates requires the existence of gravity, the gauge symmetry between electrons and neutrinos requires the existence of the photon and the W and Z messenger particles. Once again, when the symmetry is put first, the laws practically write themselves. A similar phenomenon, with symmetry dictating the presence of new particle fields, repeats itself with the strong nuclear force.

QUARK, QUARK, GROUP

Protons and neutrons, the particles that make up the atomic nucleus, are not "elementary." They are composed of elementary building blocks called *quarks.* The name *quark* was chosen by particle physicist Murray Gell-Mann in 1963. He settled on a word that combines a dog's bark with a seagull's squawk, coined by the famous Irish novelist James Joyce in *Finnegans Wake:*

Three quarks for Muster
Mark!
Sure he hasn't got much of
A bark
And sure any he had it's all
Beside the mark.

Quarks come in six "flavors" that were given the rather arbitrary names: up, down, strange, charm, top, and bottom. Protons, for instance, are made of two up quarks and one down quark, while neutrons consist of two down quarks and one up quark. Other than ordinary electric charge, quarks possess another type of charge, which has been fancifully called *color,* even though it has nothing to do with anything we can see. In the same way that the electric charge lies at the root of electromagnetic forces, color originates the strong nuclear force. Each quark flavor comes in three different colors, conventionally called red, green, and blue. There are, therefore, eighteen different quarks.

The forces of nature are color blind. Just as an infinite chessboard would look the same if we interchanged black and white, the force between a green quark and a red quark is the same as that between two blue quarks, or a blue quark and a green quark. Even if we were to use our quantum mechanical "palette" and replace each of the "pure" color states with a mixed-color state (e.g., "yellow" representing a mixture of red and green or "cyan" for a blue-green mixture), the laws of nature would still take the same form. The laws are symmetric under any color transformation. Furthermore, the color symmetry is again a gauge symmetry—the laws of nature do not care if the colors or color assortments vary from position to position or from one moment to the next.

We have already seen that the gauge symmetry that characterizes the electroweak force—the freedom to interchange electrons and neutrinos—dictates the existence of the messenger electroweak fields (photon, W, and Z). Similarly, the gauge color symmetry requires the presence of eight *gluon* fields. The gluons are the messengers of the strong force that binds quarks together to form composite particles such as the proton. Incidentally, the color "charges" of the three quarks that make up a proton or a neutron are all different (red, blue, and green), and they add up to give zero color charge or "white" (equivalent to being electrically neutral in electromagnetism). Since color symmetry is at the base of the gluon-mediated force between quarks, the theory of these forces has become known as *quantum chromodynamics.* The marriage of the electroweak theory (which describes the electromagnetic and weak forces) with quantum chromodynamics (which describes the strong force) produced the *standard model*—the basic theory of elementary particles and the physical laws that govern them.

If you are starting to feel a bit dizzy from all of these different elementary particles, you are not alone. The famous physicist Enrico Fermi (1901–54), who was considered the "last universal scientist" (meaning that he knew all the areas of physics), is quoted to have once said, "If I could remember the names of all these particles [far fewer were known at his time], I'd be a botanist." Some of the exotic properties of elementary particles have made it even into popular culture. Physicist and author Cindy Schwarz has compiled an entire collection of prose and poetry about elementary particles, written by students at Vassar College. One such poem, by Vanessa Pepoy, is entitled "Chromodynamics":

Rouge vert bleu
Trinity of color.
Fundamental.
Organizational.
Principle.

Contained within
A particle,
White light
Invisible.

You may have noticed that particles that are involved in gauge symmetries tend to form families of closely related kin (e.g., protons and neutrons). Historically, even before the suggestion that protons and neutrons are both composed of three gluon-exchanging quarks, physicists noticed striking similarities between these two intranuclear neighbors. Not only are they very close in mass, but also the strong force between them is indifferent to whether it is acting between two neutrons, a neutron and a proton, or any two mixed states of the two. With the advent of high-energy particle accelerators in the 1950s, an entire particle zoo seemed to have emerged. In an attempt to put some order into the rapidly proliferating menagerie, Murray Gell-Mann and the Israeli physicist Yuval Ne'eman noticed that the protons and neutrons looked very similar to six other particles. They also identified other such extended families of eight or ten members. Gell-Mann called this symmetry the "eightfold way," alluding to the eight principles in the Buddhist path of self-development that are supposed to lead to the end of suffering. The realization that symmetry is the key to the understanding of the properties of subatomic particles led to an inevitable question: Is there an efficient way to characterize all of these symmetries of the laws of nature? Or, more specifically, what is the basic theory of transformations that can continuously change one mixture of particles into another and produce the observed families? By now you have probably guessed the answer. The profound truth in the phrase I have cited earlier in this book revealed itself once again: "Wherever groups disclosed themselves, or could be introduced, simplicity crystallized out of comparative chaos." The physicists of the 1960s were thrilled to discover that mathematicians had already paved the way. Just as fifty years earlier Einstein learned about the geometry tool-kit prepared by Riemann, Gell-Mann and Ne'eman stumbled upon the impressive group-theoretical work of Sophus Lie. Lie's ideas have become so central to high-energy physics that a few words about this outstanding mathematician are in order.

Sophus Lie (figure 95) arrived at mathematics in a somewhat roundabout way. At the Royal Fredrik's University of Christiania (today's

Oslo) he demonstrated neither a particular passion for nor an unusual ability in mathematics, even though he did study the works of Abel and Galois. One of his teachers, Ludvig Sylow (1832–1918), himself a famous mathematician, confessed later that he never would have guessed that the young Lie would become one of the greatest mathematical minds of the century. Yet, after a few years of hesitation, during which he had been haunted by suicidal tendencies, Lie turned his interests more and more toward mathematics. In 1868 he finally concluded that "there was a mathematician hidden inside me."

Figure 95

During trips to Berlin and Paris in 1869 and 1870, Lie met and befriended Felix Klein. In Paris he also met Camille Jordan (1838–1922), and the latter convinced him that group theory could play a crucial role in the study of geometry. The combined efforts of Lie and Klein in this arena provided the seeds for Klein's celebrated Erlangen Program on the group-theoretical characterization of geometry.

In 1870, political events complicated the continued collaboration between the two young mathematicians. The outbreak of the Franco-Prussian War forced Klein to leave Paris for Berlin. Lie tried to hike his way to Italy, but he only made it as far as Fontainbleau, where he was arrested. To French army officials the dense mathematical papers of the Norwegian surely looked like the encoded messages of a German spy. Fortunately for Lie, the French mathematician Gaston Darboux intervened and released him from prison. Two years later, the University of Christiania did not repeat the mistake it had made with Abel. The faculty and officials recognized Lie's unusual talents and created a mathematics chair for him. Lie continued to collaborate with Klein on and off until 1892, when an ugly dispute erupted between the two. Partly this had to do with Lie's perception that he had not received the appropriate recognition for his role in the development of the Erlangen Program. In 1893 Lie issued a statement that publicly attacked Klein and declared, "I am no pupil of Klein nor is the opposite the case, although this might be closer to the truth." Klein did not help the situation by calling attention (supposedly in "defense" of Lie's actions) to the mental problems from which Lie had suffered during the late 1880s. None of these events take anything away from Lie's genius.

The two Norwegian giants of the end of the nineteenth century, Lie and Sylow, fully acknowledged their intellectual debt to the shining star of Norwegian mathematics—Abel. For a period of eight years they undertook the painstaking task of preparing and publishing Abel's complete works. Around the same period, Lie started to work on groups of continuous transformations (such as translations and rotations in ordinary space). This project culminated with the publication of an extensive theory and a detailed catalog of such groups between 1888 and 1893 (in collaboration with the German mathematician Friedrich Engel). The members of the class of continuous groups studied by Lie have later become known as *Lie groups.*

Lie groups were precisely the instruments Gell-Mann and Ne'eman needed to characterize the underlying pattern of the newly discovered zoo of particles. Much to their delight, the two physicists found out that the German mathematician Wilhelm Killing (1847–1923) and the French mathematician Elie-Joseph Cartan (1869–1951) had made their job even easier. Recall that for his proof on the solvability of equations Galois defined some special subgroups called normal subgroups (chapter 6). When a group has no normal subgroups (other than the two trivial subgroups, one composed only of the identity and the other being the group itself), it is called *simple.* Simple groups are the basic building blocks of group theory in the same sense that prime numbers (divisible only by themselves and 1) are the building blocks of all the integer numbers. In other words, all the groups can be constructed from simple groups, and the simple groups themselves cannot be decomposed any further by the same process. Killing outlined the classification of the simple Lie groups in 1888; the classification was completed and perfected by Cartan in 1894. There are four infinite families of simple Lie groups, and five *exceptional* (or *sporadic*) simple groups that do not fit into any of the four families. Gell-Mann and Ne'eman discovered that one such simple Lie group, called "special unitary group of degree 3," or SU(3), was particularly well suited for the "eightfold way"—the family structure the particles were found to obey. The beauty of the SU(3) symmetry was revealed in full glory via its predictive power. Gell-Mann and Ne'eman showed that if the theory were to hold true, a previously unknown tenth member of a particular family of nine particles had to be found. The extensive hunt for the missing particle was conducted in an accelerator experiment in 1964 at Brookhaven National Lab on Long Island. Yuval Ne'eman told me some years later that, upon hearing that half of the data had already been scrutinized without discovering the anticipated particle, he was contemplating leaving physics altogether. Symmetry triumphed at the end—the missing particle (called the omega minus) was found, and it had precisely the properties predicted by the theory.

All the symmetries that characterize the standard model (e.g., the symmetry of color exchange among quarks) can be represented as a product of simple Lie groups. The pioneering attempt to describe such physical symmetries mathematically was carried out by physicists Chen Ning Yang and Robert Mills in 1954. Appropriately, the equations that describe the weak force (in analogy with Maxwell's equations, which describe electromagnetism) are known as Yang-Mills equations. Through the works of Weinberg, Glashow, and Salam on the electroweak theory and the elegant framework developed by physicists David Gross, David Politzer, and Frank Wilczek for quantum chromodynamics, the characteristic group of the standard model has been identified with a product of three Lie groups denoted by U(1), SU(2), and SU(3). In some sense, therefore, the road toward the ultimate unification of the forces of nature has to go through the discovery of the most suitable Lie group that contains the product $U(1) \times SU(2) \times SU(3)$.

The experience with special and general relativity and the standard model of elementary particles can lead to only one conclusion. Symmetry and group theory have an uncanny way of directing physicists to the right path. This may seem somewhat surprising at first, since the requirement of symmetry imposes rather rigid constraints. As we have seen, once a pattern extending to infinity in one dimension is confined to obeying the rigid-motions symmetries, only seven distinct strip patterns are allowed. Even in two dimensions, one can prove that repeating "wallpaper" patterns are limited to seventeen. Similar restrictions are forced upon any theory that incorporates symmetry. Don't these constraints inhibit the freedom the theory might have otherwise had? They do, and this inhibition is a desirable outcome. Physicists are searching for *one* theory that explains the universe, not for many, all doing the job equally well. Had I presented you with twenty-three different theories of Galois's death in chapter 5, all fully consistent with the available evidence, you would probably not have been very satisfied. Symmetry helps us not only to avoid the false starts and the blind alleys, but also to cross the most difficult parts, the "decisions, decisions" phases that characterize choices.

The Bible tells us that when the Israelites left Egypt, they were led in the desert by "a pillar of fire by night, to give them light." Symmetry has been the scientists' pillar of fire, leading toward general relativity and the standard model. Can it also lead to a unification of the two?

THE HARMONY OF THE STRINGS

Historians like to point out that some social revolutions have been mistakes, when judged retrospectively. By contrast, the two scientific revolutions of the twentieth century have been unquestionable successes. General relativity predicted the bending of light by astronomical objects, the existence of the collapsed objects we call black holes, and the expansion of the universe, all of which have been observationally confirmed. Quantum theory has been confirmed in electrodynamics to an astonishing precision, and its crown jewel—the standard model—has successfully captured and predicted all the properties of the known subatomic particles. Here, however, lies the problem. We have a hugely successful theory for the largest astronomical scales (stars, galaxies, the universe) and another one for the smallest subatomic scales (atoms, quarks, photons). This might have been all right had the two worlds never had to meet. But in a universe that started to expand from a "big bang"—an extremely compact and ferociously hot state—it was unavoidable that the paths of general relativity and quantum mechanics would cross. Many pieces of evidence, such as the formation of the elements of the periodic table, point to the fact that even the large was once small. Furthermore, some entities, such as black holes, live in both the astronomical and the quantum domains. Consequently, in the wake of Einstein's unsuccessful attempts to unite general relativity with electromagnetism, many physicists have engaged in the greatest unification effort of them all—of general relativity with quantum mechanics.

The biggest stumbling block that has traditionally plagued all the unification endeavors has been the simple fact that on the face of it, general relativity and quantum mechanics really appear to be incompatible.

Recall that the key concept of quantum theory is the uncertainty principle. When you try to probe positions with an ever-increasing magnification power, the momenta (or speeds) start oscillating violently. Below a certain minuscule length known as the *Planck length,* the entire tenet of a smooth spacetime is lost. This length (equal to 0.000 . . . 4 of an inch, where the 4 is at the thirty-fourth decimal place) determines the scale at which gravity has to be treated quantum mechanically. For smaller scales, space turns into an ever-fluctuating "quantum foam." But the very basic premise of general relativity has been the existence of a gently curving spacetime. In other words, *the central ideas of general relativity and quantum mechanics clash irreconcilably when it comes to extremely small scales.*

The current best bet for a quantum theory of gravity appears to be some version of *string theory.* According to this revolutionary theory, elementary particles are not pointlike entities with no internal structure, as the standard model would have you believe, but tiny loops of vibrating strings. These infinitely thin, rubber-band-like loops are so small (on the order of the Planck length; about a hundred billion times smaller than the proton) that to the resolving power of present-day experiments they appear as points. The beauty of the principal idea of string theory is that all the known elementary particles are supposed to represent merely different vibration modes of the same basic string. Just as a violin or a guitar string can be plucked to produce different harmonics, different vibrational patterns of a basic string correspond to distinct matter particles, such as electrons and quarks. The same applies to the force carriers as well. Messenger particles such as gluons or the W and the Z owe their existence to yet other harmonics. Put simply, all the matter and force particles of the standard model are part of the repertoire that strings can play. Most impressively, however, a particular configuration of vibrating string was found to have properties that match precisely the *graviton*—the anticipated messenger of the gravitational force. This was the first time that the four basic forces of nature have been housed, if tentatively, under one roof.

You might have thought that an achievement of this magnitude—the Holy Grail of modern physics—would be immediately hailed by the entire physics community. Yet the reaction in the mid-1970s was rather different. Years of frustration with the attempts to unify general relativity with quantum mechanics have acted to build a thick wall of skepticism. The claim by physicists John Schwarz of the California Institute of Technology and Joël Scherk of the École normale supérieure in France that string theory finally unites gravity with the strong force was universally ignored. This situation persisted for over a decade. During that period, almost every step forward was followed immediately by the discovery of some subtle difficulty, resulting in nine-tenths of a step backward. The breakthrough finally occurred in 1984, when physicists Michael Green, then at Queen Mary College, and John Schwarz demonstrated that string theory might indeed provide the ultimate unification everyone was looking for. A frenzy of activity ensued as some of the best theoretical minds engaged in the hunt for what appeared to be the sought-after "theory of everything"—the ultimate foundation on which the rest of physics can be built. As is often the case in science, however, the burst of enthusiasm (dubbed the "first superstring revolution") soon gave way to a phase of frustration-rich hard work. Unlike in the case of SU(3), where all the mathematical tools had been in place, waiting for the physicists to make use of them, string theorists had to develop some of the mathematics as they went along. Nevertheless, as we shall see in the next section, groups still provided the right language to describe the underlying patterns.

So, how does string theory propose to resolve the fundamental conflict between the smooth geometry of general relativity and the violent fluctuations of quantum mechanics? By imparting some fuzziness even to spacetime, similar to the one quantum mechanics imparts to the positions and motions of particles.

Imagine that you want to draw a cloud. If the cloud you pick out in the sky for modeling is relatively distant, near the horizon, you can probably reproduce the shape that you observe quite accurately. If, on the other hand, the cloud is relatively close by, it becomes increasingly difficult to capture every twist and turn of its tiny wisps. Zooming farther in, down to the submolecular scale, will make any reproduction attempt hopeless. String theory asserts that by treating the elementary particles and the force messengers as dimensionless pointlike objects, physics attempted to probe the universe on scales that are below the limit that makes any sense. In other words, since strings, the most basic constituents of the universe, are extended objects with sizes on the order of the Planck length, sub–Planck-length distances are outside the realm of physics. By concentrating only on super–Planck-length scales, one can eliminate the violent fluctuations and avoid conflict. Not surprisingly, the fuzziness in the string-theoretical framework changes the nature of events in spacetime. While in the standard model every interaction between two particles occurs at a precisely well-defined point in spacetime, agreed upon by all observers, the situation in string theory is different (figure 96). Due to the strings' extended nature, we cannot say precisely when and where two strings interact. Both the location and the time of the interaction are "smeared out." The situation may be likened (only superficially) to our inability to predict when and where a wishbone pulled apart from both ends will break.

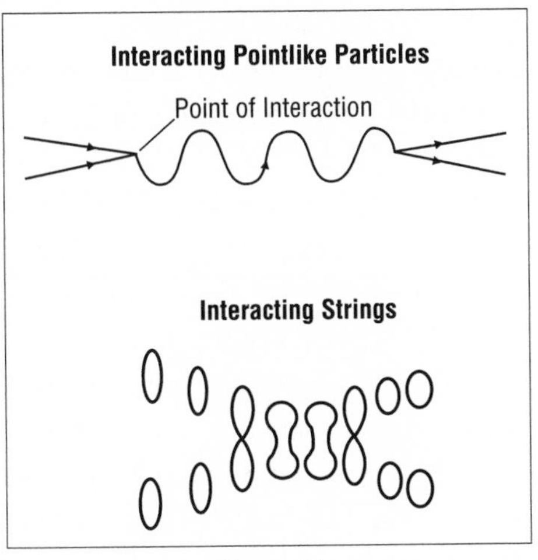

Figure 96

Having barely recovered from the revolution in the understanding of spacetime introduced by Einstein's relativity, physicists had to readjust to the new concepts introduced by the string revolution. Fortunately, one familiar concept has not only survived the revolution, but has reached its pinnacle through string theory.

NOT JUST SYMMETRY–SUPERSYMMETRY

The laws of nature do not depend on where, from which angle, or when we use them. They are symmetric under translations, rotations, and the passing of time. They are also identical for all observers, irrespective of whether these are moving at constant velocities or accelerating. This is the essence of Einstein's principle of general covariance. Just as the uniformly moving observers can declare themselves at rest, with everything around them being in motion, so can the accelerating observers. The latter are fully justified in claiming that the extra forces they feel are due to a gravitational field (according to the equivalence principle). By 1967 physicists thought that no other symmetries that are associated only with changing our vantage point in space and time could exist. In fact, there even existed a theorem that claimed to prove that this was the case.

To many physicists' surprise, intensive research during the subsequent four years led to the discovery that quantum mechanics allows for one additional symmetry. This unexpected symmetry was dubbed *supersymmetry.*

Supersymmetry is a subtle symmetry based on the quantum mechanical property spin. Recall (from chapter 1) that the spin of the electron is an intrinsic property, much like its electric charge, that resembles in some respects classical angular momentum—as if the electron were spinning around its axis. Unlike classically spinning bodies, such as tops, however, where the spin rate can assume any value fast or slow, electrons always have only one fixed spin. In the units in which this spin is measured quantum mechanically (called *Planck's constant*) the electrons have half a unit, or they are "spin-½" particles. In fact, all the matter particles in the standard model—electrons, quarks, neutrinos, and two other types called muons and taus—all have "spin ½." Particles with half-integer spin are known collectively as *fermions* (after the Italian physicist Enrico Fermi). On the other hand, the force carriers—the photon, W, Z, and gluons—all have one unit of spin, or they are "spin-1" particles in the physics lingo. The carrier of gravity—the graviton—has "spin 2," and this was precisely the identifying property that one of the vibrating strings was found to possess. All the particles with integer units of spin are called *bosons* (after the Indian physicist Satyendra Bose). Just as ordinary spacetime is associated with symmetry under rotations, the quantum-mechanical spacetime is associated with a supersymmetry that is based on spin. The predictions of supersymmetry, if it is truly obeyed, are far-reaching. In a universe based on supersymmetry, *every known particle in the universe must have an as-yet-undiscovered partner* (or "superpartner"). The matter particles with spin ½, such as electrons and quarks, should have spin 0 superpartners. The photon and gluons (that are spin 1) should have spin-½ superpartners called *photinos* and *gluinos* respectively. Most importantly, however, already in the 1970s physicists realized that the only way for string theory to include fermionic patterns of vibration at all (and therefore to be able to explain the constituents of matter) is for the theory to be supersymmetric. In the supersymmetric version of the theory, the bosonic and fermionic vibrational patterns come inevitably in pairs. Moreover, supersymmetric string theory managed to avoid another major headache that had been associated with the original (nonsupersymmetric) formulation—particles with imaginary mass. Recall that the square roots of negative numbers are called imaginary numbers. Before supersymmetry, string theory produced a strange vibration pattern (called a *tachyon*) whose mass was imaginary. Physicists heaved a sigh of relief when supersymmetry eliminated these undesirable beasts.

Needless to say, all the underlying symmetries and patterns of the current versions of string theory are described by groups. One version, for instance, known by the intimidating name the *Heterotic type* $E_8 \times E_8$, is based on one of the sporadic Lie groups.

The next critical step in the confirmation or refutation of string theory will be, of course, to be able to discover the predicted supersymmetric particles. Physicists hope that this is within the reach of the Large Hadron Collider, or LHC, at CERN. Around 2007, this world's largest accelerator is expected to reach energies that are almost eight times higher than those achievable today. If the superpartners are indeed found, their properties will provide crucial clues as to what the ultimate theory might be. If they are not found, this could be an indication that the theory is going in the completely wrong direction.

String theory progresses at such an incredible pace that anyone outside the circle of its day-to-day practitioners finds it very difficult to follow in detail. The current research continues to be spearheaded by Edward Witten of the Institute for Advanced Study at Princeton and many others too numerous to name here. The mathematics used in these studies is becoming progressively more and more advanced. Not only are ordinary numbers replaced by an extended class of numbers known as *Grassmann numbers* (after the Prussian mathematician Hermann Grassmann), ordinary geometry is also being superseded by a special branch known as *noncommutative geometry* that has been developed by the French mathematician Alain Connes.

In spite of the cutting-edge tools that have become the theory's hallmark, string theory is in fact in its infancy. One of the string theory pioneers, the Italian physicist Daniele Amati, characterized it as "part of the 21st century that fell by chance into the 20th century." Indeed, there is something about the very nature of the theory at present that points to the fact that we are witnessing the theory's baby steps. Recall the lesson learned from all the great ideas since Einstein's relativity—*put the symmetry first.* Symmetry originates the forces. The equivalence principle—the expectation that all observers, irrespective of their motions, would deduce the same laws—*requires* the existence of gravity. The gauge symmetries—the fact that the laws do not distinguish color, or electrons from neutrinos—*dictate* the existence of the messengers of the strong and electroweak forces. Yet supersymmetry is an output of string theory, a consequence of its structure rather than a source for its existence. What does this mean? Many string theorists believe that some underlying grander principle, which will necessitate the existence of string theory, is still to be found. If history is to repeat itself, then this principle may turn out to involve an all-encompassing and even more compelling symmetry, but at the moment no one has a clue what this principle might be. Since, however, we are only at the beginning of the twenty-first century, Amati's characterization may still turn out to be an astonishing prophecy.

As you have seen in this chapter, physicists have exalted symmetry to the position of *the* central concept in their attempts to organize and explain an otherwise bewildering and complex universe. This raises a few intriguing questions. First, why do we find symmetry so attractive? Second, and perhaps more difficult, are the symmetry-based group-theoretical explanations truly inevitable? Or is the human brain somehow tuned to latch onto only the symmetric aspects of the universe? In order to understand why symmetry appeals so strongly to us, we must understand how it affects the human mind.

– EIGHT –

Who's the Most Symmetrical of Them All?

For how long do you think you could bear to have a civil conversation with the man in figure 97 before the lopsided appearance of his glasses would drive you nuts? Or suppose you enter somebody's house and you discover that the pictures hung on the walls display an "arrangement" such as the one in figure 98. Wouldn't you instinctively want to adjust each and every one of them? How and why did this craving for bilateral symmetry develop in the human mind? One of the goals of evolutionary psychology is to answer precisely these types of questions.

Evolutionary psychology is a science that attempts to combine the best of two worlds—evolutionary biology and cognitive psychology. In this view, the human mind is really a collection of numerous special-purpose modules that were designed and shaped by natural selection to solve very specific adaptive problems. An adaptive problem is any challenge posed by the environment, to which the human ancestors' minds needed to rise in order for these two-legged creatures to survive and reproduce successfully. In other words, according to evolutionary psychology pioneers Leda Cosmides and John Tooby, the human mind is a bit like a Swiss army knife, with many different "gadgets," each one designed for a specific task. Evolutionary psychologists

Figure 97

Figure 98

reject ideas about more general-purpose processes in the mind. They argue convincingly that all the problems hominids have ever had to face were always specific in nature rather than general.

Clues coming from a variety of areas, ranging from biology and anthropology to archaeology and paleontology, suggest what the most crucial adaptive problems might have been. In broad terms these include escaping from predators, identifying the right food, forming alliances, supporting offspring and close kin, communicating with other humans, and selecting mates. Where does symmetry come into all of this?

FEARFUL SYMMETRY

Few could compete with Oscar Wilde in the battle of one-liners. In *The Picture of Dorian Gray* he declares, "A man cannot be too careful in the choice of his enemies." Jokes aside, from an evolutionary point of view this is a very perceptive observation. Genes cannot fulfill one of their main tasks—getting themselves passed intact to the next generation—if their carrier manages to get him- or herself devoured by a predator. Any genes that somehow help an animal to escape from predators would therefore inevitably be favored by natural selection. Such genes would participate in the evolutionary construction of mental "predator-avoidance modules." The tasks of these modules are pretty obvious. First and foremost, potential predators have to be detected. Without early detection, no action can be taken, and the consequences can be catastrophic. Only in subsequent stages do other functions need to be activated—real dangers have to be distinguished from false alarms, and responses have to be triggered accordingly. Consequently, the predator-avoidance modules have to be primarily predator-detection devices.

Numerous experiments show that the perceptual systems of many creatures, from honeybees and pigeons to humans, are highly sensitive to bilateral symmetry. Symmetric patterns are both detected faster than asymmetrical ones and are easier to learn and retrieve from memory. Could these cross-species capabilities be somehow related to the predator-avoidance needs? What was the precise adaptive problem that the perceptual hardware/software was trying to solve? A clue to the answer may be gleaned from asking the question differently: In a world devoid of churches, cars, airplanes, and other human-made artifacts, what looks bilaterally symmetric? The answer is as plain as the nose on one's face—animals and humans! In fact, while the rear end of a lion is also bilaterally symmetric, the symmetry there is not nearly as striking as that of its front view. In other words, the detection of bilateral symmetry translates for an animal more or less into "I am being watched." The watcher's intentions need not necessarily be malicious—he, she, or it might simply be enjoying the view or selecting a mate. Yet there is no question that early detection of bilateral symmetry could mean the difference between life and death to the subject of attention.

Neuroscientist Joseph LeDoux of the Center for Neural Science at New York University is one of the pioneers of the study of emotions as purely physiological, as opposed to behavioral, phenomena. LeDoux is not interested in complex feelings, such as the mingling of love and compulsion or the conscious struggle invoked by the interplay between desire and jealousy. Rather, he studies the brain circuitry that leads to the emotion of fear. LeDoux finds that the response to fear is a cognitive unconscious that does not involve "the higher processing systems of the brain." Put plainly, the brain's predator-detection module faces the same dilemma encountered by any designer of burglar alarm systems. On one hand the designers want the system to be able to respond instantaneously to any break-in attempts, but on the other, they want to minimize the number of false alarms. On balance, however, a delayed response could prove much more costly and dangerous than a few false alarms. Not surprisingly, therefore, LeDoux finds that the brain operates through two separate neural pathways. One shorter "quick and dirty" route allows animals to respond to potentially dangerous stimuli even before the brain has fully analyzed the stimuli. The other "high road" passes through the sensory cortex and benefits from more extensive processing.

Central to the immediate emotion (rather than the conscious feeling) of fear is the *amygdala*—a small almond-shaped structure in the forebrain (*amygdala* is Latin for "almond"). LeDoux used chemicals that stain neurons to trace the brain circuitry in rats and to map the precise path that fear takes. This is a significant step beyond the mere "pulling habits out of rats" approach that has characterized much earlier, purely behavioral studies. LeDoux found that as soon as one rat sounds the first alarm (in the form of high-pitched screams), the signal received by other rats goes straight from their sensory thalamus (the dual-lobe gray matter that relays sensory signals) to the amygdala. The amygdala in turn, upon reception of a powerful stimulus, triggers the entire defense system. The response can be either in the form of freezing—to avoid being seen—or in the heart racing and hormones flooding the bloodstream. These hormones help provoke the appropriate course of action—the rat either runs for its life or prepares to fight the predator.

The amygdala seems to govern the fear response in all species that have this structure, including humans. Research showed that a woman with a brain lesion in the amygdala entirely lost her ability to detect and recognize any facial expressions related to fear.

Clearly, the "quick and dirty" mechanism is likely to trigger quite a few false alarms and unnecessary panic attacks. However, the thalamus also sends information to the more accurate signal-processing center—the sensory cortex. This slower pathway eventually provides the amygdala with a more reliable representation of the actual stimulus and stops the animal from overreacting.

As we have just seen, the bare detection of bilateral symmetry could at times set off the siren that puts the entire (cognitive unconscious) fear machinery into motion. Bilateral symmetry can also act, under different circumstances, as an antipredator defense mechanism in itself. Many animals (known collectively as *aposematic animals*) use various signals, such as distinctive odors, sounds, and color patterns, to advertise their risk or distastefulness to predators. Some butterflies, for instance, have large, conspicuous eyespots that are concealed at rest but are exposed when a potential predator is detected. The sudden appearance of a pair of "eyes" often confuses the predator sufficiently so as to give the butterfly an opportunity to flee. Among the various visual warning signals that aposematic creature use, bilaterally symmetric ones have proven to be the most effective. Specifically, fascinating experiments that have subjected artificial paper "butterflies" with different wing patterns to predation from domestic chicks have shown that the protective value of such visual warning displays is enhanced by large and symmetric pattern elements. In the experiment conducted by Swedish researchers, the paper butterflies (figure 99) were affixed under plastic petri dishes, and food crumbs were placed inside each dish. In each treatment, forty-five monochrome black butterflies with palatable crumbs and forty-five aposematically signaling butterflies with unpalatable quinine-treated crumbs were placed on the floor. The aposematic butterflies had either a symmetric or asymmetric warning pattern, and each group of chicks was exposed to one type (symmetric large, symmetric small, or asymmetric) of unpalatable signaling butterfly. The experimental results suggested that asymmetry in the patterns impairs the efficacy of the aposematic signals. The researchers concluded that this was probably due to the fact that deviations from symmetry elicit a weaker neural response and thereby make the signal more difficult for the chicks to detect, remember, or associate with unpalatability. Collectively, the findings of this and similar research lead to an interesting conclusion: Prey species possessing warning coloration may be subjected to a natural selection for large and bilaterally symmetric patterns.

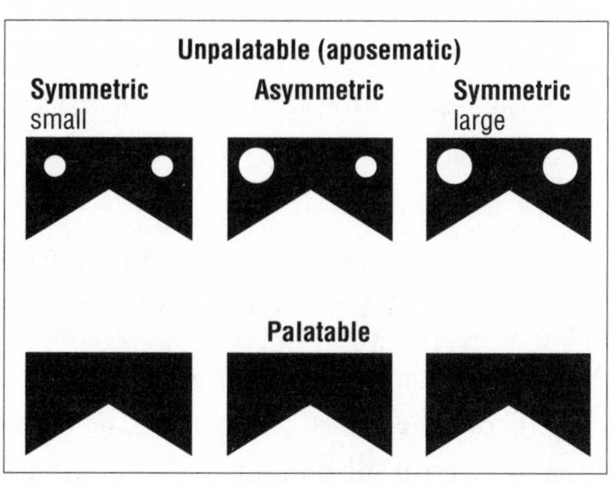

Figure 99

The British statesman and philosopher Edmund Burke (1729–97) said once that "no passion so effectually robs the mind of all its powers of acting and reasoning as fear." This is probably true, but the cognitive unconscious response to fear, occasionally triggered by the detection of symmetry, may sometimes be all that is needed to avoid a predator. Similarly, in the signaling arena, the provocative display of symmetric, aposematic signals provides in some cases a protective shield against potential predators.

The role of symmetry in both types of predator-avoidance mechanisms (detection and signaling) is a negative one. Symmetry acts in some sense as a repelling agent, if an important one. Can it also provoke an encouraging, inviting stimulus? The process of mate selection shows that indeed it can, perhaps even more so than you might expect.

BIRDS DO IT, BEES DO IT, EVEN EDUCATED FLEAS DO IT

Avoiding predators and eating the right food are very important for survival. From the genes' perspective, however, survival is only a means to an end. Even if all humans had successfully implemented the ideas in comedian George Burns's book *How to Live to Be 100—Or More,* this in itself would be of no use whatsoever to the genes, unless these humans also had offspring. Reproduction and the passing of genes to the next generation is what genes are really all about. As evolutionary biologist and author Richard Dawkins has put it, "An organism is just a gene's way of making more genes."

In some species, individuals can reproduce by themselves—they divide in two, and each part becomes a new creature. Nature clearly decided that this asexual process is less fun, because most of the 1.7 million or so species on Earth engage in sexual reproduction. More seriously, sexual reproduction must offer an adaptive benefit to species, or it would not be so prevalent. The obvious difference (between the sexual and asexual routes) is that offspring produced sexually can benefit from the swapping of their parents' genes. The new and improved genetic makeup can limit damage caused by harmful mutations and can enhance the fitness of the offspring. In order to fully exploit the advantages of sex, however, individuals have to select the most appropriate mates. "Most appropriate" from the genes' point of view means a mate with characteristics that increase the chances of survival and reproduction of the offspring. This translates into two main traits: high quality of genes (in terms of fitness), and capability of parental care. Here I shall concentrate on the first of these properties, since it is the one most directly related to symmetry.

ably preferred, resulting in a sensory bias for symmetry. Enquist and Arak used artificial neural networks as models of recognition systems. Neural networks are computer systems loosely based on the operation of the brain that are able to learn from experience in order to improve their performance. In the experiments by Enquist and Arak, the preference for symmetry was definitely a sensory exploitation—a consequence of the need to recognize signals—and it had nothing to do with the assessment of genetic quality. Similar results were obtained in a separate artificial neural network experiment by Cambridge biologist Rufus Johnstone. Again the implication was that mating preferences for symmetry evolve as a simple by-product of selection for mate recognition, rather than because of the relation between the degree of fluctuating asymmetry and mate quality.

From the perspective of the discussion here, however, it does not really matter whether the preference for symmetry in mate selection in the animal kingdom is a result of a search for quality or for recognition. Preferences for symmetry might have evolved for a variety of reasons. The important point, however, is that *there is a preference for symmetry*—symmetry plays a crucial role in animal mate selection.

WHAT'S LOVE GOT TO DO WITH IT?

Humans are very complex animals. An inseparable mixture of evolutionary psychology, culture and ethnicity, various beliefs, and personal interests and traits determines what humans find attractive. Yet, deep down, the genes' desire to procreate is still one of the powerful forces within the human mind. In the search for a healthy, fertile mate, our minds are programmed no differently from those of our Stone Age ancestors. Beauty may be in the eye of the beholder, but as evolutionary psychologist David Buss has put it, "Those eyes and the minds behind the eyes have been shaped by millions of years of human evolution." The sense of what is attractive is largely determined by an adaptive decision-making machinery that has evolved at least partially for mate selection.

If you think that attractiveness is unimportant, think again. Anna Kournikova was ranked around seventieth in women's tennis throughout most of 2003, yet she made millions of dollars more in endorsements than players ranked significantly higher. In case you wonder why, here is a hint—she was also featured twice on the cover of *Maxim* magazine. The creators of the ABC News program *20/20* conducted an experiment to gauge how often attractive men and women get preferential treatment. In one test in Atlanta, two actresses dressed alike were each made to stand helplessly next to a car that had run out of gas. For the more average looking of the two, a few pedestrians stopped, but only to point her to the nearest gas station. For the more attractive actress, no fewer than a dozen cars stopped, and six drivers actually went to get her gas!

In a second experiment, *20/20* hired two men to apply for a job. The two candidates had similar education and work experience, and even the small differences that did exist in their résumés were deliberately ironed out. There was, however, one noticeable difference between the two men—one was very attractive while the other was more ordinary looking. Believe it or not, the interviewer was eager to have the more attractive man return as soon as possible for a tryout day, while the more plain-looking man got a "don't call us, we'll call you" reply.

Even the area in the brain that responds to beauty has been identified. Researchers Hans Breiter, Nancy Etcoff, Itzhak Aharon, and their collaborators used magnetic resonance imaging (MRI) to investigate the activity in men's brains when they were shown pictures of particularly attractive women. They found that beauty triggers the same area in the brain that is triggered by food (when the person is hungry) or by other subjects of addiction (e.g., when a compulsive gambler sees a roulette wheel).

For a long time it has been assumed that the criteria for beauty are largely cultural, and therefore learned rather than innate. More recent studies by University of Texas at Austin psychologist Judith Langlois have totally overturned this conventional wisdom. Langlois first had adults rank pictures of both white and black females for attractiveness. Then the pictures were shown in pairs (one more attractive than the other) to infants in two age groups—two to three months and six to eight months old. Infants in both age groups were found to gaze longer at the faces ranked more attractive. Similarly, one-year-old infants were found to play for a significantly longer time with facially attractive dolls.

Other studies tested for changes in taste across cultures. Psychologist Michael Cunningham found an incredible consensus in the judgment of facial attractiveness of women of different races by men of different races. The agreement persisted even when different degrees of exposure to Western mass media were considered. Studies that were performed across geographical and ethnic boundaries (e.g., with Chinese, Indian, South African, and North American men) produced very similar results. Taken together, all of these studies seem to indicate that there do exist some universal criteria for attractiveness, and that attractive faces enjoy a far-extending appeal that emerges very early in life and is consistent across cultures. The beauty detectors may not quite be innate, but the human mind may have innate basic rules from which templates of attractiveness are constructed.

So maybe there is a bias for "lookism," but what is it that men and women find attractive? Biologist Randy Thornhill, psychologist Steve Gangestad, and ethologist Karl Grammer have amassed a large body of evidence that shows that symmetry is a key factor. Thornhill, Gangestad, and their colleagues measured symmetry in close to a thousand students on different facial features (placement of eye corners, pupils, cheekbones, edges of mouth, and so on) and body features (foot breadth, hand breadth, elbow breadth, ear length, length of second and fifth fingers, and so on) to develop an overall index of asymmetry. When Thornhill and Gangestad correlated these data with independent ratings of attractiveness, they found that less symmetrical people in either body or face were considered less attractive.

In a separate study, Grammer and biologist Anja Rikowski found a relation even between symmetry and attractive body odor. In a study that involved sixteen males and nineteen females, each subject wore a T-shirt on three consecutive nights under controlled conditions. Immediately after use, the T-shirts were deep frozen, and just before the evaluation of odor they were reheated to body temperature. Fifteen subjects of the opposite sex then rated the smell for sexiness on a seven-point scale. Twenty-two other men and women evaluated portraits of the subjects for attractiveness, and indices of symmetry of the subjects were calculated based on seven traits. The results showed that facial attractiveness and sexy body odor go hand-in-hand for female subjects. Moreover, the males found that the more symmetric the body of a woman, the sexier her smell. Interestingly, women found the smell of more symmetric men to be more attractive only when the women were in the most fertile phase of their menstrual cycle.

Most surprisingly perhaps, Thornhill and Gangestad discovered a relationship between symmetry and women's orgasms. The researchers reasoned that if the women's orgasms are in fact an adaptation designed for securing healthy genes for their offspring, then women should experience more orgasms with more symmetrical mates. Conducting a study with eighty-six heterosexual student couples, the researchers found that indeed the women whose partners were most symmetrical experienced a significantly higher frequency of orgasms. Somewhat unexpectedly, the researchers did not find any correlation between female orgasm during sex and the level of romantic attachment or the sexual experience of the partners. Before any female reader rushes to find a symmetrical guy, I should note that studies also show that the most symmetrical men invest the least in their relationships and cheat more often on their mates. Female orgasm seems to be less about bonding with a great person than about a cold Stone Age evaluation of the mate's genetic endowment.

Independent research by psychologists Todd Shackelford and Randy Larsen showed that symmetry in the human face correlates very well with other fitness indicators, on both the physiological and the psychological sides. In particular, men with asymmetric faces were found to be more likely to suffer from depression, anxiety, headaches, difficulties in concentrating, and even stomach problems. Women with facial asymmetry were also found to be of poorer health and more prone to emotional instability and depression. Furthermore, symmetry is also another cue to youth, because the older people get, the less symmetrical their faces become.

The picture that emerges is very suggestive. Just as in the animal kingdom the process of mate selection may have identified symmetry as a good fitness indicator, for humans too, bilateral symmetry has been equated with developmental stability, youth, and resistance to various debilitating pathogens. The result, in terms of animal/human "magnetism," was inevitable—*symmetric* has become almost synonymous with *attractive.*

I do not want to leave you with the impression that symmetry is the only quality that affects attractiveness. Psychologist Judith Langlois and her collaborators emphasize averageness in the face as being most attractive. Langlois generated computer composites of four, eight, sixteen, and thirty-two faces. To her surprise she found that the composite faces were uniformly judged to be more attractive than the individual faces from which the composites were made. Sixteen-face composites were ranked above four- or eight-face composites, and the thirty-two-face composite was found most attractive. While composite faces tend, by construction, to also be more symmetric, Langlois found that even after the effects of symmetry have been controlled, averageness was still judged to be attractive. These findings argue for a certain level of prototyping in the mind, since averageness might well be coupled with a prototypical template.

Cognitive scientist David Perrett of the University of St. Andrews in Scotland found that faces we find attractive are often appealing because they look like our own, or like the faces of our parents. Intrigued by these results, I called him up during a visit to St. Andrews, to find out why he thought that was an adaptive choice. He first emphasized that for the mind to be able to help with mate selection it has to be a learning system. "Specifically," he added, "the mind needs to have the ability to lock onto things that are relevant from the immediate environment—like symmetry or averageness. Finding someone with resemblance [to you or to your parents] attractive may also make sense, since your family has already succeeded in surviving through the evolutionary path."

Other factors that affect mate selection are related to indicators of fertility, resources, and capacity and willingness for parental care. For instance, studies by psychologist Devendra Singh show that almost universally, men prefer women with the classical "hourglass" figure characterized by a waist-to-hip ratio of 0.7. The adaptive reason behind this preference may be the fact that this ratio was also found to be a good fertility indicator. A potentially related preference was also found for breast symmetry. Other surveys find that women generally prefer men who are somewhat older than themselves, probably because of the female preference for males with resources.

Even the brief description of results and ideas from evolutionary psychology that I have presented in this chapter seems to lead to an inescapable conclusion. Either because of mate selection, cognition, predator avoidance, or a combination of all three, *our minds are attracted to and are finely tuned to the detection of symmetry.* The question of whether symmetry is truly fundamental to the universe itself, or merely to the universe as perceived by humans, thus becomes particularly acute.

DOES SYMMETRY REALLY RULE?

Imagine what would have happened if the human eye were sensitive only to blue light. Prior to the development of any other light detectors, scientists would have naturally concluded that everything in the universe is blue (even just the thought gives me the blues). Similarly, a pest control company that manufactures humane mousetraps that are three inches long might conclude that all mice are shorter than three inches, because all the mice actually trapped would be of such lengths. These are simple examples of observational *selection effects*—filterings of physical reality introduced by unrecognized biases either in the methods of observation or the observational tools. Could our mind's preference for symmetry introduce a similar bias in our perception of what is truly fundamental in the universe?

I need to emphasize here again that I am focusing on the symmetries of the laws of nature and their description using group theory, not on the symmetry of any particular structures in nature. Perfect crystals are examples of the latter. They look precisely the same when we move within the crystal by certain amounts in various directions. Crystallography is the science studying the structures and properties of assemblies made of very large numbers of identical units. The units themselves may be composed of atoms, molecules, or, in a more abstract context, even pieces of computer code. A typical question in crystallography might be, how can a large number of identical units be arranged in space so that each unit will "see" identical surroundings? Group theory is the bread and butter of crystallography—the attempts to answer the above question resulted in a proof that there exist only 230 different types of spatial symmetry groups (just as there are only 7 different symmetry groups of linear strip patterns; see chapter 7).

Symmetry principles manifest themselves also in the structure of a variety of biological molecules and organisms, from crystallized proteins and DNA to viruses. All of these symmetries are obviously important, since they represent stable (minimum-energy) systems, which in turn form minerals and living things. However, these are not symmetries that underlie the basic laws of nature.

When it comes to the laws, there is absolutely no doubt that symmetry and group theory are extremely *useful* concepts. Without the introduction of symmetry and the language of groups into particle physics the description of the elementary particles and their interactions would have been an intricate nightmare. Groups truly flesh out order and identify patterns like no other mathematical machinery.

In an interview in 1985, Harvard mathematician Andrew Gleason said, "Of course mathematics should work in physics! It is designed to discuss exactly the situation that physics confronts; namely, that there seems to be some order out there—let's find out what it is." In addition to its usefulness, symmetry removes redundancies from the description of both real and abstract systems. For instance, imagine that a certain system is represented symbolically by the string of characters

XYZXYZXYZXYZXYZ.

We can use the translational symmetry of the symbols to remove the redundancy and reduce the description to the much more compact form 5*(XYZ), reading as "repeat the substring XYZ five times." Similarly, in the string

UVWXYZZYXWVU

we can use the reflection symmetry to reduce the string to *SYM*(UVWXYZ), where the operator *SYM* indicates this type of reflection. The real question is, therefore, whether symmetry is indeed embedded in nature's fabric or if it only represents a convenient way for us to build a dialogue with physical reality. This is not an easy question. In certain steps along the road toward the ultimate theory of the universe, symmetry appears to be more fundamental than in others. The basic symmetry between any two observers that underlies relativity, for instance, is an exact symmetry that appears indeed to characterize nature's ways. On the other hand, one of the early models for atomic nuclei, known as the *Elliott model,* was described by a symmetry (and an associated group) even though that symmetry was known to be only approximate, and almost certainly not fundamental.

One potential problem with some of the gauge symmetries assumed to underpin the standard model is that of *symmetry breaking.* Let me explain this concept briefly. Examine the top view of a dinner table in figure 100, where the small plates are for bread. All the seats around the table are identical, and from the viewpoint of any person sitting at the table, left and right are indistinguishable. The configuration is therefore symmetric both under rotation (through integer multiples of

$360 \div 8 = 45$ degrees) and under reflection (about eight axes). As soon, however, as bread is served and the first person puts it on a plate (to her left, I am told), the symmetry is "spontaneously broken." Left and right become distinct, and rotational invariance is lost.

Recall that in the electroweak theory, electromagnetism and the weak force are two sides of the same coin (chapter 7). The force carriers—the photon, W, and Z—are interchangeable. A question that immediately emerges is, why then do these two forces have such different manifestations (e.g., one is a hundred thousand times stronger than the other) in today's universe? The standard model puts the blame on symmetry breaking. According to the most popular scenario, shortly after the moment our universe came into existence (the event we call the "big bang"), there was perfect symmetry between electromagnetism and the weak force. At the huge temperatures that characterized this phase, photons and W and Z particles were truly indistinguishable. As the universe expanded and cooled down, however, it underwent a phase transition—not unlike the freezing of a liquid—in which symmetry breaking occurred. This is supposed to have happened when the universe was a tiny fraction (about 10^{-12}) of a second old. The liquid analogy can, in fact, be carried one step further. A liquid looks the same however you turn it—there is no preferred direction. This symmetry is lost, however, when the liquid freezes. The crystalline structure that emerges has some preferred axes. The breaking of the symmetry between the electromagnetic and weak forces that was associated with the cosmic "freezing" is believed to have generated the differences we observe today. The W and Z were endowed with masses while the photon remained massless. The range of the weak force is limited to distances of the order of the size of the nucleus only because of its sluggish, heavyweight carriers.

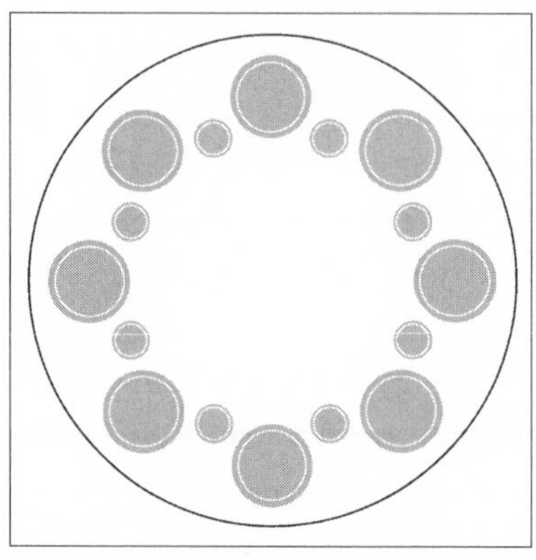

Figure 100

To the uninitiated the above description may sound a bit like an imaginative fairy tale. Such a person may think that particle physicists have invented a symmetry that is supposed to characterize the basic forces of nature, and when the present-day universe was found not to obey that symmetry, they concocted a very convenient symmetry-breaking scenario. Actually, the status of the theory is much more solid than the above description suggests. Many predictions of the standard model have been spectacularly confirmed experimentally (chapter 7). Even more important, experimental tests of the entire scheme of symmetry breaking will become feasible soon. In the same way that the freezing points of liquids can be estimated from atomic masses and the energies that bind atoms together, the known parameters of the standard model can be used to estimate the energy at the point of symmetry breaking. The required energies are either already within the reach of a large particle accelerator—such as the Tevatron at the University of Chicago's Fermilab—or will be achievable by CERN's Large Hadron Collider around 2007. At the very least, these experiments are expected to tell us whether the theoretical ideas of symmetry breaking are on the right track. The same experiments could also test the predictions of supersymmetry. Recall that if the real world obeys supersymmetry, then a whole host of new particles is waiting to be discovered. The spin-½ electron should have a spin-0 partner (called a "selectron"), the spin-1 photon should have a spin-½ "photino" partner, and similar partners are predicted to exist for every particle in the standard model.

Strictly speaking, however, even an experimental confirmation of symmetry breaking and supersymmetry will not prove unambiguously that symmetry is fundamental, as opposed to useful. As we have seen, supersymmetry is still only a feature of string theory, rather than its source. The underlying principle of the theory is still to be unveiled, and it may or may not prove to be a symmetry principle.

There is another reason why we should exercise some caution before we hail symmetry as the main mover in the genesis and workings of the universe, and group theory as its primary language. This reason can perhaps be best demonstrated using the example of the Kariera kinship-marriage rules. Recall that these rules of an Aboriginal tribe were shown to form a group that has the same structure as the famous Klein four-group. There is no doubt, however, that the Kariera did not intend their rules to represent any particular mathematical structure. We are therefore faced with a situation where we have identified a mathematical tool that provides a perfect description of the reality, but where the true reasons for that reality remain unknown. The actual motivation that has led the Kariera to choose this particular set of maxims may have relatively little to do with the order we have recognized in it, even though a deeper analysis might reveal that these rules provide for a stable society.

While I was struggling with this question of how fundamental symmetry really is, I decided to conduct a small survey among some of the world's top physicists and mathematicians to find out what their thoughts on the subject were. Steve Weinberg, Nobel Prize laureate in physics in 1979 and one of the key players in the development of the standard model, agreed that symmetry might not be the most fundamental concept in the ultimate theory. He added, "I suspect that at the end the only firm principle will be that of mathematical consistency." Ed Witten, recipient of the Fields Medal in mathematics in 1990 and the person who brought about the second string revolution, also stressed that "there are still missing or unknown ingredients in string theory" and that "some concepts, such as Riemannian geometry in general relativity, may prove to be more fundamental than symmetry." Sir Michael Atiyah, who received the Fields Medal in 1966 and the Abel Prize in 2004, alluded to human-mind-driven selection effects. "We come to describe nature with certain spectacles," he said. "Our mathematical description is accurate, but there may be better ways. The use of exceptional Lie groups may be an artifact of how we think of it." The last phrase, in particular, reminded me of another interesting statement by the famous mathematician and philosopher Bertrand Russell (1872–1970): "Physics is mathematical not because we know so much about the physical world, but because we know so little; it is only its mathematical properties that we can discover." In other words, Russell saw even our description of the universe through mathematics as being dangerously close to some sort of selection effect. Freeman Dyson, one of the principal figures in the development of quantum electrodynamics and recipient of the Wolf Prize in physics in 1981, offered, as always, his unique perspective: "I feel that we are not even at the beginning of understanding why the universe is the way it is." After a few seconds of reflection, he added, "Even such simple things as our ability to tell whether a line is perfectly straight, or to distinguish between a circle and an ellipse, are mysteries in themselves." Concerning symmetry, he confessed that he does not much like the word "fundamental" and prefers to use "fruitful" when referring to symmetry as the source of forces (as in the case of the gauge symmetries of the electroweak theory). Finally, he noted that symmetry and group theory have become much more powerful descriptors since the introduction of quantum mechanics.

What can we conclude from all of these insights in terms of the role of symmetry in the cosmic tapestry? My humble personal summary is that we don't know yet whether symmetry will turn out to be the most fundamental concept in the workings of the universe. Some of the symmetries physicists have discovered or discussed over the years have later been recognized as being accidental or only approximate. Other symmetries, such as general covariance in general relativity and the gauge symmetries of the standard model, became the buds from which forces and new particles bloomed. All in all, there is absolutely no doubt in my mind that symmetry principles almost always tell us something important, and they may provide the most valuable clues and insights toward unveiling and deciphering the underlying principles of the universe, whatever those may be. Symmetry, in this sense, is indeed fruitful.

In *The Feynman Lectures on Physics,* a book based on a course given by the famous physicist Richard Feynman during the academic year 1961–62, Feynman concludes his discussion of symmetry thus:

> So our problem is to explain where symmetry comes from. Why is nature so nearly symmetrical? No one has any idea why. The only thing we might suggest is something like this: There is a gate in Japan, a gate in Neiko, which is sometimes called by the Japanese the most beautiful gate in all Japan; it was built in a time when there was great influence from Chinese art. This gate is very elaborate, with lots of gables and beautiful carving and lots of columns and dragon heads and princes carved into the pillars, and so on. But when one looks closely he sees that in the elaborate and complex design along one of the pillars, one of the small design elements is carved upside down; otherwise the thing is completely symmetrical. If one asks why this is, the story is that it was carved upside down so that the gods will not be jealous of the perfection of man. So they purposely put an error in there, so that the gods would not be jealous and get angry with human beings. We might like to turn the idea around and think that the true explanation of the near symmetry of nature is this: that God made the laws only nearly symmetrical so that we should not be jealous of His perfection!

Symmetries associated with the laws of nature are not the only topic in which Galois's legacy has generated and continues to generate new ideas. We can get at least a taste of this incredible heritage by examining a few simple examples, spanning a range of artistic and intellectual activities from music to modern algebra.

WHAT PASSION CANNOT MUSIC RAISE AND QUELL?

The title of this section is taken from "A Song for St. Cecilia's Day," by the famous English poet and dramatist John Dryden (1631–1700). The feast of St. Cecilia (November 22) commemorated the legend that this patron saint of music invented the organ. The theme of this poem is a tribute to the power of music. Indeed, few art forms are as allied to both emotional states and the rhythm of the human body as music is. Our breathing and heartbeat, for instance, are intimately correlated with the level and nature of our activities and with the intensity of our excitement or fear. Many pieces of music, none perhaps more so than Ravel's celebrated *Boléro,* provide a direct reflection of these rhythms of life. In fact, in Blake Edwards's 1979 movie *10,* the *Boléro* was declared to be the perfect soundtrack for making love. As I noted already in chapter 1, to say that symmetry plays a major role in music is to state the obvious. Consequently, it was only to be expected that group theory would describe musical structures and patterns beautifully.

The notes on a piano keyboard provide the simplest example of a groups-music relationship. The pitch of a tone is characterized by the number of vibrations per second (e.g., of a string), the *frequency.* Frequency is measured in vibrations per second or hertz (denoted by Hz), after the German physicist Heinrich Rudolf Hertz. For example, the frequency of middle C (or "do" in the major scale) on the piano keyboard (figure 101) is about 261.6 Hz. The frequency of A_4 (or "la") is 440 Hz. The *octave* is defined so that the ratio of frequencies is precisely equal to 2. One octave higher than middle C has a frequency of $261.6 \times 2 = 523.2$ Hz, and one octave lower has a frequency of $261.6 \div 2 = 130.8$ Hz. Notes that are separated by a precise whole number of octaves have the same name and they sound alike. In the "equally tempered system" popularized by Bach in his impressive collection of preludes and fugues, all keys have equal status. The ratio between the frequencies of any two adjacent keys is the same and equal to 1.05946. This number (equal to the

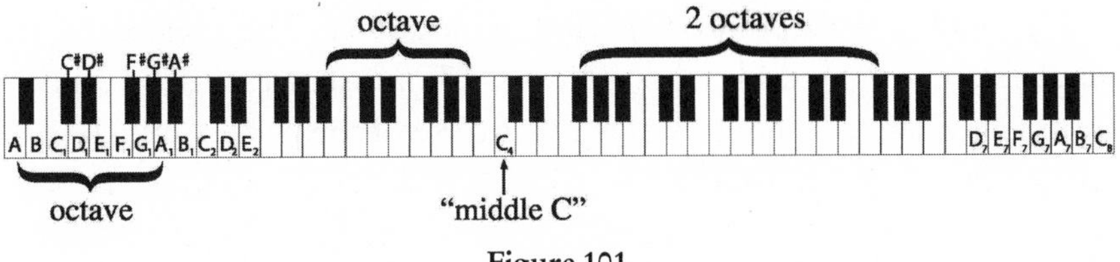

Figure 101

twelfth root of 2) is obtained simply by the requirement that when raised to the power 12 (there are twelve semitones or half steps in the octave) it gives a ratio of 2, corresponding to the tone one octave higher.

The Greek mathematician Pythagoras is traditionally credited with the discovery that two notes that correspond to frequencies whose ratio is equal to the ratio of two simple whole numbers (such as 3:2) yield harmonious ("consonant") and pleasing sounds. A perfect fifth, for instance, is characterized by a frequency ratio of 3:2, which corresponds to a separation by seven semitones (the seventh power of 1.05946 is very close to 1.5). A perfect fourth corresponds to a frequency ratio of 4:3 and five semitones.

Since there are twelve semitones in the octave, we can conveniently represent them on a clock face, as in figure 102. We can now move from any note to any other note by performing precisely the same operation as when we calculate the hour in the day. That is, when we want to know what will be the time 9 hours after 7:00 p.m., we calculate 7 + 9 = 16 = 4:00 a.m. (because 12 is considered also as 0). Adding numbers in this fashion is called in the mathematical lingo *addition modulo 12.* For instance, 8 + 7 = 15 = 3 (modulo 12), and 10 + 2 = 12 = 0 (modulo 12). The semitones of the equally tempered system obey the same rules. If you want to know which note is 10 semitones above D# (figure 102), you calculate 3 + 10 = 13 = 1 (modulo 12) = C#. The set of numbers {0, 1, 2, 3, 4, 5, 6, 7, 8, 9, 10, 11} or the corresponding notes on the musical scale form a group under the operation of addition modulo 12. You can easily check closure—e.g.,

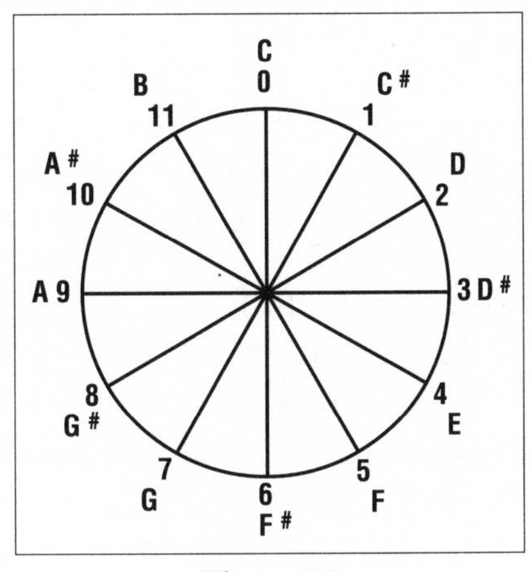

Figure 102

9 + 4 = 13 = 1 (modulo 12)—and associativity. The identity is the number 0, and any number has an inverse. For instance, the perfect fifth (corresponding to 7 semitones) is the inverse of the perfect fourth (corresponding to 5 semitones), since 7 + 5 = 12 = 0 (modulo 12). This makes good sense even from a purely musical perspective, since when these two intervals are combined, this corresponds to a frequency ratio of $\frac{3}{2} \times \frac{4}{3} = 2$, which is precisely an octave—which gives the same sound. In fact, very appropriately, musicians call two intervals that combine to give an octave "inversions" of each other. Another example of two such inversions (figure 103) is the minor third (ratio of 6:5; 3 semitones) and the major sixth (ratio of 5:3; 9 semitones), since 3 + 9 = 12 = 0 (modulo 12).

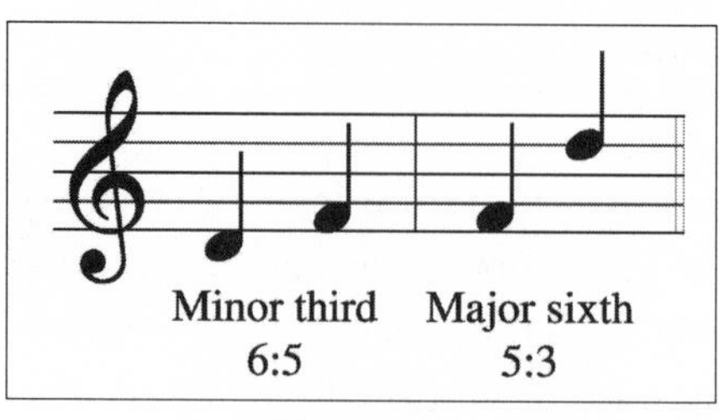

Figure 103

Groups show up not only in the musical scale, but also in the structure of certain forms of music. A simple case is that of the round—a type of short canon in which each voice enters in turn to sing the same melody, as in the familiar "Frère Jacques" (figure 104).

If we denote the four different phrases by A, B, C, D respectively (figure 104), then the structure is represented by AABBCCDD (each phrase is repeated), and the round for four voices takes the form

1: A A B B C C D D A A B B C C D D A A B B C C D D
2: _ _ A A B B C C D D A A B B C C D D A A B B C C
3: _ _ _ _ A A B B C C D D A A B B C C D D A A B B
4: _ _ _ _ _ _ A A B B C C D D A A B B C C D D A A

Note that if a fifth voice were to enter, it would simply be repeating or doubling the first voice. In fact, starting with any voice, if we were to continue to add voices, then four voices down the line would result in doubling. We could now denote by *a* the instruction "come in two

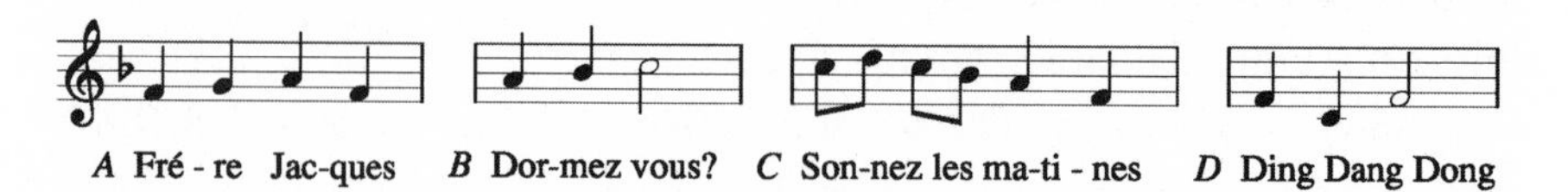

Figure 104

bars later." This moves us from one voice to the next. Symbolically, a^2 (or $a \circ a$) would denote "come in four bars later," a^3 ($a \circ a \circ a$) would be "come in six bars later," and a^4 ("come in eight bars later") would result in doubling of the same voice, or the identity. You can easily verify that the four instructions I, a, a^2, a^3 (where I is the identity) form a group under the operation "multiply" (e.g., a and a^3 are each other's inverses since $a \circ a^3 = a^4 = I$).

Clearly, neither Bach nor any of the other classical composers had group theory in mind when they composed their music. Group theory inevitably finds its way into the description of musical patterns simply because of its very nature as a language of symmetries. Some twentieth-century composers, most notably Arnold Schoenberg, Alban Berg, and Anton Webern from the Second Viennese School, were said to have flirted more deliberately with mathematically based music. In particular, in the "method of composing with twelve tones" used in such pieces as Berg's *Lyric Suite,* or Schoenberg's Piano Concerto, all harmonies are based on a "twelve-tone row" that is in fact a permutation of the common twelve chromatic notes. A twelve-tone row could be used either in its original order (selected by the composer), or it could be further transformed by some operations. Three basic operations used by the Viennese composers were *row inversion, retrogression,* and *retrograde inversion.* In row inversion, descending intervals were replaced by ascending intervals, and vice versa. For instance, if the original row started with C and rose a perfect fourth to F, then the inverted row fell a perfect fourth to G (figure 102). The retrogression reversed the order of the melodic jumps. If the last jump in the original row was up a major third, then this would be the first jump of the new row. Finally the retrograde inversion applied both the row inversion and the retrogression simultaneously. You can easily convince yourself that these three transformations together with the identity ("do absolutely nothing") form a group under the operation "followed by." In particular, each member of this group is its own inverse.

Many people, including avid concertgoers, feel discomfort with Schoenberg's atonal music, as they do with similarly experimental pieces by Igor Stravinsky, Aaron Copland, Pierre Boulez, Luciano Berio, and many others. Members of this anti-atonality audience would probably argue that the use of mathematics by these composers (if indeed deliberate) did not help the quality of the music. However, irrespective of what one's opinion on atonal music may be, there is no denial of the fact that Schoenberg's and even more so Webern's "mathematical" experimentation opened the door to interesting avant-garde New Music and was the inspiration for *serialism.* This revolution in composition replaced all the traditional rules and conventions by a structural series of notes that governs the entire development of the composition. Fascinating music by such composers as Olivier Messiaen and Milton Babbitt originated from this radical change of tenets.

Music represents an art form in which only the very basic concepts of groups have been implicated. The development of group theory itself, however, did not stop at the beginning of the twentieth century. Rather, a proof in group theory that was completed only in August of 2004 is in some respects the most complex proof in the history of mathematics.

THE "THIRTY YEARS' WAR," OR THE TAMING OF THE MONSTER

Scientific endeavors are often a search for the most basic building blocks. Regarding the structure of matter, this centuries-long quest has led to the discovery of molecules and atoms, then of protons and neutrons, then of the elementary particles of the standard model (quarks, electrons, neutrinos, muons, taus), and finally to the suggestion of strings. In the vast expanses of space, astronomers are now searching for the first stars and clusters of stars to have formed in the universe—the building blocks of today's giant galaxies. In group theory, the hunt has been for a classification of all the simple groups (which have no nontrivial normal subgroups) from which all other groups can be constructed. As we have seen in chapter 7, the landmark classification of the simple Lie groups was carried out at the end of the nineteenth century by Wilhelm Killing and Elie Cartan. Lie groups are the groups of continuous transformations (such as rotations in three dimensions) that were defined by Sophus Lie in 1874. By their very nature, Lie groups have an infinite number of elements (e.g., there is an infinite number of possible rotation angles). Still, it suffices to specify a finite number of parameters to fully characterize any Lie group. For instance, the elements of the group of rotations of a circle in the plane, usually denoted by SO(2) or U(1), are fully determined by specifying one parameter—the angle of rotation. The dimension of this group is therefore 1. The group of rotations of a sphere in three-dimensional space can be characterized by three parameters—two angles that identify the axis of rotation and one angle for the rotation itself. This group, denoted by SO(3), therefore has dimension 3. Killing and Cartan managed to find four infinite families of Lie groups (traditionally known as A_m, B_m, C_m, D_m, for values of $m = 1, 2, 3, \ldots$), and five sporadic groups that were individual "one-offs" that did not fit into any of the families. These sporadic groups are normally called G_2, F_4, E_6, E_7, and E_8, and they have dimensions of 14, 52, 78, 133, and 248 respectively. As I have described in chapter 7, the simple Lie groups play a crucial role in the standard model and may prove to be an essential tool in string theory.

The classification of the finite simple groups turned out to be a much more daunting task than its Lie-group equivalent. By the end of the nineteenth century there were six infinite families and five sporadic (exceptional) finite simple groups known. One of those families was defined by none other than Galois himself as he was struggling with the insolubility of the quintic. Recall that in an even permutation of a set of objects, there is an even number of reversals from the natural or original order (chapter 6), while in an odd permutation the number of reversals is odd. For instance, 1324 represents an odd permutation of 1234, because it involves only one reversal (3 appears before 2), but 4321 represents an even permutation because you can check that it involves six reversals. We already know (chapter 6) that the collection of permutations of n objects forms a group with $n!$ elements. In fact, Cayley's theorem states that every group has the same structure as a group of permutations. The set of even permutations of any number of objects also forms a group—a subgroup of the full group of permutations. This is easy to understand: If one permutation involving an even number of reversals is followed by a second even permutation, then clearly the total number of reversals is also even, implying closure. The groups of even permutations are known as the *alternating groups.* Galois showed that the alternating groups obtained from permutations of more than four elements are all simple, and this was precisely the property he used to prove the insolvability of the quintic by a formula.

A second family of simple groups that was known to mathematicians at the end of the nineteenth century was of the type we have encountered with the musical scale. In the same way that the numbers zero to eleven form a group under the operation of addition modulo 12, the numbers zero to $n - 1$ form a group under addition modulo n for any value of n.

Groups of this type are known as *cyclic groups,* and cyclic groups with a prime number of elements are simple. The four other families of finite simple groups were equivalent in many ways to corresponding families of Lie groups. In 1955, the French mathematician Claude Chevalley (1909–84) discovered new families of simple groups. In fact, the sporadic Lie groups were found to be the source of families of finite simple groups. Eventually, eighteen families of simple groups were identified.

The story of the sporadic simple groups started with the French mathematician Émile Léonard Mathieu (1835–90). Between 1860 and 1873, while studying finite geometries, Mathieu discovered the first five sporadic simple groups that were later named after him. The smallest of these has 7,920 elements, and the largest 244,823,040. An entire century passed before the next sporadic simple group was discovered by Yugoslav mathematician Zvonimir Janko in 1965. This and several other simple groups had been predicted to exist before they were actually "discovered." Just as the SU(3) symmetry predicted the existence of the omega minus particle, Janko managed to prove that if a simple group with certain properties was to exist, it absolutely had to consist of 175,560 elements. After pages and pages of calculations, Janko's search bore fruit and he succeeded in constructing the simple group now called J_1. Janko's discovery ended a century of hibernation and marked the beginning of a decade of discovery. Between 1965 and 1975 no fewer than twenty-one sporadic simple groups were constructed, bringing the total to twenty-six (in addition to the eighteen families). The largest of the twenty-six exceptional groups, usually referred to as the "monster," contains the staggering number of

808,017,424,794,512,875,886,459,904,961,
710,757,005,754,368,000,000,000

elements! For the prime-number aficionados, this number is equal to

$$2^{46} \times 3^{20} \times 5^{9} \times 7^{6} \times 11^{2} \times 13^{3} \times$$
$$17 \times 19 \times 23 \times 29 \times 31 \times 41 \times 47 \times 59 \times 71.$$

The monster was predicted to exist by the German mathematician Bernd Fischer and the American Robert Griess (independently) in 1973, and it was constructed by Griess in 1980. Fischer discovered in addition four other sporadic groups, as did Janko in Australia and Germany. In England, John Conway discovered three more.

The identification of eighteen families and twenty-six sporadic simple groups was just the starting point for what turned out to be one of the most impressive and challenging projects in the history of mathematics. The goal was clear: to prove unequivocally that this classification truly exhausted all the possibilities of finite simple groups. In other words, prove that every finite simple group is either a member of one of the eighteen families or is one of the twenty-six sporadic groups. The man who took charge of this awe-inspiring project, Daniel Gorenstein, later called it the "thirty years' war" because much of the classification effort was achieved during the three decades between 1950 and 1980.

Daniel Gorenstein (1923–92) grew up in Boston, studied at Harvard, and became interested in finite groups during his undergraduate days. During World War II, he taught mathematics to the military as part of the war effort. After the war he returned to Harvard for graduate school, completing his doctorate in 1950. Following a few years in which he worked primarily in the field of algebraic geometry, he returned to finite groups in 1957 and became involved with the classification of finite simple groups in the academic year 1960–61.

In addition to the actual discoveries of the twenty-one sporadic simple groups, two other events were instrumental in setting the stage for the massive assault on the classification problem. One was a lecture delivered in Amsterdam in 1954 by the German-American mathematician Richard Brauer (1901–77). In this seminal lecture, Brauer proposed a method of classification that relied on the identification of small "nuclei" of the simple groups that resembled in their properties the parent groups themselves. Brauer's idea was to use these nuclei as the first step in checking whether any arbitrary group can indeed be identified with one of the known simple groups.

The second crucial element for the classification war was an important theorem that was proved in 1963 by University of Chicago mathematicians Walter Feit and John Thompson. The theorem basically states that every finite simple group (that is not cyclic) must have an even number of elements. While the correctness of this statement had been anticipated already in 1906 by the British mathematician William Burnside (1852–1927) and was known as the *second Burnside conjecture*, the actual 1963 proof by Feit and Thompson filled an entire issue (255 pages) of the *Pacific Journal of Mathematics.* The impact of this proof was enormous. Both the ideas and the methods introduced in the paper became the foundation for the classification effort. As Gorenstein described in 1989, "Largely under the impetus of the odd order theorem [the Feit-Thompson theorem states equivalently that finite groups with an odd number of elements are solvable], there was an awakening interest in finite group theory. Throughout the next decade and a half a long list of gifted young mathematicians, who were to play a prominent role in the classification proof, were attracted to the field." Armed with Brauer's insights and with the Feit-Thompson theorem, Gorenstein outlined in 1972 a bold sixteen-step plan to complete the classification proof. He expressed cautious optimism that the full proof could be achieved by the end of the twentieth century.

Given that the proof turned out to involve about a hundred mathematicians who produced some fifteen thousand pages of proof in some five hundred journal articles, Gorenstein's original estimate for the time necessary to complete the proof certainly did not seem excessive. In fact, Ohio State mathematician Ron Solomon, one of the leaders of the endeavor, wrote in 1995, "Not a single leading group theorist besides Gorenstein believed in 1972 that the classification could be completed in this century." As is often the case in mathematics, however, one person can make a big difference. For the classification theorem, that person was Caltech mathematician Michael Aschbacher. Through a series of lightning assaults he cracked a few of the major stumbling blocks, blasting through much of the proof. In Gorenstein's words:

> There were a great many other group theorists who made significant contributions to the classification proof. But it was Aschbacher's entry into the field in the early 1970s that irrevocably altered the simple group landscape. Quickly assuming a leadership role in a single minded pursuit of the full classification theorem, he was to carry the entire "team" along with him over the following decade until the proof was completed.

Indeed, to everyone's amazement, the proof was thought to have been completed as early as 1983. Still, because of the almost unmanageable length of the proof, Gorenstein, Solomon, and mathematician Richard Lyons joined forces in 1982, launching a revision project whose goal was to produce a shorter, more coherent version of the proof. In the years that followed, a few significant gaps have been identified in the main proof. The last of these was finally closed in August 2004, in a two-volume work by Aschbacher and University of Illinois mathematician Stephen Smith. The Gorenstein-Lyons-Solomon revision project is also proceeding well, with six monographs already published or in press. Nevertheless, at least five more years will be needed to complete this monumental undertaking.

The study of finite groups in recent years is intricately connected with a rich variety of other areas of mathematics, from topology to graph theory. Some suspected but not yet fully explored potential connections to quantum field theory may also exist.

Galois introduced the group concept and constructed the first family of finite simple groups with a modest goal in mind—to prove which equations are solvable by a formula and which are not. He surely would have been delighted to see what those humble beginnings have yielded. Ron Solomon described the results of the "thirty years' war" beautifully: "The eruption of mathematics during the heyday of the study of simple groups generated amazing insights into the structure of finite groups and uncovered several of the most fascinating objects in the mathematical firmament."

– NINE –

Requiem for a Romantic Genius

Of the many thousands of mathematicians to have lived since ancient Babylon, who have been the most influential? Mathematician and author Clifford Pickover conducted an informal survey asking precisely this question, and he presented the list of the top ten names in his entertaining book *Wonders of Numbers.* Évariste Galois is on that distinguished roster (at number eight), even though this tormented romantic died at twenty. What is it that makes certain individuals so creatively superior to all others? And how is it possible that such an abundantly overflowing creativity would manifest itself at such a young age? If I could actually provide precise answers to these questions, I am sure that many psychologists, biologists, educators, and corporations would be very appreciative. Since I can't, however, I will instead briefly present some of the current thoughts on these topics and examine if and how they apply to Galois.

First, let me clarify that by extraordinary creativity I mean a process that has a significant cultural impact—an idea or act that brings about a meaningful change. Obvious examples include Sigmund Freud's founding of psychoanalysis and Newton's formulation of the laws of motion.

University of Chicago psychologist Mihaly Csikszentmihalyi has insightfully pointed out that by its very nature, creativity is not just something that happens inside someone's head. To be able to declare any idea or accomplishment "creative," we must compare it to some existing criteria and standards. For instance, we can say without any qualification that Einstein's general relativity is one of the most creative theories of all time only after we judge it against the background of all other physical theories of the universe. Creativity therefore always involves relations among at least three components: the creative person; the domain in which the creative act occurs (e.g., mathematics or some part thereof, music, literature); and the field of players or practitioners who act as gatekeepers and judges (e.g., other mathematicians, museum curators, literature readers, and critics). By any standards, Galois was astonishingly creative. This young man's ideas changed mathematics in a profound way. The new domain that he established—group theory—has expanded far beyond the boundaries of pure mathematics into the realms of visual arts, music, physics, and wherever symmetries can be found.

As I noted above, understanding how creativity works is something that intrigues not only cognitive scientists, neurologists, and educators. Large companies and corporations are scrambling to find ways to foster creativity and innovation among their employees. Many millions of dollars are being spent every year on seminars, retreats, brainstorming sessions, and special courses, all designed with the specific purpose of producing the next Bill Gates. But can the sources of creativity be identified? Or are creative ideas merely sparked by chance and stray bits of knowledge cleverly snatched from loosely related disciplines?

THE SECRETS OF A CREATIVE MIND

The English poet Owen Meredith (the pseudonym of Edward Robert Bulwer-Lytton, earl of Lytton) said once, "Genius does what it must, and Talent does what it can." This is an interesting quote, since it combines and contrasts two terms that may occasionally overlap with creativity, but which shouldn't be confused with it—"talent" and "genius." Over the centuries, there have certainly been many talented painters and inventors, but very few (if any) who could match Leonardo da Vinci for creativity. On the other hand, to be creative—that is, to bring about a paradigm shift—one does not necessarily have to be a genius. In particular, many studies show that beyond a certain level of IQ, probably around 120, there is no clear correlation between intelligence and creativity. In other words, true creativity probably requires some degree of intelligence, but there is absolutely no guarantee that a person with an IQ of 170 will be any more creative than one with an IQ of 120. One of the main reasons why there is no "explanation" of creativity is precisely the fact that all humans are creative at some level. When you cannot open a jar and you grab a towel to prevent your hand from slipping, you have come up with a creative solution. When a kid at school writes a friend's phone number on the back of his hand, he responds creatively to an urgent need. At the end of the day, even the most creative people ever to have lived still had to use a human mind.

Another point to remember is that creative outbursts in different domains do not admit easy comparisons. As Harvard cognition and education researcher Howard Gardner has observed, "Creative breakthroughs in one realm cannot be collapsed uncritically with breakthroughs in other realms; Einstein's thought process and scientific achievements differ from those of Freud, and even more so from those of Eliot [the poet T. S. Eliot] or Gandhi. A single variety of creativity is a myth." In spite of these caveats, in an almost desperate attempt to get to the bottom of creativity, researchers (including Gardner himself) have often relied on trying to identify common traits in many creative individuals. The hope has been that characteristics shared by most represent potential sources for outstanding creativity. Qualities that have been examined include physiological features in the brain, personality traits, various cognitive characteristics (such as the ability to make remote associations), and societal circumstances in both the immediate (e.g., family and close friends) and more global (e.g., ethnic, political) environments. We can get at least a taste of the degree to which various creativity models work from a simple exercise that is based on the concepts of the scientific method. The latter represents the organized approach to explain a collection of observed facts with a model. This idealized process can be summed up by three words: induction, deduction, verification. More explicitly, the scientific method begins with the gathering of experimental or observational facts. On the basis of those facts a model, a scenario, or sometimes a complete theory is constructed. Finally, the model or theory is tested against new experiments, observations, or the collection of new facts that had not been used in the formulation of the model itself.

We can follow a simple version of this general philosophy by examining how Galois measures up against some agreed-upon "template" of personality traits of the creative mind, as long as Galois himself has not been used in the creation of that "template." The latter requirement turned out to be easy to satisfy—I did not find Galois's name on any of the lists compiled by creativity researchers. The first thing I should note is that there is a very good reason why I have put the word "template" in quotation marks—no such "template" truly exists! Even if someone has the genetic predisposition for creativity as a painter, unless this person also has access to the proper training and boasts some connections in the art world, chances are we would never hear of her or him. Moreover, not all creators, not even those within a given domain, are alike. As Csikszentmihalyi has put it, "Michelangelo was not greatly fond of women, while Picasso couldn't get enough of them." Similarly, in chapter 3 we have seen that Cardano flamboyantly burned the candle at both ends, while dal Ferro, who contributed to the solution of precisely the same mathematical problems, was reclusive and modest. Nevertheless, as Boston College psychologist Ellen Winner has noted about gifted children, "For those who *do* [the emphasis is mine] make it into the roster of creators, a certain set of personality traits proves far more important than having a high general IQ, or a high domain-specific ability, even one at the level of prodigy. Creators are hard-driving, focused, dominant, independent risk-takers." While researchers have not been able to discern with any certainty whether personal characteristics can indeed be the direct causes of creativity, there is little doubt that some qualities are intimately involved in the creative process. So what are these traits? Psychologists John Dacey and Kathleen Lennon emphasize tolerance of ambiguity—the ability to think, operate, and remain open-minded in situations where the rules are unclear, where there are no guidelines, or where the usual support systems (e.g., family, school, society) have collapsed. Indeed, without the competence to function where there are no rules, Picasso would have never invented cubism and Galois would not have come up with group theory. Tolerance of ambiguity is a necessary condition for creativity.

Psychologist Csikszentmihalyi concentrates on a somewhat related quality, which he refers to as "complexity." *Complexity* means to be able to harbor tendencies that normally appear to be at opposite extremes. For instance, most people are somewhere in the middle of the continuum between being rebellious or highly disciplined. Very creative individuals can alternate between the two extremes almost at the drop of a hat. Csikszentmihalyi interviewed many dozens of creative people from a wide range of domains, stretching from the arts, humanities, and sciences to business and politics. Based on these interviews, he compiled a list of ten dimensions of complexity—ten pairs of apparently antithetical characteristics that are often both present in the creative minds. The list includes:

1. Bursts of impulsiveness that punctuate periods of quiet and rest.
2. Being smart yet extremely naïve.
3. Large amplitude swings between extreme responsibility and irresponsibility.
4. A rooted sense of reality together with a hefty dose of fantasy and imagination.
5. Alternating periods of introversion and extroversion.
6. Being simultaneously humble and proud.
7. Psychological androgyny—no clear adherence to gender role stereotyping.
8. Being rebellious and iconoclastic yet respectful to the domain of expertise and its history.
9. Being on one hand passionate but on the other objective about one's own work.
10. Experiencing suffering and pain mingled with exhilaration and enjoyment.

Interestingly, psychologist Ellen Winner finds that child prodigies usually exhibit only one extreme of the spectrum of characteristics—they tend to be intense, driven, and introverted. We should remember, however, that gifted children are still in the soaking-up knowledge mode, rather than in the creative mode. The reality that most prodigies do not become particularly creative in their adult life may reflect (among other things) the fact that only a small fraction of the wunderkinder actually possess the capacity for complexity.

Even though Csikszentmihalyi's list is clearly only suggestive at best, it actually describes Galois astonishingly well. Galois was, in many ways, the epitome of contradictions and complexity. Take, for instance, his letter of May 25 to Auguste Chevalier: "How can I console myself when I have exhausted in one month the greatest source of happiness a man can have?" Can one imagine larger mood swings? Or examine the following description in one of Raspail's letters from prison. Galois's behavior oscillates between calm and eruption:

> He was wandering around the prison yard, one day, deep in thought, as if he was daydreaming. He had the sickly look of a man who was barely physically present on earth, and who was kept alive only by his thoughts.
>
> Our bully boys shouted out: "Hey, you may be only twenty, but you are an old man! You cannot take your drink, can you? Drinking frightens you, doesn't it?" He then marched straight to confront the danger, emptied down his throat an entire bottle, all at once, and threw it at his teaser.

Being smart but naïve, realistic yet imaginative, simultaneously rebellious and respectful toward mathematics and mathematicians, are combinations of traits that could have been invented to literally describe Galois. How else would you characterize his experiences with the entrance examinations to the École polytechnique, his acrimonious exchanges with his school principal, his paranoiac interactions with the mathematical establishment, and his confrontations with the law?

Psychological androgyny—being on one hand very sensitive and more "feminine" and on the other aggressive and offensive—was another obvious Galois trait. Consider the following letter, which he wrote from prison to his aunt, Céleste-Marie Guinard:

> My dear aunt, I have been told that you are sick and bedridden. I feel the need to let you know how sorry I am, and this feeling is further aggravated by the fact that I am deprived of the pleasure of seeing you, since I am confined to my room and cannot visit anybody. You were kind enough to think of sending me presents. It is very pleasant to receive reminders of the living, while being in a tomb. I hope you will be in good health when I leave the prison. My first visit will be to you.

Hard to believe, but this is the same person about whom mathematician Sophie Germain had written the following to her friend and colleague Guglielmo Libri Carucci dalla Sommaja: "Having returned home, he [Galois] continued his habit of insult, a taste of which he gave you after your best lecture at the Academy. The poor woman [Galois's mother] fled her home, leaving just enough for her son to live on."

Dacey and Lennon identify a few additional traits that in their opinion contribute to tolerance of ambiguity and to its role in promoting creativity. One of these—*stimulus freedom*—is what we might call the ability to think outside the box. To a large extent, the very essence of creativity is the capacity to break out of common assumptions and to escape any preexisting mind-sets. Let me give a very simple example of this type of stimulus freedom. You are given six matches of equal length, as in figure 105, and the objective is to use them to form exactly four triangles, in which all the sides of all the four triangles are equal. Try this for a few minutes, but be aware that the solution requires an unconventional approach. In case you have not succeeded, don't despair; most people have trouble with this problem. The solution is shown in appendix 10. Galois's proof concerning which equations are solvable by a formula (chapter 6) is the embodiment of thinking outside the box—to answer a question about algebraic equations he invented a whole new domain in mathematics.

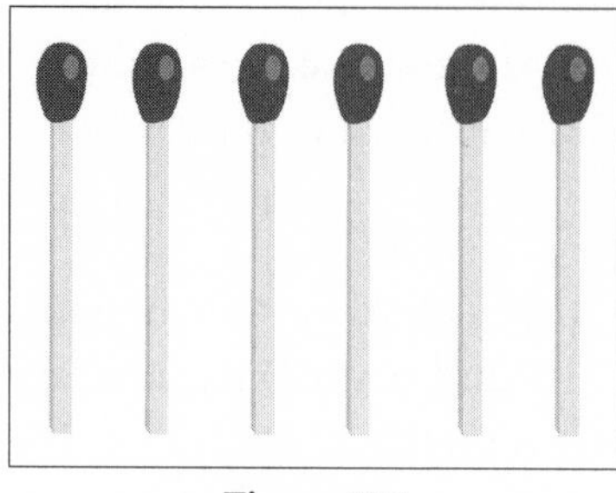

Figure 105

There is another characteristic that appears to be shared by many creative individuals (especially creative men), and that applies to Galois as well—the loss of a father early in life. Among nearly a hundred creative interviewees, Csikszentmihalyi actually found that no fewer than three out of ten men and two out of ten women were orphaned by the time they reached their teens.

How can the loss of a father stimulate creativity? Life deals youngsters who have lost their fathers a complex hand that is a mixture of burden and opportunity. On one hand, there is the huge psychological burden of having to live up to the perceived expectations of the missing father. On the other, such youngsters have the immense opportunity of truly inventing themselves. The French philosopher Jean-Paul Sartre (1905–80) observed in his autobiographical *Les mots* (*Words*): "The death of Jean Baptiste [Sartre's father] was the big event of my life: it sent my mother back to her chains and gave me freedom. . . . Had my father lived, he would have lain on me full length and would have crushed me. As luck had it, he died young." This is surely an overly cynical view. While some creators, including possibly Abel and Galois, may have been driven toward independence and curiosity by the death of their fathers, many others thrived on the support they had received from their families. There are cases, for instance, where the father and son were both

Nobel Prize laureates. Niels Bohr received the physics prize in 1922, and his son Aage Bohr received it in 1975. An even more impressive example is that of William Henry Bragg and his son William Lawrence Bragg. The father-son team won the Nobel Prize in physics together in 1915, when Lawrence was only twenty-five.

Galois accomplished all of his brilliantly seminal work on group theory before the age of twenty-one, Abel's genius dazzled the mathematical world before this poor mathematician was twenty-seven. Should we be surprised? Not really. Some of the most creative mathematicians, lyric poets, and composers of music were extraordinarily young when they produced their best work. Most painters, novelists, and philosophers, on the other hand, continue to create and are often at their peak well into old age. Music critic and novelist Marcia Davenport (1903–96) expressed this reality beautifully: "All the great poets died young. Fiction is the art of middle age. And essays are the art of old age."

I asked Sir Michael Atiyah, the recipient of the 2004 Abel Prize, why he thought mathematicians were so insightful early in life. He answered immediately:

> In mathematics, if you are of quick mind, you can get to the "front line" of cutting-edge research very quickly. In some other domains you may have to read entire thick volumes first. Moreover, if you have been for too long in a certain domain, you get conditioned to think like everybody else. When you are new, you are not compelled to the ideas of the people around you. The younger you are, the more likely you are to be truly original.

Psychologist Howard Gardner makes a similar distinction between mathematicians and scientists on one hand and artists on the other:

> It is important to note here a decisive difference from creation in the sciences or mathematics. Individuals in these latter areas begin to be productive at an early age and certainly have the option of making numerous innovations during their early years. However, unlike the arts, these domains progress and accumulate at a rapid rate, stimulated by the discoveries of the most creative individuals; tools fashioned earlier in life may become irrelevant or dysfunctional.

Creative minds in mathematics can even be distinguished from those in the other sciences in that they often don't obey what Gardner calls the "ten-year rule." This is the observation that many creative individuals make a breakthrough after ten years of work in their domain. Both Abel and Galois had the guts to attack the quintic while still in high school! They gave the definitive answer to its solubility at or before their early twenties, well ahead of when the ten-year rule would have applied.

There is one other aspect of Galois's personality that fits current thinking about creativity—the fact that he exhibited strong symptoms of paranoia. His continual delusions of being persecuted and haunted by mediocrity certainly went beyond normal. Genius has often been linked to mental disorder. Already in ancient times, the Roman philosopher Seneca wrote that "no great genius has ever existed without some touch of madness." In 1895, psychiatrist W. L. Babcock published an article entitled "On the Morbid Heredity and Predisposition to Insanity of the Man of Genius" in which he claimed that like proneness to early death, genius was a characteristic of inferior genetic makeup. On a more solid basis, recent research supports the general association of creativity with psychopathology. For instance, psychologist Arnold Ludwig examined the lives of more than a thousand creative individuals and found that about 28 percent of the prominent scientists experienced at least some sort of mental disturbance. The fraction increased to a staggering 87 percent among outstanding poets. Psychologist Donald MacKinnon, then of the Institute for Personality Assessment and Research at the University of California, Berkeley, conducted an extensive psychometric evaluation of many creative mathematicians, architects, and writers. The findings showed that the creative individuals consistently scored higher on dimensions that are indicative of various affective disorders such as schizophrenia, depression, and paranoia. The conclusion from these and numerous similar studies is, as University of California, Davis psychologist Dean Keith Simonton puts it, "The genius-madness link may be more than myth." I should note that, as in Galois's case, the levels of the disorder were rarely found to be so high as to debilitate the creative individual. Galois and many other creative geniuses possessed enough ego-strength and other mental resources to help contain their psychopathology. Yet the evidence for this Faustian bargain that creative minds often have to negotiate is quite compelling. The English essayist Sir Max Beerbohm (1872–1956) expressed his own experience with this phenomenon: "I have known no man of genius who had not to pay, in some affliction or defect either physical or spiritual, for what the gods had given him."

As compelling as the case may be for Galois fitting the profile of a creative genius, we have to wonder, was there also something distinctively special about his brain?

THE STORY OF TWO BRAINS

Albert Einstein died on April 18, 1955, at Princeton Hospital in New Jersey. Thomas S. Harvey, the pathologist who performed the autopsy, removed the great scientist's brain, dissected it into 240 pieces, and embedded the pieces in a plastic-like substance called celoidin.

Évariste Galois died on May 31, 1832, at the Cochin Hospital in Paris. The pathologist opened his skull and conducted a thorough examination of his brain. This is truly astonishing, given that Galois was shot in the stomach and died of peritonitis. More than half of the autopsy report is devoted to the brain.

For more than two decades, no one, not even Einstein's family, knew that Einstein's brain was being kept in jars at Harvey's home. In 1978, Steven Levy, then a reporter for the *New Jersey Monthly,* tracked Harvey down at his home in Wichita, Kansas. After a long conversation with the reporter, Harvey admitted that he had the brain. Out of a box labeled "Costa Cider," he pulled the two Mason jars that contained the brain that had brought about a revolution in science.

Since then, Harvey has allowed three teams to examine parts of the brain. University of California, Berkeley anatomist Marian Diamond and her colleagues published a paper on Einstein's brain in 1985. They found that the ratio of neurons to glial cells (the cells that support and protect neurons) in one part of Einstein's brain was smaller than the ratios in eleven normal brains. While the authors concluded that the larger number of glial cells per neuron might indicate that Einstein's neurons worked harder—needed more energy—than normal, this interpretation was later questioned by other researchers. A second paper, by Britt Anderson of the University of Alabama at Birmingham, was published in 1996. Anderson and Harvey showed that while Einstein's brain weighed less than the average (2 pounds 11.4 ounces compared to 3 pounds 1.4 ounces for the average; 1,230 grams compared to

1,400 grams) it packed more neurons in a given area. Finally, in 1999, McMaster University neuropsychologist Sandra Witelson and her colleagues discovered what was hailed as a potential key to Einstein's genius. The inferior parietal region that is thought to be used for mathematical reasoning was found to be 15 percent wider than normal. In addition, a groove (sulcus) was found to be partially missing in that area. The researchers argued that the absence of that fissure could have resulted in more effective communication among neurons. Although interesting, all of this research could not be regarded as conclusive. After all, even though Witelson's study used thirty-five brains as a control group, it had only one brain in the experimental group—Einstein's.

The remaining pieces of Einstein's brain were eventually brought by Harvey to their final resting place—the Pathology Department at Princeton Hospital. When asked why he took the brain in the first place (Einstein's body was cremated), Harvey explained that he felt obligated to salvage the precious gray matter for posterity.

The autopsy report on Galois's brain reads:

> Stripped of its envelope, the skull presents the two pieces that form the coronal in young children, being joined at an obtuse angle. This has at most a width of one-fifth of an inch. At the edge where the coronal sutures the parietal bones, one can see a deep, flat, circular depression, which follows the joint between the two bones; the parietal humps are very developed, wide apart from one another; the development of this portion is remarkable, by comparison to the occipital bone . . .
>
> Once the skull is opened, the inner walls of the frontal sinuses are very close; the remaining space is less than one-fifth of an inch; in the middle of the skull's dome, two depressions correspond to the humps described above . . .
>
> The brain is heavy, its convolutions large, its crevices deep, especially on the lateral parts; there are protuberances matching the cavities of the skull; one in front of each anterior lobe, two on top of the upper face; the cerebral substance is generally soft; the ventricular cavities are small, empty of any serous fluids; the pituitary gland is voluminous and contains gray granulations; the cerebellum is small; the weight of the brain and the cerebellum together is three pounds, two ounces, less one-eighth of an ounce.

Why did the pathologist examine Galois's brain so thoroughly when the cause of death was obvious? The first sentence in the report may provide a hint: "Young Galois Évariste, 21 years of age, a good mathematician, known primarily for his ardent imagination, has just succumbed in 12 hours to acute peritonitis, caused by a bullet shot from 25 paces." My hunch is that the pathologist was driven by the same curiosity that caused Harvey to take Einstein's brain. The pathologist was aware of both Galois's reputation as a mathematician and of his fiery, passionate imagination, and he felt compelled to examine the brain for potential clues as to the origin of these attributes. As in Einstein's case, the autopsy did not reveal any clear "smoking gun." Still, this was probably a worthwhile effort, since its goal was to unveil the mind of a person who stood, in both mathematics and politics, at the heart of revolutionary romanticism.

INDIVISIBLE

Unlike in most other sciences, in mathematics ideas have a lasting value. Aristotle's views of the universe are interesting historical curiosities but nothing more. The theorems in Euclid's *Elements*, on the other hand, are as valid, as correct, and as immortal today as they were in 300 BC. This is not to say the mathematics is stagnant. Far from it. Just as new generations of telescopes expand our horizons without necessarily invalidating previous findings in the nearby universe, mathematics continually reveals new vistas while building on existing knowledge. The perspective may change but the truths do not. Mathematician and author Ian Stewart expressed this reality beautifully: "In fact, there is a word in mathematics for previous results that are later changed: they are called 'mistakes.' "

Galois's ideas, with all their brilliance, did not appear out of thin air. They addressed a problem whose roots could be traced all the way back to ancient Babylon. Still, the revolution that Galois had started grouped together entire domains that were previously unrelated. Much like the Cambrian explosion—that stunning burst of diversification in life forms on Earth—the abstraction of group theory opened windows into an infinity of truths. Fields as far apart as the laws of nature and music suddenly became mysteriously connected. The Tower of Babel of symmetries miraculously fused into a single language.

Web designer Brenda C. Mondragon manages an inviting website entitled "Neurotic Poets." Her first line on the romantic English poet Percy Bysshe Shelley (1792–1822) reads: "The spirit of revolution and the power of free thought were Percy Shelley's biggest passions in life." One could use precisely the same words to describe Galois. On one of the pages that Galois had left on his desk before leaving for that fateful duel, we find a fascinating mixture of mathematical doodles, interwoven with revolutionary ideas (figure 106). After two lines of functional analysis comes the word "indivisible," which appears to apply to the mathematics. This word is followed, however, by the revolutionary slogans "unité; indivisibilité de la république" ("unity; indivisibility of the republic") and "Liberté, égalité, fraternité ou la mort" ("Liberty, equality, brotherhood, or death"). After these republican proclamations, as if this is all part of one continuous thought, the mathematical analysis resumes. Clearly, in Galois's mind, the concepts of unity and indivisibility applied equally well to mathematics and to the spirit of the revolu-

Figure 106

tion. Indeed, group theory achieved precisely that—a unity and indivisibility of the patterns underlying a wide range of seemingly unrelated disciplines.

There are two other phrases that catch the eye among Galois's scribbles. One, "Pas l'ombre," almost certainly refers to the phrase "pas l'ombre d'un doute" ("without the shadow of a doubt"). Again, Galois would have had such convictions about both the correctness of his mathematical proofs and of his republican ideals. The second phrase, "une femme" ("a woman"), is a sad reminder of the annoyingly trivial circumstances that were about to cause his untimely death only a few hours later.

The famous Indian poet Rabindranath Tagore (1861–1941) wrote that "death is not extinguishing the light. It is putting out the lamp because dawn has come." This was certainly true in Galois's case. His insights announced the dawn of a new era in mathematics. He belongs to that very exclusive club of those who are genuinely immortal.

Generations of young mathematicians moved by Galois's tragic story and pointless death have found consolation in his incredible legacy. Through this gratification, they were spared the fate of some of the more impressionable youth who a few decades before Galois's time had read Goethe's masterpiece *The Sorrows of Young Werther.* The romantic agony of Goethe's sensitive protagonist touched a universal chord. The story was so powerful that it inspired a series of youthful suicides all across Europe. Incidentally, one might have thought that such passion has long since disappeared from a much more cynical world. Yet the spontaneous outpouring of grief that followed Princess Diana's death has demonstrated that romanticism isn't quite dead yet. Galois's story continues simultaneously to sadden and inspire even today, and the spirit of his work permeates much of modern mathematics. I can find no better words to describe this contrast between the perishability of the flesh and the endurance of the ideas than those in Emily Dickinson's poem:

Death is a Dialogue between
The Spirit and the Dust.
"Dissolve" says Death—The Spirit "Sir
I have another Trust"—

APPENDIX 1

Card Puzzle

A solution to the card puzzle on page 22. The goal is to arrange the jacks, queens, kings, and aces in a square so that no suit or value would appear twice in any row, column, or the two main diagonals.

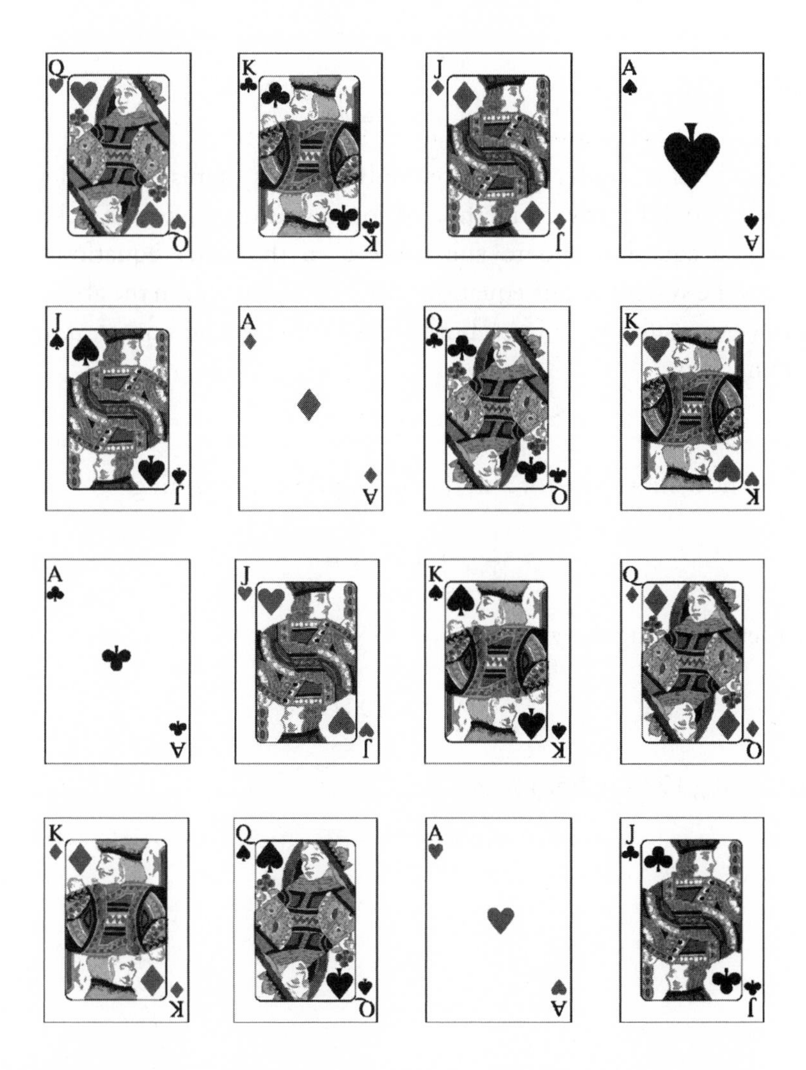

APPENDIX 2

Solving a System of Two Linear Equations

On page 52 we encountered the ancient Babylonian system of equations:

$$\frac{1}{4}y + x = 7$$

$$x + y = 10.$$

This is a brief reminder of how one solves such systems of equations. A relatively straightforward method of solution is to isolate one unknown from one equation and to substitute it in the other equation. This reduces the system to one equation with one unknown. In the above system, we can subtract y from both sides of the second equation, transforming it to

$$x = 10 - y$$

We can now substitute for x in the first equation, obtaining

$$\frac{1}{4}y + 10 - y = 7$$

or, after collecting the y terms

$$-\frac{3}{4}y + 10 = 7$$

Subtracting 10 from both sides:

$$-\frac{3}{4}y = -3$$

Multiplying both sides by $\left(-\frac{4}{3}\right)$ we obtain:

$$y = 4.$$

Now, substituting the value of y in $x = 10 - y$ above, we get: $x = 6$. The solution is therefore length = 6; width = 4.

APPENDIX 3

Diophantus's Solution

This represents Diophantus's solution to problem 28 from the first book of *Arithmetica* (mentioned on page 59).

We need to find two numbers such that their sum and the sum of their squares are given numbers. Suppose that the sum is 20 and the sum of the squares is 208. Diophantus does not designate the numbers by x and y, but rather by $10 + x$ and $10 - x$, taking advantage of the fact that the sum has to be 20. The equation he obtains for the sum of the squares is therefore

$$(10 + x)^2 + (10 - x)^2 = 208.$$

Now, since:
$$(10 + x)^2 = (10 + x)(10 + x) = 100 + 20x + x^2$$
$$(10 - x)^2 = (10 - x)(10 - x) = 100 - 20x + x^2$$

the equation reads (by collecting all the terms):

$$200 + 2x^2 = 208.$$

Subtracting 200 from both sides: $2x^2 = 8$.

Dividing by 2: $x^2 = 4$.

Taking the positive square root: $x = 2$.

Therefore, the two numbers sought are 12 and 8.

APPENDIX 4

A Diophantine Equation

We need to find whole-number (such as 1, 2, 3, . . .) solutions to the equation (page 59)

$$29x + 4 = 8y.$$

We can subtract 4 from both sides to obtain

$$29x = 8y - 4.$$

Taking out the common factor 4 from both terms on the right-hand side results in

$$29x = 4(2y - 1).$$

Since x must be a whole number, the left-hand side is divisible by 29, and so must be the right-hand side. However, 29 is a prime (divisible only by 1 and by itself); therefore $2y - 1$ must be divisible by 29. In particular, we could take:

$$2y - 1 = 29 \text{ and } x = 4 \text{ (for the equality to hold true).}$$

Adding 1 to both sides of $2y - 1 = 29$ and dividing by 2, we get $y = 15$. A solution is therefore $x = 4$, $y = 15$.

APPENDIX 5

Tartaglia's Verses and Formula

Tartaglia's rules for solving the three forms of the cubic were put into verses (page 68 of the text). The translation by Ron G. Keightley reads:

In cases where the cube and the unknown
Together equal some whole number, known:
Find first two numbers diff'ring by that same;
Their product, then, as is the common fame,
Will equal one third, cubed, of your unknown;
The residue of their cube roots, when shown
And properly subtracted, next will give
Your main unknown in value, as I live!
As to the second matter of this kind,
When cube on one side lonely you shall find,
The other terms together being bound:
Two numbers from that one, once they are found,
Together multiplied, swift as a bird,
Give product clear and simple, of one-third
Cubed of th'unknown; by common precept, these
You take, cube rooted; add them, if you please,
T'achieve your object in their sum with ease.
The third case, now in these our little sums,
From the second is solved; for, as it comes,
In kind it is the same, or so say I!
These things I found—O, say not tardily—
In thrice five-hundred, four and thirty more,
Of this our age; the gallant proof's in store
Where City's girt by Adriatic Shore.

When the cubic equation is of the form

$$x^3 + px = q$$

where p and q are any numbers, as in $x^3 + 6x = 20$, the dal Ferro–Tartaglia-Cardano formula for the solution is given by the somewhat intimidating expression

$$x = \sqrt[3]{\frac{q}{2} + \sqrt{\frac{p^3}{27} + \frac{q^2}{4}}} + \sqrt[3]{\frac{q}{2} - \sqrt{\frac{p^3}{27} + \frac{q^2}{4}}}\,.$$

For example, substituting $p = 6$ and $q = 20$ from the above example (and taking the positive square roots), gives the positive solution $x = 2$.

Examine, however, the following equation considered by Bombelli (page 79 in the text):

$$x^3 - 15x = 4.$$

Here, $p = -15$, $q = 4$. You can easily check that substituting these values into the formula above gives

$$x = \sqrt[3]{2 + \sqrt{-121}} + \sqrt[3]{2 - \sqrt{-121}}\,.$$

Here the intermediate step involves the square root of the negative number -121. Yet a simple inspection reveals that $x = 4$ is a solution of the original equation. While Bombelli managed to solve this specific equation using an ingenious trick, the general problem of dealing with square roots of negative numbers was solved only with the introduction of complex numbers.

Bombelli's trick was as follows: Write $\sqrt[3]{2 + \sqrt{-121}} = 2 + c\sqrt{-1}$ where the value of c needs to be determined. Raise the two sides of this equation to the third power. This would give

$$2 + \sqrt{-121} = \left(2 + c\sqrt{-1}\right)^3.$$

Since the square root of 121 is equal to 11, the left-hand side is equal to $2 + 11\sqrt{-1}$. The right-hand side can be expanded, using the identity

$$(a + b)^3 = a^3 + 3a^2b + 3ab^2 + b^3$$

to give $8 + 12c\sqrt{-1} - 6c^2 - c^3\sqrt{-1}$.

Equating the two sides and collecting terms we obtain:

$$2+11\sqrt{-1}=\left(8-6c^{2}\right)+\left(12c-c^{3}\right)\sqrt{-1}\ .$$

By inspection this equation holds true for $c = 1$. From Bombelli's substitution we therefore find that

$$\sqrt[3]{2+\sqrt{-121}}=2+\sqrt{-1}\,.$$

Using a similar substitution Bombelli found that

$$\sqrt[3]{2-\sqrt{-121}}=2-\sqrt{-1}\,.$$

Substituting both expressions into the formula for x above, Bombelli found the solution

$$x=2+\sqrt{-1}+2-\sqrt{-1}=4\,.$$

APPENDIX 6

Adriaan van Roomen's Challenge

The equation presented by van Roomen was (page 80 in the text):

$$
\begin{aligned}
x^{45} &- 45x^{43} + 945x^{41} - 12{,}300x^{39} + 111{,}150x^{37} \\
&- 740{,}459x^{35} + 3{,}764{,}565x^{33} \\
&- 14{,}945{,}040x^{31} + 469{,}557{,}800x^{29} - 117{,}679{,}100x^{27} \\
&+ 236{,}030{,}652x^{25} - 378{,}658{,}800x^{23} + 483{,}841{,}800x^{21} \\
&- 488{,}494{,}125x^{19} + 384{,}942{,}375x^{17} - 232{,}676{,}280x^{15} \\
&+ 105{,}306{,}075x^{13} - 34{,}512{,}074x^{11} + 7{,}811{,}375x^{9} \\
&- 1{,}138{,}500x^{7} + 95{,}634x^{5} - 3{,}795x^{3} + 45x = C
\end{aligned}
$$

where C is a known number. In particular, he asked for a solution when

$$C = \sqrt{\frac{7}{4} - \sqrt{\frac{5}{16}} - \sqrt{\frac{15}{8} - \sqrt{\frac{45}{64}}}}\,.$$

Viète, who already knew the formula for the sines and cosines of $n\alpha$ (where n is any integer and α is some angle), was able to use this knowledge. He recognized that the left-hand side of the equation is the expression for $2 \sin 45\alpha$, when the latter is expressed in terms of $2 \sin\alpha$. Therefore, by simply finding the value of α such that $2 \sin 45\alpha = C$, gives the solution to van Roomen's equation as $x = 2 \sin\alpha$.

APPENDIX 7

Properties of the Roots of Quadratic Equations

The most general quadratic equation has the form (page 82 in the text):

$$ax^2 + bx + c = 0.$$

Dividing by a we obtain

$$x^2 + \frac{b}{a}x + \frac{c}{a} = 0.$$

If, on the other hand, we denote the solutions by x_1 and x_2, then the equation can also be written as

$$(x - x_1)(x - x_2) = 0$$

since the product is equal to zero when $x = x_1$, or $x = x_2$. Multiplying through, we get

$$x^2 - (x_1 + x_2)x + x_1x_2 = 0.$$

Comparing this expression to the previous form, we see that the solutions have to satisfy

$$x_1 + x_2 = -\frac{b}{a}$$

$$x_1x_2 = \frac{c}{a}.$$

Let us now examine the expression

$$\frac{1}{2}\left[(x_1 + x_2) \pm \sqrt{(x_1 + x_2)^2 - 4x_1x_2}\right].$$

Given that:

$$(x_1 + x_2)^2 = x_1^2 + 2x_1x_2 + x_2^2$$
$$(x_1 - x_2)^2 = x_1^2 - 2x_1x_2 + x_2^2$$

we see that

$$\pm\sqrt{(x_1 + x_2)^2 - 4x_1x_2} = \pm(x_1 - x_2)$$

and therefore

$$\frac{1}{2}\left[(x_1 + x_2) \pm \sqrt{(x_1 + x_2)^2 - 4x_1x_2}\right] = \frac{1}{2}\left[(x_1 + x_2) \pm (x_1 - x_2)\right]$$

giving x_1 (when the "+" sign is chosen) and x_2 (when the "−" sign is chosen).

APPENDIX 8

The Galois Family Tree

On Évariste's paternal side I have uncovered only the following, starting with Évariste's grandfather:

Jacques Olivier Galois (Évariste's grandfather)
Born 1742 at Ozouer-le-Voulgy (Seine-et-Marne)
Married Marie-Jeanne Deforge (Évariste's grandmother)
Died at Bourg-la-Reine on May 12, 1806

Évariste's grandparents had six children:

Marie Anne Olivier Galois
Born November 3, 1768
Married Joseph Martin Blondelot

Marie Antoinette Galois
Born October 20, 1770
Married Denis François Le Guay

Théodore Michel Galois
Born March 14, 1774
Married Victoire Antoinette Grivet

Nicolas-Gabriel Galois (Évariste's father)
Born December 3, 1775
Married Adélaide Marie Demante (Évariste's mother)
Died July 2, 1829

Maria Pauline Galois
Born September 7, 1778
Married André Robert Hyard

Jacques Antoine Raphaël Galois
Born 1781

Auffray (2004) lists another son—Jean Baptiste Olivier—but I did not find his name in the Bourg-la-Reine lists. He also spells the middle name of the next-to-last daughter as Apolline (instead of Pauline).

Nicolas Gabriel Galois and Adéläide Marie Demante had three children:

Nathalie Théodore Galois
Born December 26, 1808
Married Benoît Chantelot

Évariste Galois
Born October 25, 1811
Died May 31, 1832

Alfred Galois
Born December 18, 1814
Married Pauline Chantelot

The following generations look as follows:

Natalie (1808–)   Évariste (1811–32)   Alfred (1814–)
| |
Pauline (1833–1901)   Elisabeth (1843–55)
Married Guinard Felix
|
Nathalie (–1877)

The *direct* family tree on Évariste's maternal side is:

Michel de Mante
Married Barbe de Criquebeuf
|
Pierre de Mante (1590–1670)
Married Anne Bréard
Had ten children, of which the tenth follows
|
François Demante (1645–1711)
Married Marguerite de Gruchy
Had fourteen children, of which the thirteenth follows
|
Michel Demante (1692–1766)

Married Anne Marguerite Leclerc
Had fourteen children, of which the sixth follows:

|

François Demante (1723–90)
Married Marie-Madeleine Martin
Had two children; the daughter died at a young age

|

Thomas François Demante (1752–1823)
Married Marie Thérèse Élisabeth Durand

Adéläide Marie Demante (1788–1872)	Antoine-Marie Demante (1789–1856)	Céleste-Marie Demante (1804–60)
Évariste's mother	Married Anne Delaporte	Married Étienne-Charles Guinard
Married Nicolas-Gabriel Galois	Had seven children	Had seven children
Married the second time Jean François Loyer		

I have extensive information on the generations that followed Antoine-Marie Demante and Céleste-Marie Demante, but I do not present it, since it is not directly related to Galois.

APPENDIX 9

The 14–15 Puzzle

The original configuration in Samuel Loyd's 14–15 puzzle (page 160 in the text):

1	2	3	4
5	6	7	8
9	10	11	12
13	15	14	

can be changed into the following configuration:

	1	2	3
4	5	6	7
8	9	10	11
12	13	14	15

using forty-four moves. The following numbers indicate which square (in order) should be slid into the vacant spot: 14, 11, 12, 8, 7, 6, 10, 12, 8, 7, 4, 3, 6, 4, 7, 14, 11, 15, 13, 9, 12, 8, 4, 10, 8, 4, 14, 11, 15, 13, 9, 12, 4, 8, 5, 4, 8, 9, 13, 14, 10, 6, 2, 1.

APPENDIX 10

Solution to the Matches Problem

With six matches of equal length (figure 105), we need to form four triangles in which all the sides are equal. The naïve tendency is to attempt to solve the problem in two dimensions (with the matches lying on a desktop), where no solution exists. The "outside the box" solution is to construct a tetrahedron in three dimensions (as in the figure below). This automatically forms four triangles with equal sides.

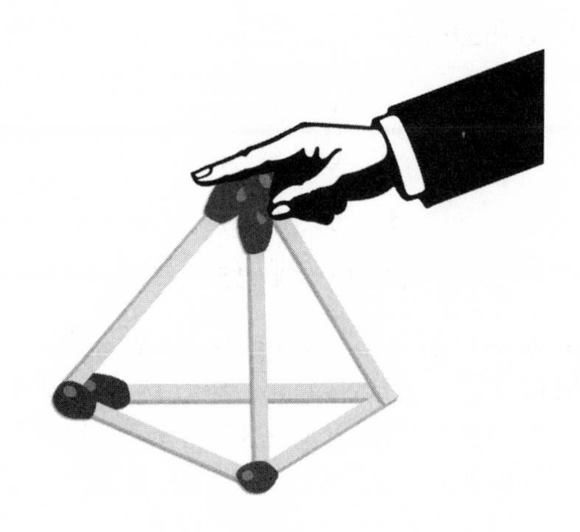

Notes

CHAPTER 1: SYMMETRY

Two excellent popular books on symmetry in general are by Stewart 2001 and Stewart and Golubitsky 1992. A highly recommended popular book on symmetry in physics is by Zee 1986. Somewhat more technical on symmetry in science are Rosen 1995 and Icke 1995. Symmetry in chemistry is nicely described in Heilbronner and Dunitz 1993. An extremely well-documented book on symmetries in different cultures is by Washburn and Crowe 1988. The book by Evans 1975 examines symmetry in West European ornaments. Two invaluable collections of more technical articles on symmetry are Hargittai 1986 and Hargittai 1989. An extensive technical discussion of symmetry in science and art can be found in Shubnikov and Koptsik 1974. A book with many examples is Walser 2000. Finally, Weyl 1952 remains a classic reference.

PAGE

1 *famous Rorschach test:* Loftus 2001; Wood, Nezworski, Lilienfeld, and Garb 2003.

1 *Botticelli's* Birth of Venus: Gombrich 1995.

3 *The word* symmetry *has ancient roots:* Nagy 1995; *Oxford English Dictionary* 1978.

4 *Roman architect Vitruvius:* Vitruvius ca. 27 BC; Osborne 1952.

4 *modern meaning of symmetry:* Clear descriptions can be found in Rosen 1975 and in Stewart and Golubitsky 1992.

4 *Examine for example the verses:* Shubnikov and Koptsik 1974.

4 *Phrases with this property are called* palindromes: Bergeron 1973; Gardner 1979.

5 *"Able was I ere I saw Elba":* Fayen 1977.

5 *The Y's full genome sequencing:* Skaletsky et al. 2003; Willard 2003; Rozen et al. 2003; Pagán Westphal 2003.

5 *bilateral symmetry that characterizes the animal kingdom:* Excellent discussions can be found in Weyl 1952; Gardner 1990; Gregory 1997; Corballis and Beale 1976.

8 *Chicago entertainer:* "Senator" Clarke Crandall, described in Gardner 1990.

8 *A computer-based superintelligence:* Excellent arguments are presented in Kurzweil 1999.

10 *"The Snow Storm":* Emerson 1847.

10 *The famous astronomer Johannes Kepler:* Kepler 1966.

10 rotational symmetry: Discussions can found, for example, in: Weyl 1952; Boardman, O'Connor, and Young 1973.

10 *the painter James McNeill Whistler:* Whistler 1890.

11 *Clive Bell:* Bell 1997.

11 *Aesthetics theorist Harold Osborne:* Osborne 1952, 1986.

12 *psychologists Peter G. Szilagyi and John C. Baird:* Szilagyi and Baird 1977.

12 *George David Birkhoff:* Wilson 1945; O'Connor and Robertson 2001b.

13 Aesthetic Measure: Birkhoff 1933.

16 *A mammoth-ivory bracelet:* mentioned in Wolfram 2002.

16 *the drawings of the fantastic:* A wonderful description of Escher's works is Schattschneider 2004.

16 *William Morris:* Three good books on his life and work are, for instance, Parry 1996; MacCarthy 1995; Menz 2003.

17 The Beauty of Life: Hares-Stryker 1997.

18 *crystal physicist G. V. Wulff:* Shubnikov and Koptsik 1974.

18 *symmetry under translation in music:* Wilson 1986.

18 *Mozart's association with objects of mathematics:* There are many references including: Hyatt King 1944; Tovey 1957; Mozart 1966; Hyatt King 1976; Putz 1995.

19 *Johann Sebastian Bach:* Detailed descriptions of his life and music, as well as connections to mathematics, can be found in Schweitzer 1967; Altschuler 1994; Wolff 2001; Wilson 1986; Smith 1996.

21 glide reflection: For instance in Washburn and Crowe 1988.

22 Latin squares: Popular descriptions can be found in Peterson 2000; Gardner 1959a; more technical in Ball and Coxeter 1974.

22 *Permutations feature in such diverse circumstances:* J. Rosen 1995; Boardman, O'Connor, and Young 1973.

23 *Charles Lamb:* Lamb 1823.

24 *People playing roulette:* Fabricand 1989.

24 *The expectation to win:* Suppose you bet on some red number. Then, on *average,* you can expect to win 18 times out of every 38 times you bet. In the other 20 times you lose. If all the bets are for $1, then in over 38 games the expectation is to lose $2 (there are 18 wins vs. 20 losses). The net return is therefore a *loss* of $2 for a $38 bet, and the *expectation* (the loss for each dollar bet over many games) is $2/38, or 5.3¢.

24 *Blackjack is a card game:* Mezrich 2002; Fabricand 1989.

26 *Pauli exclusion principle:* A brief description can be found in Gamow 1959.

27 *color transformation:* Rosen 1975; Loeb 1971.

28 *Escher himself was never quite sure:* The quote appears in the preface to MacGillavry 1976.

CHAPTER 2: EYE S'DNIM EHT NI YRTEMMYS

PAGE

30 *picture taken with the Hubble Space Telescope:* Brown et al. 2003.

30 stereoscopic vision: Julesz 1960; Brindley 1970; Pinker 1997; Goldstein 2002; Wheatstone 1838.

32 *In two remarkable books:* Kepler's books *Astronomiae Pars Optica* (published in 1604) and *Dioptrice* (published in 1611) are a part of his *Collected Works;* see Caspar and Hammer 1937.

32 *Charles Wheatstone:* Bowers 2001.

33 Gestalt psychology: Description of the principles can be found in Wertheimer 1912; Palmer 1999; Goldstein 2002; Barry 1997.

34 *Garner and Palmer in particular:* Garner 1974; Palmer 1991.

34 structural information theory: Leeuwenberg 1971; Buffart and Leeuwenberg 1981; van der Helm and Leeuwenberg 1991.

36 *The biblical prophet Amos:* Amos 3:3.

36 common region, connectedness, and synchrony: Summarized nicely in Palmer 1999.

37 *Oscar Wilde once said:* Wilde 1892.

37 *role of symmetry in perception:* Fox 1975; Howe 1980; Palmer and Hemenway 1978.

37 *psychologists Jennifer Freyd and Barbara Tversky:* Freyd and Tversky 1984.

37 *Ioannis Paraskevopoulos:* Paraskevopoulos 1968.

37 *(MRI) to map the areas in the brain:* Tyler 2002; Tootell et al. 1998; Mendola et al. 1999.

38 *interrelation between symmetry and orientation:* Detailed descriptions can be found in Rock 1973; Marr 1982; Corballis 1988; Tarr and Pinker 1989; Palmer 1999.

39 *philosopher Ernst Mach:* Mach 1914.

39 *who also invented the kaleidoscope:* An excellent, if technical, article on the mathematics of kaleidoscopes is Goodman 2004.

40 *computer-generated* autostereograms: Tyler 1983, 1995; N. E. Thing Enterprises 1995.

42 *Biologist Thomas Henry Huxley:* Huxley 1868.

43 *A symmetry of the laws:* Popular discussions of symmetries of the laws of nature can be found, for instance, in Weinberg 1992; Zee 1999; Livio 2000; Greene 2004; Kane 2000; Lederman and Hill 2004.

45 group theory: There are many textbooks on group theory but relatively few that are popular or semi-popular accounts. Brief popular explana-

tions can be found, for instance, in Stewart 1995 and Devlin 1999, 2002. An excellent, comprehensive description that is still at a quite elementary (but not popular) level can be found in Budden 1972. A simple discussion of groups and symmetry is presented in Farmer 1996. Other elementary-level books include: Gardner 1966; Maxwell 1965; McWeeny 2002. See notes to chapter 6 for more advanced books.

49 *historian of mathematics:* The quote is from Bell 1951.

CHAPTER 3: NEVER FORGET THIS IN THE MIDST OF YOUR EQUATIONS

PAGE

51 *"Science and Happiness":* The quote appears in Calaprice 2000.

52 *Babylonians did not truly use the concept of algebraic equations:* Calinger 1999; van der Woerden 1983; Boyer 1991; O'Connor and Robertson 2001a; Kline 1972.

53 *Egyptian mathematics comes from the fascinating Ahmes Papyrus:* Gillings 1972; Calinger 1999; O'Connor and Robertson 2000a, b, c; Newman 1956.

54 *Problem 79 reads:* Wells 1997; Gillings 1972.

54 Mother Goose *collection:* Newman 1956.

55 *Chinese collection* Nine Chapters: van der Woerden 1983.

55 *member of Parliament Tony McWalter explained:* In the session held on June 26, 2003.

55 *The distressed man went to Rabbi Huna:* Gandz 1940a.

56 *The Greek historian Polybius:* Gandz 1940b.

57 *problem 2 in tablet 13901 in the British Museum:* Berriman 1956.

58 *geometric algebra:* This popular mathematical genre is described in van der Woerden 1983; Heath 1956, 1981; Gandz 1937.

58 *One of the most original thinkers of the Alexandrian school was Diophantus:* Some of his fascinating mathematics can be found in Turnbull 1993; Crossley 1987; Gow 1968; van der Woerden 1983; Vogel 1972.

58 *Regiomontanus could not curb his admiration:* Calinger 1999.

59 *consider problem 28 from the first book:* van der Woerden 1983.

59 *Fermat's Last Theorem:* There are a few excellent books about Fermat's last theorem. A beautiful description of the history that had led to Andrew Wiles's proof is given in Singh 1997 and in Aczel 1996; a brief but clear explanation of a few elements of the proof can be found in Devlin 1999; more detailed, technical expositions are Edwards 1996; Mozzochi 2004.

61 *mathematician and astronomer Brahmagupta:* van der Woerden 1983; Calinger 1999.

61 *The man who literally gave algebra its name was Muhammad ibn Musa*

al-Khwarizmi: van der Woerden 1985; Crossley 1987; O'Connor and Robertson 1999a.

62 *Abraham bar Hiyya Ha-nasi:* Levey 1954; O'Connor and Robertson 1999b.

63 *Omar Khayyam:* Yardley 1990; Amir-Moéz 1994.

63 *Maestro Benedetto . . . Maestro Biaggio and Antonio Mazzinghi:* van der Woerden 1985; Calinger 1999.

63 *Maestro Dardi:* van der Woerden 1985; Calinger 1999.

63 *No wonder that mathematician and author Luca Pacioli:* Franci and Toti Rigatelli 1985; Taylor 1942; Livio 2002.

64 *a mathematician from Bologna named Scipione dal Ferro:* van der Woerden 1985; Cardano 1993; Bortolotti 1947; Crossley 1987; Dunham 1991; Rose 1975; Masotti 1972.

64 *A collection of lecture notes from the University of Bologna:* Bortolotti 1947.

65 *mathematicians were interested in such confrontations:* al-Nadim 1871–72; Crossley 1987.

66 *Niccolò Tartaglia:* Crossley 1987; Rose 1975; van der Woerden 1985; Bortolotti 1933; Di Pasquale 1957a, b, 1958; Schultz 1984; Masotti 1972.

66 *Rumors of Tartaglia's claim:* Some historians of mathematics, such as Moritz Cantor (1829–1920) in his *Lectures on the History of Mathematics,* have expressed skepticism as to Tartaglia's ability to rediscover dal Ferro's formula. Cantor suggests that Tartaglia may have simply managed to obtain the formula secondhand. Other historians, such as Gustav Eneström, disagree with Cantor's speculation.

68 *the physician, mathematician, astrologer, gambler, and philosopher Gerolamo Cardano:* Detailed descriptions of his life and work can be found in Cardano 1993; Fierz 1983; Ore 1953; Crossley 1987; van der Woerden 1985; Gliozzi 1972; Hale 1994.

69 *"Even if gambling were altogether":* Ore 1953.

71 *Ludovico Ferrari (1522–65) takes center stage:* Candido 1941; Di Pasquale 1957a, b, 1958; Jayawardene 1972.

75 *The history of the solutions:* The story of the cubic and the quartic equations is described nicely also in Gindikin 1988.

77 *Cardano published horoscopes:* According to one legend that is almost certainly false, Cardano committed suicide only to vindicate his own horoscope.

79 *It started with another Bolognese, Rafael Bombelli:* van der Woerden 1985; Crossley 1987; Boyer 1991; Rose 1975; Pesic 2003.

80 *The French lawyer François Viète:* Ritter 1895; Crossley 1987; Pesic 2003.

80 *The embarrassed king called upon Viète:* While the trigonometric method of solution outlined in appendix 6 is correct, it is hard to believe that Viète was able to actually find the solutions within minutes.

80 *the Scot James Gregory:* O'Connor and Robertson 2000d; Whiteside 1972.

81 *the German Count Ehrenfried Walther von Tschirnhaus:* O'Connor and Robertson 1997; Hofmann 1972; Ayoub 1980.

81 *The Frenchman Étienne Bézout:* O'Connor and Robertson 2001a; Tignol 2001.

81 *Leonhard Euler:* Many sources depict the life and work of Euler. The following represent a collection of sources in which different elements are emphasized. Youschkevitch 1972a; O'Connor and Robertson 1998; Wells 1997; Boyer 1991; Bell 1937; James 2002; Ayoub 1980; Tignol 2001.

82 *the Swede Erland Samuel Bring:* Youschkevitch 1972b; O'Connor and Robertson 1996a.

83 *the Frenchman Alexandre-Théophile Vandermonde (1735–96) and the Englishman Edward Waring:* van der Woerden 1985; Ayoub 1980; Tignol 2001.

83 *Lagrange . . . was born in Turin:* There are many references. The following emphasize aspects relevant for the discussion here: van der Woerden 1985; Bell 1937; James 2002; Ayoub 1980; Kiernan 1971; Nový 1973; Stubhaug 2000; Tignol 2001.

85 *Gauss's genius was recognized:* Among the many excellent references I note: Fine and Rosenberger 1997; Tignol 2001; Bell 1937; Dörrie 1965; Gray 2004.

85 *Argand's proof:* Fehr 1902.

85 *he also expressed his skepticism:* Gauss 1876.

85 *Jean Étienne Montucla:* Ayoub 1980.

86 *Paolo Ruffini:* The best reference is Ayoub 1980; interesting information can also be found in van der Woerden 1985; Carruccio 1972; Wussing 1984; Pesic 2003.

87 *the French mathematician and astronomer Jean-Baptiste Joseph Delambre:* Ayoub 1980.

88 *Cauchy, generally reserved with compliments, writes:* Ayoub 1980.

CHAPTER 4: THE POVERTY-STRICKEN MATHEMATICIAN

There exist a few good Abel biographies: A relatively recent and very well researched biography is Stubhaug 2000. The first major biography was that by Bjerknes 1880 (in Norwegian), which appeared in expanded form in French in 1885. An excellent biography is Ore 1954 (in Norwegian), which appeared in English translation in 1957. Shorter biographies can be found in Mittag-Leffler 1904 (in Norwegian), which also appeared in French, Mittag-Leffler (1907); James 2002; de Pesloüan 1906; Ore 1972; Pesic 2003; a romantic account is Bell 1937. In addition, there are a few articles in a memorial volume published on the occasion of the centenary of Abel's birth (Holst, Stormer, and Sylow 1902) and at the Abel Bicentennial (Laudal and Piene 2004). Abel's publications can be found in Sylow and Lie 1881.

PAGE

90 *Mittag-Leffler (1846–1927) described Abel's mathematical achievements:* Steiner 2001.

90 *Emil Artin . . . wrote about Galois:* Tignol 2001.

97 *Abel's proof is too technical:* Sylow and Lie 1881; Kiernan 1971; Kline 1972; Nový 1973; Gårding and Skau 1994: Very clear, if technical proofs are presented in M. I. Rosen 1995; Dörrie 1965. An excellent, somewhat less technical explanation of the proof can be found in Pesic 2003. On the quintic equation see also (the more technical) Shurman 1997; Spearman and Williams 1994.

102 *Abel took up lodging:* Stubhaug 2000; Auffray 2004.

110 *Abel Prize in Mathematics:* Battersby 2003; Thomas 2001.

CHAPTER 5: THE ROMANTIC MATHEMATICIAN

There exist a few Galois biographies: The most recent and well-researched biography (in French) is Auffray 2004. The first, excellent, biography was Dupuy 1896. Other well-documented biographies include: Dalmas 1956; Sarton 1921; Rothman 1982a, b; Taton 1972; Toti Rigatelli 1996; Astruc 1994; Verdier 2003. Shorter biographies and biographical studies include: Chevalier 1832; Bell 1937; Davidson 1938; Barbier 1944; Kollros 1949; Malkin 1963; Hoyle 1977; James 2002. Fictionalized biographies include: Infeld 1948; Petsinis 1998; Berloquin 1974; Mondor 1954. Interesting material on the Web can be found, for instance, at Gales 2004, Bychan 2004, and O'Connor and Robertson 1996b.

Galois's mathematical work can be found in many sources, including Liouville 1846; Picard 1897; Tannery 1906, 1907, 1908; Verriest 1934; Bourgne and Azra 1962; Toti Rigatelli 1996; Verdier 2003; Auffray 2004.

PAGE

112 *Évariste Galois was born:* Documents on birth and family tree provided by Philippe Chaplain at Bourg-la-Reine Town Hall.

114 *Lycée Louis-le-Grand:* Donnay 1939.

117 *"it is not the only striking analogy:* Bourgne and Azra 1962.

118 *Richard proved to be for Galois:* Terquem 1849.

119 *solutions that leave the relations unchanged:* For the equation given as an example, one can prove that the relation remains correct, for instance, under the permutation:

$$\begin{pmatrix} x_1 & x_2 & x_3 & x_4 \\ x_2 & x_4 & x_1 & x_3 \end{pmatrix}$$

119 *Cauchy wrote the following apologetic letter:* Taton 1947, 1971, 1983; Rothman 1982a, b.

121 *the examination took place just one month:* Dupuy 1896; Bertrand 1899; Verdier 2003.

122 *there were no* arithmetical *logarithms:* The examiner, Dinet, might have referred to a concept developed by Gauss in 1801. These objects, which Gauss called *indices,* really represent a special class of roots. These "indices" obey all the rules of logarithms, and therefore could be thought of as "arithmetical logarithms."

124 *issuing on July 26 an infamous series of ordinances:* Bérard 1834; Blanc 1841–1844; Burnand 1940; Chenu 1850; Girard 1929.

124 Liberty Leading the People: An excellent description can be found in De Hureaux 1993.

130 *The famous mathematician Sophie Germain:* Dahan-Dalmédico 1991; Sampson 1990–1991; James 2002; O'Conner and Robertson 1996c.

130 *She added a sad comment:* Henry 1879.

130 *In Dumas's words:* Dumas 1863–1865.

131 *The legal proceedings opened:* Dalmas 1956; Verdier 2003.

133 *a lukewarm report:* Bertrand 1899.

135 *his friend Ernest Duchatelet:* Auffray 2004; Verdier 2003.

137 Letters on the Prisons of Paris: Raspail 1839.

138 *De Nerval's description:* De Nerval 1841, 1855.

138 *The preface starts by mocking:* Bourgne and Azra 1962.

140 *Galois was transferred on March 13:* Astruc 1994; Auffray 2004.

140 *to a convalescent home:* Some of the Galois biographers took the "*maison de santé*" ("house of health") to mean a professional medical clinic. However, at the time, the expression really referred to homes where people would come to rest or even spend a vacation. Usually these homes did indeed provide some medical supervision and had a resident doctor who treated both physical and mental problems.

140 *Few love affairs in history:* Bibliothèque de l'Institut, Manuscrits d'É. Galois, f°59v°; Infantozzi 1968; Auffray 2004; Dalmas 1956; Astruc 1994; Toti Rigatelli 1996; Rothman 1982a, b.

143 *The known facts concerning Galois's activities:* Dupuy 1896; Bourgne and Azra 1962; Rothman 1982a, b; Toti Rigatelli 1996; Auffray 2004; Barbier 1944.

143 *Christ on the cross:* Luke 23:34.

144 *The third and most important letter: Revue encyclopédique* 1832.

146 *Galois refused the services of a priest:* Communicated by Gabriel Demante (Galois's cousin) to Galois's first biographer, Paul Dupuy. Gabriel's brother Victor was a priest, and he was probably the one whose services Galois had refused.

146 *the death certificate was signed:* Reconstitution Des Actes de L'État Civil de Paris; Archives de la Seine.

146 *The* Bulletin de Paris *of May 31:* Archives Nationales f^{O} 7. 3886.

148 *the police chief Henri-Joseph Gisquet:* The political events surrounding Galois's death are described in Hodde 1850; Gisquet 1840–1844.

152 *fully consistent with their characters:* Auffray thinks that Galois would not have referred to his young friend Duchatelet (born on May 19, 1812) as one of two "men" or two "patriots" (e.g., in his letters "to all republicans" and to N.L. and V.D.). However, Galois's style is in fact fully consistent with the fact that he felt compelled (by Faultrier's authority) to keep the identity of his opponents secret.

152 *Dumas has been known to be wrong:* Astruc (1994) refers to him as "often boastful."

152 *I humbly* propose: The suggestion that the opponent may have been Duchatelet had already been made by Dalmas 1956, and it was adopted by Rothman 1982a in his later discussion, which appeared on the Web. However, Toti Rigatelli advanced her very different scenario in 1996, and this has dominated the subsequent discussions. The important realization that one needs to identify two adversaries was made by Auffray 2004. My suggestion is based on *all* of the existing material.

152 *he did not stand fully sideways:* The precise practices and history of duels is discussed, for instance, in Baldick 1965.

152 *Riquet à la Houppe:* Appeared first in a story by Charles Perrault in 1697, which was inspired by the novel *Inès de Cordoue* by Catherine Bernard (appeared in 1696).

154 *He writes in his memoirs:* Gisquet 1840–1844.

154 *the funeral of the general:* The event is described in Hugo 1862.

154 *drawn from memory by Alfred:* The picture appeared in the *Magasin Pittoresque,* vol. 16, July 1848, pp. 227–228.

155 *Lie concluded:* Verdier 2003.

155 *the placement of a commemorative plaque:* Described in the newspaper *La Banlieue,* No. 24, 20 June 1909.

155 *Tannery finished with a moving mea culpa:* Tannery 1909.

CHAPTER 6: GROUPS

Interesting descriptions of the development and history of group theory include Kiernan 1971; Tignol 2001; Nový 1973; Wussing 1984; Chandler and Magnus 1982. A brief summary is given in Kline 1972. Of the many (more advanced) treatments of groups and Galois theory, I recommend in particular the following books: Rotman 1990, 1995; Stewart 2004; Tignol 2001; Fraleigh 1989; Garling 1960; Edwards 1984. For a brief but brilliant application of groups to science see Eddington 1956.

PAGE

159 *The Sefer yetzira:* Kaplan 1990; Katz 1989.

159 *examine the operation that permutes:* A similar example appears in Verdier 2003.

159 *The 14–15 puzzle:* Gardner 1959b; Singh 1997.

162 *Rubik invented the cube in 1974:* Rubik, Varga, Kéri, Marx, and Vekerdy 1987.

162 *In honor of the inventor:* Goudey 2001–2003.

162 *the solution to the cube's puzzle:* Frey and Singmaster 1982.

162 *Mathematician David Joyner:* Joyner 2002.

166 The collection of all the permutations of three objects: Groups are called "commutative" or "Abelian" (after Abel) if, for every two group members, the order in which they are combined (by the group operation) does not matter. That is, $x \circ y = y \circ x$ for any group members x and y. For instance, the group of all integer numbers with the operation of ordinary addition is Abelian, since the sum of any two integers is the same, irrespective of the order (e.g., $5 + 3 = 3 + 5 = 8$). From the table on page 165 we can see that the group of permutations of three objects is non-Abelian (e.g., $t_1 \circ t_2 = s_1$ while $t_2 \circ t_1 = s_2$).

167 Every group is cast in the same mold: Rigorous proofs of Cayley's theorem can be found, for instance, in Birkhoff and Maclane 1953 and in Patterson and Rutherford 1965.

169 *The proof contains three crucial ingredients:* A simple exposition of the logic of the proof is presented in Rothman 1982b. Rigorous proofs can be found, for instance, in: Edwards 1984; Rotman 1990; Tignol 2001; Stewart 2004. Galois's original version is reproduced in Toti Rigatelli 1996.

170 *the combination that is the sum of the two solutions:* Generally, determining the Galois group for equations of a higher degree is not easy.

173 *finding a partner for marriage:* Smith 1997; Cooper 1997. The problem is also discussed in Cresswell 2003.

176 *The biblical story of the judge Gideon:* Judges 7:2.

176 *the 1970 draft lottery:* Starr 1997; Fienberg 1971; Rosenbaum 1970.

178 *A recent study by statisticians:* Diaconis, Holmes, and Montgomery 2004; Peterson 2004; Keller 1986.

180 *"Premature abstraction falls on deaf ears":* Kline and Alder 2000.

180 *Cayley's life:* Bell 1937; Crilly 1995; Forsyth 1895; Gray 1995; O'Connor and Robertson 1996d.

181 *Start with four operations:* Budden 1972.

183 *An extremely complex kinship:* Fletcher 1967; Fox 1967; Verdier 2003.

186 *Claude Lévi-Strauss's extensive analyses:* Lévi-Strauss 1949, 1958.

186 *universal grammar:* This topic was introduced and discussed extensively in works of the linguist Noam Chomsky. Seminal works include: Chom-

sky 1966, 1968, 1975; Chomsky and Halle 1968. To get just a glimpse of Chomsky's work in general, see Maher and Groves 1996.

186 *this does not mean that all languages:* Pennisi 2004.

186 *novelist and philosopher Umberto Eco:* See, for instance, the collection of essays Eco 2004.

189 *the Jesuit Giovanni Girolamo Saccheri:* Angelelli 1995; Emch 1933; Pascal 1914.

189 *the fifth postulate is in fact replaced:* Bonola 1955; Trudeau 1987; Gray 1979; Sommerville 1960; O'Connor and Robertson 1996a; Coxeter 1998.

189 *János Bolyai (1802–60):* Szénássy 1992; Mayer 1982; Bier 1992.

190 *Nikolai Ivanovich Lobachevsky:* Daniels 1975; Halsted 1895; Vucinich 1962.

190 *Eugenio Beltrami:* Chandrasekhar 1989; Bryan 1901.

190 *Georg Friedrich Bernhard Riemann:* Interesting popular books are Derbyshire 2003; Du Sautoy 2003; Sabbagh 2003. More technical descriptions of Riemann's work in geometry can be found in Portnoy 1982; Zund 1983; Scholz 1992.

192 *In a seminal lecture entitled:* Yaglom 1988; Birkhoff and Bennett 1988; James 2002.

193 *These ideas have been later expanded upon:* A spectacular presentation of some of the symmetries to have emerged from Klein's work is Mumford, Series, and Wright 2002.

194 *Leopold Kronecker:* Bell 1937; Edwards 1987; James 2002.

195 *Charles Hermite:* Bell 1937; Belhoste 1996; James 2002; Darboux 1906.

196 *The philosophy behind Klein's:* Klein 1884.

CHAPTER 7: SYMMETRY RULES

There are quite a few popular science books that discuss the role of symmetry in the laws of nature. The most extensive discussion is in the excellent book by Zee 1999. Other interesting books are Icke 1995; Lederman and Hill 2004. Shorter, but very clear, expositions of some of the main points include Weinberg 1992; Greene 1999, 2004; Penrose 2004; Kaku 1994, 2004; Gell-Mann 1994; Barrow 2003; Webb 2004.

PAGE

198 *"Nature and Nature's laws":* Epitaph intended for Sir Isaac Newton.

198 *In his scientific masterpiece:* See in Chandrasekhar 1995; Motte 1995; Gleick 2003.

200 *the speed of light always comes out:* Excellent popular accounts of special and general relativity can be found, for instance, in Kaku 2004; Greene 2004; Penrose 2004; Galison 2004; Davies 1977; Bodanis 2000; Wheeler 1990; Hawking and Penrose 1996; Deutsch 1997. For a recent textbook

see Schwarz and Schwarz 2004. From the father of relativity himself: Einstein 1953, 2001, 2004.

203 *confirmed that the traveling muons lived longer:* Rossi and Hall 1941; Bailey et al. 1977.

203 *Physicists Joseph Carl Hafele and Richard Keating:* Hafele and Keating 1972a, b.

204 *Einstein's attitude toward pure mathematics:* To get a better understanding of Einstein the man, beautiful accounts include: Overbye 2000; Pais 1982; Miller 2001; Frank 1949; Fölsing 1997.

206 *there are only seven distinct strip patterns:* Budden 1972; Farmer 1996; Barrow 1995; Stevens 1996.

209 The force of gravity and the force resulting from acceleration: In addition to the popular books on general relativity mentioned above, it is always illuminating to examine Einstein's original papers, e.g. in Stachel 1989, or in the monumental *Collected Papers of Albert Einstein,* published by Princeton University Press. A by-now-classic textbook on general relativity is Misner, Thorne, and Wheeler 1973.

209 *"I was sitting in the patent office":* From a lecture given in Kyoto in 1922. The notes were taken by Yon Ishiwara, and were translated in August 1932 in *Physics Today* by Y. A. Ono. Quoted in Calaprice 2000.

210 *Newton was fully aware of the fact:* See Chandrasekhar 1995.

211 *later became known as* Ehrenfest's paradox: A technical discussion can be found in Hill 1946.

212 *In an uncharacteristically helpless tone:* Quoted in Kaku 2004.

212 *In an address to the Prussian Academy of Sciences:* The address was entitled "Geometry and Experience." Quoted in Calaprice 2000.

215 *the change in the wavefunction that results from shifting:* See also Weinberg 1992; Feynman, Leighton, and Sands 1965. For a general exposition of quantum mechanics, an excellent popular account is Lindley 1996.

216 *Emmy Noether was born in Erlangen:* James 2002; Osen 1974; Kimberling 1972; Weyl 1935.

217 *Noether proved the theorem:* For a relatively simple proof see Baez 2002, and "Noether's theorem" on the Wikipedia website.

220 *They are composed of elementary building blocks:* There are many popular books on particle physics and theories of the fundamental forces. A few excellent ones are: Gell-Mann 1994; Barrow 1991; Davies 1984; Glashow 1988; Guth 1997; Weinberg 1992; Lederman 1993; Ferris 1997.

221 *Physicist and author Cindy Schwarz has compiled an entire collection:* Schwarz 2002.

222 *Lie's ideas have become so central:* For a relatively gentle introduction to Lie groups see Lipkin 2002; Hall 2003.

222 *Sophus Lie . . . arrived at mathematics:* Stubhaug 2002; O'Connor and Robertson 2002.

225 *repeating "wallpaper" patterns are limited to seventeen:* e.g. Budden 1972. Interestingly, of the seventeen classes, thirteen can be found in the Alhambra in Granada, Spain; Grünbaum, Grünbaum, and Shephard 1986.

226 *The Bible tells us that when the Israelites:* Exodus 13:21.

226 *But in a universe that started to expand from a "big bang":* The list of excellent recent books in popular cosmology includes: Rees 1997; Guth 1997; Silk 2000; Kirshner 2002; Goldsmith 2000; Chown 2002; Tyson and Goldsmith 2004; and of course I have to mention Livio 2000.

226 *general relativity and quantum mechanics really appear:* Excellent popular accounts of string theory and other potential paths to combining gravity with quantum mechanics are: Greene 1999, 2004; Smolin 2002; Kaku 1994; Davies 1995, 2001; Gribbin 1999. For brief overviews see Witten 2004a, b.

228 *The claim by physicists John Schwarz . . . and Joël Scherk:* Somewhat similar or related claims were made independently by other physicists, e.g., the Swedish Lars Brink, and the Japanese Tamiaki Yoneya.

228 *The breakthrough finally occurred in 1984:* For a popular account see Green 1986.

228 *how does string theory propose to resolve:* For a recent textbook on string theory see Zweibach 2004.

229 *In fact, there even existed a theorem:* Proven in 1967 by physicists Sidney Coleman and Jeffrey Mandula; Coleman and Mandula 1967.

230 *Supersymmetry is a subtle symmetry:* Kane 2000; Greene 1999.

231 *known as* Grassmann numbers: The algebra associated with Grassmann is known as *exterior algebra.*

CHAPTER 8: WHO'S THE MOST SYMMETRICAL OF THEM ALL?

PAGE

233 *Evolutionary psychology is a science:* Excellent books on evolutionary psychology include: Dawkins 1986, 1989; Barkow, Cosmides, and Tooby 1992; Pinker 1994; Buss 1999. The idea of the modularity of the mind is discussed, for instance, in Fodor 1983. Other excellent reviews include: Cosmides and Tooby 1987; Tooby and Cosmides 1990. A very brief overview is presented in Evans and Zarate 1999.

235 *What was the precise adaptive problem:* Braitenberg 1984; Dennett 1988.

235 *the brain circuitry that leads to the emotion of fear:* LeDoux 1996; Kalin 1993.

236 *Central to the immediate emotion:* Davis 1992; Fanselow 1994.

236 *conspicuous eyespots that are concealed at rest:* Lyytinen, Brakefield, and Mappes 2003; Blest 1957.

237 *In the experiment conducted by Swedish researchers:* Forsman and Merilaita 1999, 2003. More general reviews on the detection of asymmetry can be found in Palmer 1994; Møller and Swaddle 1997.

237 *The British statesman and philosopher Edmund Burke:* Burke 1757.

238 *George Burns's book:* Burns 1984.

239 *evolution is also shaped by sexual selection:* Interesting books on sexual selection include Ridley 2003; Buss 2003; Miller 2000.

240 *biologists William Hamilton and Marlene Zuk:* Hamilton and Zuk 1982.

240 *Other experiments, with barn swallows:* Møller 1992, 1994.

240 *and with zebra finches:* Swaddle and Cuthill 1994.

240 *The Israeli biologist Amotz Zahavi:* Zahavi 1975, 1991; Zahavi and Zahavi 1997.

240 *Swedish biologist Magnus Enquist and British engineer Anthony Arak:* Enquist and Arak 1994.

241 *Similar results were obtained in a separate:* Johnstone 1994.

241 *as evolutionary psychologist David Buss has put it:* Buss 1999.

242 *Even the area in the brain:* Aharon et al. 2001.

242 *More recent studies by University of Texas:* Langlois and Roggman 1990.

242 *Psychologist Michael Cunningham found:* Cunningham et al. 1995.

243 *Studies that were performed across:* Discussed, for instance, in Jackson 1992 and Jones 1996.

243 *what is it that men and women find attractive:* An excellent review is Grammer et al. 2003. See also Gangestad, Thornhill, and Yeo 1994; Grammer and Thornhill 1994; Thornhill and Gangestad, 1996; Thornhill and Grammer 1999.

243 *Grammer and biologist Anja Rikowski:* Rikowski and Grammer 1999. See also Schaal and Porter 1991; Thornhill et al. 2003.

243 *a relationship between symmetry and women's orgasms:* Thornhill, Gangestad, and Comer 1995.

244 *Independent research by psychologists:* Shackelford and Larsen 1997.

244 *Langlois generated computer composites:* Langlois and Roggman 1990; Langlois, Roggman, and Musselman 1994; Langlois et al. 2000.

245 *Cognitive scientist David Perrett:* Perrett et al. 2002. See also Perrett, May, and Yoshikawa 1994.

245 *men prefer women with the classical:* Singh 1993, 1995.

247 *In an interview in 1985:* Gleason 1990.

247 *known as the* Elliott model: Described, for instance, in Dyson 1966.

249 *In the same way that the freezing points:* For a brief, popular account of string theory see Witten 2004b.

249 *theoretical ideas of symmetry breaking are on the right track:* In particular, using the most economical implementation of the symmetry breaking, the theory predicts the existence of a hitherto unseen new particle known as the Higgs particle (after the Scottish physicist Peter

Higgs). This particle is expected to be produced at the Large Hadron Collider.

250 *"We come to describe nature:* Sir Michael Atiyah discussed closely related topics also in Atiyah 1993.

251 *Feynman concludes his discussion:* Feynman, Leighton, and Sands 1963.

253 *we can conveniently represent them on a clock face:* Groups in music are nicely discussed in Budden 1972. See also Winchel 1967; Lewin 1993. On the human physiological response to sound see Maor 1994.

255 *most notably Arnold Schoenberg:* Schoenberg 1969.

255 *atonal music:* See, e.g., Rahn 1980.

257 *One of those families was defined:* The original definition was due to Cauchy, but Galois was the first to distinguish the notion of a "simple group."

258 *The story of the sporadic simple groups:* Excellent popular accounts can be found in Devlin 1999; Odifreddi 2004. A somewhat more technical description is Solomon 1995. For the more mathematically inclined: Aschbacher 1994; Gorenstein 1982, 1986.

259 *Daniel Gorenstein (1923–92):* Aschbacher 1992.

260 *As Gorenstein described in 1989:* In response to the awarding of the Steele Prize by the American Mathematical Society.

260 *The last of these was finally closed:* The two volumes *The Classification of Quasithin Groups* appeared as volumes 111 and 112 in the American Mathematical Society series Mathematical Surveys and Monographs.

CHAPTER 9: REQUIEM FOR A ROMANTIC GENIUS

PAGE

262 *Mathematician and author Clifford Pickover:* Pickover 2001.

262 *University of Chicago psychologist:* Csikszentmihalyi 1996.

263 *The English poet Owen Meredith:* In *Last Words of a Sensitive Second-Rate Poet.*

263 *many studies show that beyond a certain level of IQ:* Summarized e.g. in Gardner 1993; Simonton 1999. See also Dartnall 2002; Ambrose, Cohen, and Tannenbaum 2003; Rothenberg and Hausman 1976. A comprehensive review of studies of creativity is Sternberg 1998.

264 *As Harvard cognition and education researcher:* Gardner 1993. An interesting compilation is Brockman 1993.

265 *I did not find Galois's name:* A brief examination of the creativity of Poincaré and Einstein is presented in Miller 1996. A mosaic of one hundred creative minds in literature and religion is presented in Bloom 2002. An interesting examination of the creativity of the members of the U.S. team to the Mathematics Olympiad is presented in Olson 2004. See also Csikszentmihalyi 1996; Gardner 1993.

265 *Psychologists John Dacey and Kathleen Lennon:* Dacey and Lennon 1998.

266 *ten pairs of apparently antithetical characteristics:* Csikszentmihalyi 1996.

266 *psychologist Ellen Winner finds:* Winner 1996.

268 *The French philosopher Jean-Paul Sartre:* Sartre 1964.

270 *In 1895, psychiatrist W. L. Babcock:* Babcock 1895.

270 *psychologist Arnold Ludwig examined the lives:* Ludwig 1995; Simonton 1999. An interesting early study of this problem is that of Lombroso 1895.

270 *The findings showed that the creative individuals:* MacKinnon 1975.

270 *The English essayist Sir Max Beerbohm:* In "No. 2. The Pines."

271 *For more than two decades:* On his website Steven Levy tells the story of how he found Einstein's brain. The story is also told in Abraham 2002 and Paterniti 2000.

271 *They found that the ratio of neurons to glial cells:* Diamond et al. 1985.

271 *this interpretation was later questioned:* Hines 1998.

271 *Anderson and Harvey showed that while Einstein's brain:* Anderson and Harvey 1996.

272 *McMaster University neuropsychologist:* Witelson, Kigar, and Harvey 1999.

272 *The autopsy report on Galois's brain:* Dupuy 1896.

272 "*the weight of the brain*": The value of the Parisian pound at the time was somewhat larger than the current value; 489.75 grams compared to 453.59 grams.

273 *Mathematician and author Ian Stewart expressed:* Stewart 2004.

274 *an inviting website entitled:* http://www.neuroticpoets.com/shelley.

275 *The famous Indian poet Rabindranath Tagore:* Quoted, for instance, at: http://www.en.wikiquote.org/wiki/Death.

275 *than those in Emily Dickinson's poem:* Can be found, for instance, at: http://www.everypoet.com/archive/index.htm/.

References

Abel, Niels Henrik. 1902. *Mémorial publié à l'occasion du centenaire de sa naissance* (Christiania: Jacob Dybwal).

Abraham, C. 2002. *Possessing Genius: The Bizarre Odyssey of Einstein's Brain* (New York: St. Martin's Press).

Aczel, A. D. 1996. *Fermat's Last Theorem: Unlocking the Secret of an Ancient Mathematical Problem* (New York: Four Walls Eight Windows).

Aharon, I., Etcoff, N., Ariels, D., Chabris, C. F., O'Connor, E., and Breiter, H. C. 2001. "Beautiful Faces Have Variable Reward Value: fMRI and Behavioral Evidence." *Neuron,* 32(3), 537.

al-Nadim, I. 1871–72. *Kitab al-Fihrist.* J. Roediger and A. Mueller, eds. (Leipzig: FCW Vogel).

Altschuler, E. L. 1994. *Bachanalia* (Boston: Little, Brown and Company).

Ambrose, D., Cohen, L. M., and Tannenbaum, A. J., eds. 2003. *Creative Intelligence: Toward Theoric Integration* (Cresskill, NJ: Hampton Press).

Amir-Moéz, A. R. 1994. "Khayyam, Al-Biruni, Gauss, Archimedes, and Quartic Equations." *Texas Journal of Science,* 46, no. 3, 241.

Anderson, B., and Harvey, T. 1996. "Alterations in Cortical Thickness and Neuronal Density in the Frontal Cortex of Albert Einstein." *Neuroscience Letters,* 210, 161.

Angelelli, I. 1995. "Saccheri's Postulate." *Vivarium,* 33(1), 98.

Aschbacher, M. 1992. "Daniel Gorenstein (1923–1992)." *Notices of the AMS,* 39(10), 1190.

Aschbacher, M. 1994. *Sporadic Groups* (Cambridge: Cambridge University Press).

Astruc, A. 1994. *Évariste Galois* (Paris: Flammarion).

Atiyah, M. 1993. "Mathematics: Queen and Servant of the Sciences." *Proceedings of the American Philosophical Society,* 137, no. 4, 527.

Auffray, J.-P. 2004. *Évariste 1811–1832, le roman d'une vie* (Lyon: Aléas).

Ayoub, R. G. 1980. "Paolo Ruffini's Contributions to the Quintic." *Archive for History of Exact Sciences,* 23, 253.

Babcock, W. L. 1895. "On the Morbid Heredity and Predisposition to Insanity of the Man of Genius." *Journal of Nervous and Mental Disease,* 20, 749.

Baez, J. 2002. http://www.math.ucr.edu/home/baez/noether.html/.

Bailey, J., et al. 1977. "Measurements of Relativistic Time Dilation for Positive and Negative Muons in a Circular Orbit." *Nature,* 268, 301.

Baldick, R. 1965. *The Duel* (London: Chapman & Hall).

Ball, W. W. R., and Coxeter, H. S. M. 1974. *Mathematical Recreations and Essays,* 12th ed. (Toronto: University of Toronto Press).

Barbier, A. 1944. Un Météore: Évariste Galois, Mathématicien 1811–1832 (manuscript in the Bourg-la-Reine collection).

Barkow, J., Cosmides, L., and Tooby, J. 1992. *The Adapted Mind: Evolutionary Psychology and the Generation of Culture* (New York: Oxford University Press).

Barrow, J. D. 1991. *Theories of Everything: The Quest for Ultimate Explanation* (Oxford: Clarendon Press).

Barrow, J. D. 1995. *The Artful Universe* (Boston: Back Bay Books).

Barrow, J. D. 2003. *The Constants of Nature: From Alpha to Omega—The Numbers that Encode the Deepest Secrets of the Universe* (New York: Pantheon).

Barry, A. M. S. 1997. *Visual Intelligence, Perception, Image, and Manipulation in Visual Communication* (Albany: State University of New York Press).

Bartusiak, M. 2000. *Einstein's Unfinished Symphony* (Washington, D.C.: Joseph Henry Press).

Battersby, S. 2003. "Will Abel Prize for Maths Rival the Nobels?" *New Scientist,* 7 June, 12.

Belhoste, B. 1996. "Autour d'un Mémoire Inédit: La Contribution d'Hermite au Développement de la Théorie des Fonctions Elliptiques." *Revue d'Histoire des Mathématiques,* 2(1), 1.

Bell, C. 1997. "The Aesthetic Hypothesis." In *Aesthetics,* S. L. Feagin and P. Maynard, eds. (Oxford: Oxford University Press), 15.

Bell, E. T. 1937. *Men of Mathematics* (New York: Simon & Schuster); reprinted in 1986.

Bell, E. T. 1951. *Mathematics, Queen and Servant of Science* (New York: McGraw-Hill).

Bérard, A. S. L. 1834. *Souvenirs historiques sur la révolution de 1830* (Paris: Perrotin).

Bergeron, H. W. 1973. *Palindromes and Anagrams* (New York: Dover Publications).

Berloquin, P. 1974. *Un Souvenir d'Enfance d'Évariste Galois* (Paris: Balland).

Berriman, A. E. 1956. "The Babylonian Quadratic Equation." *The Mathematical Gazette,* XL, no. 333, 185.

Bertrand, J. 1899. "Sur 'La vie d'Évariste Galois' par Paul Dupuy." *Journal des savants,* July, 289.

Bier, M. 1992. "A Transylvanian Lineage." *The Mathematical Intelligencer,* 14(2), 52.

Birkhoff, G., and Bennett, M. K. 1988. "Felix Klein and His 'Erlanger Programm.' " In *History and Philosophy of Modern Mathematics (Minnesota*

Studies in the Philosophy of Science XI), W. Aspray and P. Kitcher, eds. (Minneapolis: University of Minnesota Press), 145.

Birkhoff, G., and Mac Lane, S. 1953. *A Survey of Modern Algebra* (Basingstoke Hampshire: Collier Macmillan).

Birkhoff, G. D. 1933. *Aesthetic Measure* (Cambridge, MA: Harvard University Press).

Bjerknes, C. A. 1880. *Niels Henrik Abel: En skildring af hans Liv og vitenskapelige Virksomhed* (Stockholm); translated into French in 1885 as *Niels Henrik Abel: Tableau de sa vie et de son action scientifique* (Paris: Gauthier-Villars).

Blanc, L. 1841–1844. *L'Histoire de dix ans (1830–1840)* (Paris: Paguerre).

Blest, A. D. 1957. "The function of eyespot patterns in the Lepidoptera." *Behaviour*, 11, 209.

Bloom, H. 2002. *Genius* (New York: Warner Books).

Boardman, A. D., O'Connor, D. E., and Young, P. A. 1973. *Symmetry and Its Application in Science* (New York: John Wiley & Sons).

Bodanis, D. 2000. $E = mc^2$ (New York: Walker).

Bonola, R. 1955. *Non-Euclidean Geometry: A Critical and Historical Study of Its Development* (New York: Dover).

Bortolotti, E. 1933. *I Cartelli Di Matemàtica Disfida* (Imola: Cooperation Tip. Edit. Paolo Galeati).

Bortolotti, E. 1947. *La Storia Della Matemàtica Nella Università Di Bologna* (Bologna: Nicola Zanichelli Editore).

Bourgne, R., and Azra, J. P., eds. 1962. *Écrits et mémoires mathématiques d'Évariste Galois* (Paris: Gauthier-Villars).

Bowers, B. 2001. *Sir Charles Wheatstone FRS 1802–1875* (Edison, NJ: IEE Publishing).

Boyer, C. B. 1991. *A History of Mathematics*, revised by U. C. Merzbach (New York: John Wiley & Sons).

Braitenberg, V. 1984. *Vehicles* (Cambridge, MA: MIT Press).

Brindley, G. S. 1970. *Physiology of the Retina and Visual Pathway* (Baltimore: Williams & Wilkins Company).

Brockman, J., ed. 1993. *Creativity* (New York: Touchstone).

Brown, T. M., Ferguson, H. C., Smith, E., Kimble, R. A., Sweigert, A. V., Renzini, A., Rich, R. M., and Vandenberg, D. A. 2003. "Evidence of a Significant Intermediate-Age Population in the M31 Halo from Main-Sequence Photometry." *The Astrophysical Journal*, 592, L17.

Bryan, G. H. 1901. "Eugenio Beltrami." *Proceedings of the London Mathematical Society*, 32, 436.

Budden, F. J. 1972. *The Fascination of Groups* (Cambridge: Cambridge University Press).

Buffart, H., and Leeuwenberg, E. L. J. 1981. "Structural Information Theory."

In H. G. Geissler, E. L. J. Leeuwenberg, S. Link, and V. Sarris, eds., *Modern Issues in Perception* (Berlin: Erlbaum).

Burand, R. 1943. *La vie quotidienne en France en 1830* (Paris: Hachette).

Burke, E. 1757. *A Philosophical Enquiry into the Origin of Our Ideas of the Sublime and Beautiful* (new edition published in 1998; Oxford: Oxford University Press).

Burns, G. 1984. *How to Live to Be 100—or More* (London: Robson Books).

Buss, D. M. 1999. *Evolutionary Psychology: The New Science of the Mind* (Needham Heights: Allyn & Bacon).

Buss, D. M. 2003. *The Evolution of Desire* (New York: Basic Books).

Bychan, B. 2004. "The Evariste Galois Archive." http://www.galois-group.net/.

Calaprice, A., collector and ed. 2000. *The Expanded Quotable Einstein* (Princeton: Princeton University Press).

Calinger, R. 1999. *A Contextual History of Mathematics* (Upper Saddle River, NJ: Prentice-Hall).

Candido, G. 1941. "La risoluzione della equazione di 4° grado." *Periodico di Matematiche,* ser. IV, vol. XXI, no. 1, 21.

Cardano, G. 1993. *Ars Magna or the Rules of Algebra,* T. R. Witmer, trans. and ed. (Mineola, NY: Dover Publications); original edition 1545.

Carruccio, E. 1972. In *Dictionary of Scientific Biography,* C. C. Gillespie, ed. (New York: Charles Scribner's Sons).

Caspar, F. H., and Hammer, F., eds. 1937. *Johannes Kepler: Gesammelte Werke* (Munich: C. H. Beck'sche Verlagsbuchhandlung).

Chandler, B., and Magnus, W. 1982. *The History of Combinatorial Group Theory: A Case Study in the History of Ideas* (New York: Springer-Verlag).

Chandrasekhar, T. R. 1989. "Non-Euclidean Geometry from Early Times to Beltram." *Indian Journal of History of Science,* 24(4), 249.

Chandrasekhar, S. 1995. *Newton's Principia for the Common Reader* (Oxford: Clarendon Press).

Chenu, A. 1850. *Les Conspirateurs* (Paris: Garnier Frères).

Chevalier, A. 1832. "Nécrologie Evariste Galois." *Revue encyclopédique,* 55, 744.

Chomsky, N. 1966. *Topics in the Theory of Generative Grammar* (The Hague: Mouton).

Chomsky, N. 1968. *Language and Mind* (New York: Harcourt Brace Jovanovich).

Chomsky, N. 1975. *The Logical Structure of Linguistic Theory* (New York: Plenum).

Chomsky, N., and Halle, M. 1968. *The Sound Pattern of English* (New York: Harper & Row).

Chown, M. 2002. *The Universe Next Door* (Oxford: Oxford University Press).

Coleman, S., and Mandula, J. 1967. "All Possible Symmetries of the S-matrix." *Physical Review,* 159, 1251.
Cooper, L. 1997. "Maths, love and men's best friend." *The Independent,* April 5.
Corballis, M. C. 1988. "Recognition of Disoriented Shapes." *Psychological Review,* 95, 115.
Corballis, M. C., and Beale, I. L. 1976. *The Psychology of Left and Right* (Hillsdale, NJ: Erlbaum).
Corballis, M. C., and Roldan, C. E. 1974. "On the Perception of Symmetrical and Repeated Patterns." *Perception and Psychophysics,* 16(1), 136–142.
Cosmides, L., and Tooby, J. 1987. "From Evolution to Behavior: Evolutionary Psychology as the Missing Link." In *The Latest on the Best: Essays on Evolution and Optimality,* J. Dupre, ed. (Cambridge, MA: MIT Press).
Coxeter, H. S. 1998. *Non-Euclidean Geometry* (Washington, DC: Mathematical Association of America).
Cresswell, C. 2003. *Mathematics and Sex* (Crows Nest NSW, Australia: Allen & Unwin).
Crilly, T. 1995. "A Victorian Mathematician: Arthur Cayley (1821–1895)." *Mathematical Gazette,* 79, 259.
Crossley, J. N. 1987. *The Emergence of Number* (Singapore: World Scientific).
Csikszentmihalyi, M. 1996. *Creativity* (New York: HarperCollins).
Cunningham, M. R., Roberts, A. R., Wu, C.-H., Barbeis, A. P., and Druen, P. B. 1995. "Their Ideas of Beauty Are, on the Whole, the Same as Ours: Consistency and Variability in the Cross-Cultural Perception of Female Attractiveness." *Journal of Personality and Social Psychology,* 68, 261.
Dacey, J. S., and Lennon, K. H. 1998. *Understanding Creativity* (San Francisco: Jossey-Bass Publishers).
Dahan-Dalmédico, A. 1991. "Sophie Germain." *Scientific American,* 265, 117.
Dalmas, A. 1956. *Évariste Galois, Révolutionnaire et Géomètre* (Paris: Fasquelle); reprinted in 1982 by Le Nouveau Commerce (Paris).
Daniels, N. 1975. "Lobachevsky: Some Anticipations of Later Views on the Relation between Geometry and Physics." *Isis,* 66, 75.
Darboux, G. 1906. "Charles Hermite." *La Revue du Mois,* 1, 37.
Dartnall, T., ed. 2002. *Creativity, Cognition, and Knowledge* (Westport, CT: Praeger).
Davidson, G. 1938. "The Most Tragic Story in the Annals of Mathematics." *Scripta Mathematica,* 6, 95.
Davies, P. 1977. *Space and Time in the Modern Universe* (Cambridge: Cambridge University Press).
Davies, P. 1995. *About Time* (New York: Simon & Schuster).
Davies, P. 2001. *How to Build a Time Machine* (New York: Allen Lane).
Davies, P. C. W. 1984. *Superforce* (New York: Simon & Schuster).

Davis, M. 1992. "The role of the amygdala in fear-potentiated startle: Implications for animal models of anxiety." *Trends in Pharmacological Science,* 13, 35.

Dawkins, R. 1986. *The Blind Watchmaker* (New York: W. W. Norton).

Dawkins, R. 1989. *The Selfish Gene* (Oxford: Oxford University Press).

De Hureaux, A. D. 1993. *Delacroix* (Paris: Éditions Hazau).

De la Hodde, L. 1850. *Histoire des sociétés secrètes et du parti républicain de 1830 à 1848* (Paris: Julien et Lanier).

De Nerval, G. 1841. "Mémoire d'un Parisien." *L'Artiste,* April 11.

De Nerval, G. 1855. "Mes Prisons." In *La bohème galante* (Paris: Michel Lévy).

Dennett, D. C. 1988. In *Sourcebook on the Foundations of Artificial Intelligence,* Y. Wilks and D. Partridge, eds. (Albuquerque: New Mexico University Press).

De Pesloüan, C. L. 1906. *N. H. Abel, sa vie et son oeuvre* (Paris: Gauthier-Villars).

Derbyshire, J. 2003. *Prime Obsession* (Washington, DC: Joseph Henry Press).

Deutsch, D. 1997. *The Fabric of Reality* (New York: Allen Lane).

Devlin, K. 1999. *Mathematics, the New Golden Age* (New York: Columbia University Press).

Devlin, K. 2002. *The Millennium Problems* (New York: Basic Books).

Diaconis, P., Holmes, S., and Montgomery, R. 2004. "Dynamical Bias in the Coin Toss." http://www-stat.stanford.edu/~susan/papers/headswithJ.pdf/.

Diamond, M. C., Scheibel, A. B., Murphy, G. M. Jr., and Harvey, T. 1985. "On the Brain of a Scientist: Albert Einstein." *Experimental Neurology,* 88, 198.

Di Pasquale, L. 1957a. "La equazioni di terzo gradi nei 'Quesiti et inventioni diverse' di Niccolò Tartaglia." *Periodico di Matematiche,* ser. IV, vol. XXXV, no. 2, 79.

Di Pasquale, L. 1957b. "I cartelli di matematica disfida di Ludovico Ferrari e i controcartelli di Niccolò Tartaglia." *Periodico di Matematiche,* ser. IV, vol. XXXV, no. 5, 253.

Di Pasquale, L. 1958. "I cartelli di matematica disfida di Ludovico Ferrari e i controcartelli di Niccolò Tartaglia (continuazione e fine)." *Periodico di Matematiche,* ser. IV, vol. XXXVI, no. 3, 175.

Donnay, Maurice. 1939. *Le Lycée Louis-le-Grand* (Paris: Nouvelle Revue Française).

Dörrie, H. 1965. *100 Great Problems of Elementary Mathematics* (New York: Dover Publications).

Dumas, A. 1862–1865. *Mes mémoires* (Paris: Calman-Lévy).

Dunham, W. 1991. *Journey Through Genius* (New York: Penguin Books); first published in 1990 by John Wiley & Sons (New York).

Dunnington, G. W. 1955. *Gauss, Titan of Science* (New York: Hafner Publishing); reprinted 2004 by The Mathematical Association of America, with additional material by J. Gray and F.-E. Dohse (Washington, DC).

Dupuy, P. 1896. "La Vie d'Évariste Galois." *Annales scientifiques de l'École normale supérieure;* 3rd ser., 13, 197. Reprinted as a book in 1992 by Éditions Jacques Gabay.

Du Sautoy, M. 2003. *The Music of the Primes* (New York: HarperCollins).

Dyson, F. J. 1966. *Symmetry Groups in Nuclear and Particle Physics* (New York: W. A. Benjamin).

Eco, U. 2004, *On Literature,* trans. M. McLaughlin (Orlando, FL: Harcourt).

Eddington, A. S. 1956. "The Theory of Groups." In *The World of Mathematics,* J. R. Newman, ed. (New York: Simon & Schuster), 1558.

Edwards, H. M. 1984. *Galois Theory* (New York: Springer).

Edwards, H. M. 1987. "An Appreciation of Kronecker." *The Mathematical Intelligencer,* 9, 28.

Edwards, H. M. 1996. *Fermat's Last Theorem: A Genetic Introduction to Algebraic Number Theory* (Berlin: Springer-Verlag).

Einstein, A. 1953. *The Meaning of Relativity* (Princeton: Princeton University Press).

Einstein, A. 2001. *Relativity: The Special and the General Theory* (New York: Routledge).

Einstein, A. 2004. *Einstein's 1912 Manuscript on the Special Theory of Relativity* (New York: George Braziller).

Eisenman, R., and Rappaport, J. 1967. "Complexity Preference and Semantic Differential Ratings of Complexity-Simplicity and Symmetry-Asymmetry." *Psychonomic Science,* 7(4), 147–148.

Emch, A. F. 1933. *The "Legia Demonstrativa" of Girolamo Saccheri* (Cambridge, MA: Harvard).

Emerson, R. W. 1847. *Poems.* In *Ralph Waldo Emerson: Collected Poems and Translations,* P. Kaye and H. Bloom, eds. (New York: Library of America, 1994).

Enquist, M., and Arak, A. 1994. "Symmetry, Beauty and Evolution." *Nature,* 372, 169.

Evans, D., and Zarate, O. 1999. *Introducing Evolutionary Psychology* (Cambridge, England: Icon Books).

Evans, J. 1975. *Pattern* (New York: Hacker Art Books).

Fabricand, B. P. 1989. "Symmetry in Free Markets." *Computers & Mathematics with Applications,* 17, nos. 4–6, 653.

Fanselow, M. S. 1994. "Neural Organization of the Defensive Behavior System Responsible for Fear." *Psychonomic Bulletin and Review,* 1, 429.

Farmer, D. W. 1996. *Groups and Symmetry: A Guide to Discovering Mathematics* (Providence, RI: American Mathematical Society).

Fayen, G. 1977. "Ambiguities in Symmetry-Seeking: Borges and Others." In *Patterns of Symmetry*, M. Senechal and George Fleck, eds. (Amherst: University of Massachusetts Press), 104.

Fehr, H. 1902. *Intermédiaire des mathématiciens*, 9, 74.

Ferris, T. 1997. *The Whole Shebang* (New York: Simon & Schuster).

Feynman, R. P., Leighton, R. B., and Sands, M. 1963. *The Feynman Lectures on Physics*, vol. I.

Feynman, R. P., Leighton, R. B., and Sands, M. 1965. *The Feynman Lectures on Physics*, vol. III (Reading, MA: Addison-Wesley).

Fienberg, S. E. 1971. "Randomization and Social Affairs: The 1970 Draft Lottery." *Science*, 171, 255.

Fierz, M. 1983. *Girolamo Cardano 1501–1576* (Boston: Birkhäuser).

Fine, B., and Rosenberger, G. 1997. *The Fundamental Theorem of Algebra* (Berlin: Springer-Verlag).

Fletcher, D. J. 1967. "Carry On Kariera." *Mathematics Teaching*, 43, 35.

Fodor, J. 1983. *The Modularity of Mind: An Essay on Faculty Psychology* (Cambridge, MA: MIT Press).

Fölsing, A. 1997. *Albert Einstein* (New York: Viking).

Forsman, A., and Merilaita, S. 1999. "Fearful Symmetry: Pattern Size and Asymmetry Affects Aposematic Signal Efficacy." *Evolutionary Ecology*, 13, 131.

Forsman, A., and Merilaita, S. 2003. "Fearful Symmetry? Intra-Individual Comparisons of Asymmetry in Cryptic vs. Signaling Colour Patterns in Butterflies." *Evolutionary Ecology*, 17, 491.

Forsyth, A. R. 1895. "Arthur Cayley." *Proceedings of the Royal Society of London*, 58, 1.

Fox, J. 1975. "The Use of Structural Diagnostics in Recognition." *Journal of Experimental Psychology*, 104(1), 57–67.

Fox, R. 1967. *Kinship and Marriage: An Anthropological Perspective* (Harmondsworth, England: Penguin Books).

Fraleigh, J. B. 1989. *A Fast Course in Abstract Algebra* (Reading, MA: Addison-Wesley).

Franci, R., and Toti Rigatelli, L. 1985. "Towards a History of Algebra from Leonardo of Pisa to Luca Pacioli." *Janus*, 72 (1–3), 17.

Frank, P. 1949. *Einstein: His Life and His Thoughts* (New York: Alfred A. Knopf).

Frey, A. H. Jr., and Singmaster, D. 1982. *Handbook of Cubik Math* (Hillside, NY: Enslow Publishers).

Freyd, J., and Tversky, B. 1984. "Force of Symmetry in Form Perception." *American Journal of Psychology*, 97(1), 109–126.

Gales, F. 2004. http://perso.wanadoo.fr/frederic.gales/Laviedegalois.htm/.

Galison, P. 2004. *Einstein's Clocks, Poincaré's Maps* (New York: W. W. Norton).

Gamow, G. 1959. "The Exclusion Principle." *Scientific American,* 201, no. 1, 74.

Gandz, S. 1937. "The Origin and Development of the Quadratic Equations in Babylonian, Greek and Early Arabic Algebra." *Osiris,* vol. III, 405.

Gandz, S. 1940a. "Studies in Babylonian Mathematics III: Isoperimetric Problems and the Origin of the Quadratic Equations." *Isis,* 32, 103; quotes *Talmud,* "Sotah" IV, 4.

Gandz, S. 1940b, ibid.; quotes from Heath 1956.

Gangestad, S. W., Thornhill, R., and Yeo, R. A. 1994. "Facial Attractiveness, Developmental Stability, and Fluctuating Asymmetry." *Ethology and Sociobiology,* 15, 73.

Gårding, L., and Skau, C. 1994. "Niels Henrik Abel and Solvable Equations." *Archive for the History of Exact Sciences,* 48, 81.

Gardner, H. 1993. *Creating Minds* (New York: Basic Books).

Gardner, K. L. 1966. *Discovering Modern Algebra* (Oxford: University Press).

Gardner, M. 1959a. "Mathematical Games: How three modern mathematicians disproved a celebrated conjecture of Leonhard Euler." *Scientific American,* 201 (November), 181.

Gardner, M. 1959b. *Mathematical Puzzles of Sam Loyd* (New York: Dover).

Gardner, M. 1979. *Mathematical Circus* (New York: Alfred A. Knopf).

Gardner, M. 1990. *The New Ambidextrous Universe: Symmetry and Asymmetry from Mirror Reflections to Superstrings* (New York: W. H. Freeman and Company).

Garling, D. J. H. 1960. *A Course in Galois Theory* (Cambridge: Cambridge University Press).

Garner, W. R. 1974. *The Processing of Information and Structure* (Hillsdale, NJ: Erlbaum).

Gauss, C. F. 1876. *Collected Works,* vol. 3 (in German, published in 1969 and in 1987 by Göttingen: Vandenhoeck & Ruprecht; there are also several publications by G. Olms, Hildesheim).

Gell-Mann, M. 1994. *The Quark and the Jaguar* (San Francisco: W. H. Freeman).

Gillings, R. J. 1972. *Mathematics in the Time of the Pharaohs* (Cambridge, MA: MIT Press).

Gindikin, S. G. 1988. *Tales of Physicists and Mathematicians* (Boston: Birkhäuser).

Girard, G. 1929. *Les Trois Glorieuses* (Paris: Firmin Didot).

Gisquet, H. J. 1840–1844. *Mémoires de M. Gisquet, ancien préfet de police, écrits par lui-même* (Bruxelles: Meline et Cans).

Glashow, S. 1988. *Interactions* (New York: Time-Warner Books).

Gleason, A. M. 1990. In *More Mathematical People,* D. J. Albers, G. L. Alexanderson, and C. Reed, eds. (Boston: Harcourt Brace Jovanovich).

Gleick, J. 2003. *Isaac Newton* (New York: Pantheon).

Gliozzi, M. 1972. In *Dictionary of Scientific Biography,* C. C. Gillespie, ed. (New York: Charles Scribner's Sons).
Goldsmith, D. 2000. *The Runaway Universe: The Race to Discover the Future of the Cosmos* (New York: Perseus Publishing).
Goldstein, E. B. 2002. *Sensation and Perception* (Pacific Grove, CA: Wadsworth).
Gombrich, E. H. 1995. *The Story of Art,* 16th edition (London: Phaidon Press Inc.).
Goodman, R. 2004. "Alice Through Looking Glass After Looking Glass: The Mathematics of Mirrors and Kaleidoscopes." *The American Mathematical Monthly,* 111(4), 281.
Gorenstein, D. 1982. *Finite Simple Groups: An Introduction to Their Classification* (New York: Plenum Press).
Gorenstein, D. 1985. "The Enormous Theorem." *Scientific American,* 253, 104.
Gorenstein, D. 1986. "Classifying the Finite Simple Groups." *Bulletin of the American Mathematical Society,* 14, 1.
Goudey, C. 2001–2003. http://cubeland.free.fr/infos/infos.htm/.
Gow, J. 1968. *A Short History of Greek Mathematics* (New York: Chelsea Publishing Company).
Grammer, K., Fink, B., Møller, A. P., and Thornhill, R. 2003. "Darwinian Aesthetics: Sexual Selection and the Biology of Beauty." *Biological Reviews,* 78, 385.
Grammer, K., and Thornhill, R. 1994. "Human (Homo Sapiens) Facial Attractiveness and Sexual Selection: The Role of Symmetry and Averageness." *Journal of Comparative Psychology,* 108, 233.
Gray, J. 2004. *Gauss: Titan of Science* (Cambridge: Cambridge University Press).
Gray, J. J. 1979. "Non-Euclidean Geometry—A Reinterpretation." *Historia Mathematica,* 6(3), 236.
Gray, J. J. 1995. "Arthur Cayley (1821–1895)." *The Mathematical Intelligencer,* 17(4), 62.
Green, M. B. 1986. "Superstrings." *Scientific American,* 255, 48.
Greene, B. 1999. *The Elegant Universe* (New York: W. W. Norton).
Greene, B. 2004. *The Fabric of the Cosmos* (New York: Alfred A. Knopf).
Gregory, R. 1997. *Mirrors in Mind* (New York: W. H. Freeman).
Gribbin, J. 1999. *The Search for Superstrings, Symmetry, and the Theory of Everything* (New York: Little, Brown and Company).
Grünbaum, B., Grünbaum, Z., and Shephard, G. C. 1986. In *Symmetry,* I. Hargittai, ed. (New York: Pergamon Press), 641.
Guth, A. H. 1997. *The Inflationary Universe* (Reading, MA: Helix Books).
Hafele, J. C., and Keating, R. E. 1972a. "Around-the-World Atomic Clocks: Predicted Relativistic Time Gains." *Science,* 177, 166.

Hafele, J. C., and Keating, R. E. 1972b. "Around-the-World Atomic Clocks: Observed Relativistic Time Gains." *Science,* 177, 168.

Hale, J. 1994. *The Civilization of Europe in the Renaissance* (New York: Atheneum).

Hall, B. C. 2003. *Lie Groups, Lie Algebras, and Representations* (Berlin: Springer-Verlag).

Halsted, G. B. 1895. "Biography, Lobachevsky." *American Mathematical Monthly,* 2, 137.

Hamilton, W. D., and Zuk, M. 1982. "Heritable True Fitness and Bright Birds: A Role for Parasites?" *Science,* 218, 384.

Hares-Stryker, C., ed. 1997. *An Anthology of Pre-Raphaelite Writings* (New York: New York University Press), 284.

Hargittai, I. 1986. *Symmetry: Unifying Human Understanding* (New York: Pergamon Press).

Hargittai, I. 1989. *Symmetry 2: Unifying Human Understanding* (New York: Pergamon Press).

Hawking, S., and Penrose, R. 1996. *The Nature of Space and Time* (Princeton: Princeton University Press).

Heath, T. 1956. *The Thirteen Books of Euclid's Elements* (New York: Dover Publications).

Heath, T. 1981. *A History of Greek Mathematics* (New York: Dover Publications).

Heilbronner, E., and Dunitz, J. D. 1993. *Reflections on Symmetry* (Basel: Verlag Helvetica Chimica Acta).

Henry, C. 1879. "*Manuscrits de Sophie Germain.*" *Revue philosophique de la France et de l'étranger,* 8, 619.

Hill, E. L. 1946. "A Note on the Relativistic Problem of Uniform Rotation." *Physical Review,* 69, 488.

Hines, T. 1998. "Further on Einstein's Brain." *Experimental Neurology,* 150, 343.

Hodde, L. D. L. 1850. *Histoire de sociétés secrètes et du parti républicain de 1830 à 1848* (Paris: Julien, Lanier et Cie).

Hofmann, J. E. 1972. In *Dictionary of Scientific Biography,* C. C. Gillespie, ed. (New York: Charles Scribner's Sons).

Hofstadter, D. R. 1979. *Gödel, Escher, Bach: An Eternal Golden Braid* (New York: Basic Books).

Holst, E., Stormer, C., and Sylow, L., eds. 1902. Festskrift ved hundreaarsjubiloeet for Niels Henrik Abel fosdel (Christiania).

Howe, E. S. 1980. "Effects of Partial Symmetry, Exposure Time, and Backward Masking on Judged Goodness and Reproduction of Visual Patterns." *Quarterly Journal of Experimental Psychology,* 32, 27.

Hoyle, F. 1977. *Ten Faces of the Universe* (San Francisco: W. H. Freeman and Company).

Hugo, V. 1862. *Les Misérables* (Bruxelles: A. Lacroix, Verboeckhoven & Cie.), translated by C. E. Wilbour (New York: Modern Library, 1902).

Huxley, T. H. 1868. "A Liberal Education." http://human-nature.com/darwin/huxley/chap2.html.

Hyatt King, A. 1944. "Mozart's Piano Music." *The Music Review,* 5, 163–191.

Hyatt King, A. 1976. *Mozart in Retrospect* (Westport, CT: Greenwood Press).

Icke, V. 1995. *The Force of Symmetry* (Cambridge: Cambridge University Press).

Infantozzi, C. A. 1968. "Sur la mort d'Evariste Galois." *Revue d'histoire des sciences,* 21, 1968.

Infeld, L. 1948. *Whom the Gods Love: The Story of Évariste Galois* (New York: Whittlesey House, McGraw-Hill).

Jackson, L. A. 1992. *Physical Appearance and Gender: Sociobiological and Sociocultural Perspectives* (Albany: State University of New York Press).

James, I. 2002. *Remarkable Mathematicians: From Euler to von Neumann* (Cambridge: Cambridge University Press).

Jayawardene, S. A. 1972. In *Dictionary of Scientific Biography,* C. C. Gillespie, ed. (New York: Charles Scribner's Sons).

Johnstone, R. A. 1994. "Female Preference for Symmetrical Males as a By-Product of Selection for Mate Recognition." *Nature,* 372, 172.

Jones, D. 1996. *Physical Attractiveness and the Theory of Sexual Selection* (Ann Arbor: University of Michigan Press).

Joyner, D. 2002. *Adventures in Group Theory: Rubik's Cube, Merlin's Machine and Other Mathematical Toys* (Baltimore: Johns Hopkins University Press).

Julesz, B. 1960. "Binocular Depth Perception of Computer-Generated Patterns." *Bell System Technical Journal,* 39, 1125.

Kaku, M. 1994. *Hyperspace* (New York: Oxford University Press).

Kaku, M. 2004. *Einstein's Cosmos* (New York: Atlas Books).

Kalin, N. H. 1993. "The Neurobiology of Fear." *Scientific American,* May, 94.

Kane, G. L. 2000. *Supersymmetry: Unveiling the Ultimate Laws of Nature* (New York: Perseus).

Kaplan, A. 1990. *Sefer Yetzira: The Book of Creation: In Theory and Practice* (Boston: Weiser).

Katz, V. 1989. "Historical Notes," in Fraleigh 1989.

Keiner, I. 1986. "The Evolution of Group Theory: A Brief Survey." *Mathematics Magazine,* 59(4), 195.

Keller, J. B. 1986. "The Probability of Heads." *American Mathematical Monthly,* 93, 191.

Kepler, J. 1966. *The Six-Cornered Snowflake* (Oxford: Oxford University Press; originally published in 1611).

Keyser, C. J. 1956. "The Group Concept." In *The World of Mathematics,* vol.

3, J. R. Newman, ed. (New York: Simon & Schuster; republished in 2000, Mineola, NY: Dover).

Kiernan, B. M. 1971. "The Development of Galois Theory from Lagrange to Artin." *Archive for History of Exact Sciences,* 8, 40.

Kimberling, C. 1972. "Emmy Noether." *American Mathematical Monthly,* 79, 136.

Kirshner, R. P. 2002. *The Extravagant Universe: Exploding Stars, Dark Energy and the Accelerating Cosmos* (Princeton: Princeton University Press).

Klein, F. 1884. *Lectures on the Icosahedron and the Solution of Equations of the Fifth Degree,* G. G. Morrice, trans. Published in 1956 by Dover (New York).

Kline, H. M., and Alder, M. 2000. www.marco-learningsystems.com/pages/kline/johnny/johnny-chapt7-8.html/.

Kline, M. 1972. *Mathematical Thought from Ancient to Modern Times* (New York: Oxford University Press).

Kollros, L. 1949. "Evariste Galois." *Elemente der Mathematik,* no. 7, 1 (Basel: Verlag Birkhäuser).

Kurzweil, R. 1999. *The Age of Spiritual Machines* (London: Orion Business Books).

Lamb, C. 1823. "Essays of Elia: The Two Races of Men." In *The Norton Anthology of English Literature,* 6th edition, vol. 2 (New York: W. W. Norton).

Langlois, J. H., Kalakanis, L., Rubenstein, A. J., Larson, A., Hallam, M., and Smoot, M. 2000. "Maxims or Myths of Beauty? A Meta-Analytic and Theoretical Review." *Psychological Bulletin,* 126, 390.

Langlois, J. H., and Roggeman, L. A. 1990. "Attractive Faces Are Only Average." *Psychological Science,* 1, 115.

Langlois, J. H., Roggman, L. A., and Musselman, L. 1994. "What Is Average and What Is Not Average About Attractive Faces?" *Psychological Science,* 5(4), 214.

Langlois, J. H., Roggman, L. A., and Reiser-Danner, L. A. 1990. "Infants' Differential Social Responses to Attractive and Unattractive Faces." *Developmental Psychology,* 26, 153.

Laudal, O. A., and Piene, R., eds. 2004. *The Legacy of Niels Henrik Abel: The Abel Bicentennial, Oslo, June 3–8, 2002* (Berlin: Springer).

Lederman, L., with Tersei, D. 1993. *The God Particle* (Boston: Houghton Mifflin).

Lederman, L. M., and Hill, C. T. 2004. *Symmetry and the Beautiful Universe* (Amherst, NY: Prometheus).

LeDoux, J. E. 1996. *The Emotional Brain* (New York: Simon & Schuster).

Leeuwenberg, E. L. J. 1971. "A Perceptual Coding Language for Visual and Auditory Patterns." *American Journal of Psychology,* 84(3), 307.

Levey, M. 1954. "Abraham Savasorda and His Algorism: A Study in Early European Logistic." *Osiris*, 11, 50.

Lévi-Strauss, C. 1949. *The Elementary Structure of Kinship.* Republished in 1971 (Boston: Beacon Press).

Lévi-Strauss, C. 1958. *Structural Anthropology.* Republished in 1974 (New York: Basic Books).

Levy, S. "I Found Einstein's Brain." http://www.echonyc.com/~steven/einstein.html/.

Lewin, D. 1993. *Musical Form and Transformation: 4 Analytic Essays* (New Haven: Yale University Press).

Lindley, D. 1996. *Where Does the Weirdness Go?* (New York: Basic Books).

Liouville, J., ed. 1846. "Oeuvres mathématiques d'Évariste Galois." *Journal de mathématiques pures et appliquées*, 11, 381.

Lipkin, H. J. 2002. *Lie Groups for Pedestrians* (Mineola, NY: Dover Publications).

Livio, M. 2000. *The Accelerating Universe: Infinite Expansion, the Cosmological Constant, and the Beauty of the Cosmos* (New York: Wiley).

Livio, M. 2002. *The Golden Ratio: The Story of Phi, the World's Most Astonishing Number* (New York: Broadway Books).

Loeb, A. L. 1971. *Color and Symmetry* (New York: John Wiley & Sons).

Loftus, M. J. 2001. "The Rorschach Inkblot Test." *Emory Magazine*, vol. 77, number 2.

Lombroso, C. 1895. *The Man of Genius* (London: Charles Scribner's Sons).

Ludwig, A. M. 1995. *The Price of Greatness: Resolving the Creativity and Madness Controversy* (New York: Guilford Press).

Lyytinen, A., Brakefield, P. M., and Mappes, J. 2003. "Significance of Butterfly Eyespots as an Anti-Predator Device in Ground-Based and Aerial Attacks." *Oikos*, 100, 373.

MacCarthy, F. 1995. *William Morris: A Life for Our Time* (London: Faber and Faber).

MacGillavry, C. H. 1976. *Fantasy & Symmetry: The Periodic Drawings of M. C. Escher* (New York: Harry N. Abrams).

Mach, E. 1914. *The Analysis of Sensation* (Chicago: Open Court); republished in 1959 by Dover (New York).

MacKinnon, D. W. 1975. "IPAR's Contribution to the Conceptualization and Study of Creativity," in *Perspectives in Creativity*, I. Taylor and J. W. Getzels, eds. (Chicago: Aldine Publishing).

Maher, J., and Groves, J. 1996. *Introducing Chomsky* (Cambridge, England: Icon Books).

Malkin, I. 1963. "On the 150th Anniversary of the Birth Date of an Immortal in Mathematics." *Scripta Mathematica*, 26, 197.

Maor, E. 1994. *e: The Story of a Number* (Princeton: Princeton University Press).

Marr, D. 1982. *Vision* (New York: W. H. Freeman).

Maxwell, E. A. 1965. *Gateway to Abstract Algebra* (Cambridge: Cambridge University Press).

Mayer, O. 1982. "János Bolyai's Life and Work," in *Proceedings of the National Colloquium on Geometry and Topology* (Napoca, Romania: Cluj-Napoca Technical University Press).

McWeeny, R. 2002. *Symmetry: An Introduction to Group Theory and Its Applications* (Mineola, NY: Dover).

Mendola, J. D., Dale, A. M., Fischel, B., Liu, A. K., and Tootell, R. B. H. 1999. "The Representation of Illusory and Real Contours in Human Cortical Visual Areas Revealed by fMRI." *Journal of Neuroscience,* 19, 8560.

Menz, C. 2003. *Morris & Co.* (Adelaide, Australia: Art Gallery of South Australia).

Mezrich, B. 2002. *Bringing Down the House* (New York: Free Press).

Miller, A. I. 1996. *Insights of Genius: Imagery and Creativity in Science and Art* (New York: Copernicus).

Miller, A. I. 2001. *Einstein, Picasso* (New York: Perseus Books).

Miller, G. F. 2000. *The Mating Mind* (New York: Doubleday).

Misner, C. W., Thorne, K. S., and Wheeler, J. A. 1973. *Gravitation* (San Francisco: W. H. Freeman).

Mittag-Leffler, G. 1904. "Niels Henrik Abel." *Ord Och Bild,* 12, 65, and 129.

Mittag-Leffler, G. 1907. "Niels Henrik Abel." *Revue du Mois* (Paris), 4, 5, and 207.

Møller, A. P. 1992. "Female Swallow Preference for Symmetrical Male Sexual Ornaments." *Nature,* 357, 238.

Møller, A. P. 1994. *Sexual Selection and the Barn Swallow* (Oxford: Oxford University Press).

Møller, A. P., and Swaddle, J. P. 1997. *Asymmetry, Developmental Stability and Evolution* (Oxford: Oxford University Press).

Molnar, V., and Molnar, F. 1986. "Symmetry-Making and -Breaking in Visual Art." *Computers and Mathematics with Applications,* 12B, Nos. 1–2, 291–301.

Mondor, H. 1954. "L'étrange rencontre de Nerval et de Galois." *Arts,* 7 Juillet.

Mosotti, A. 1972. In *Dictionary of Scientific Biography,* C. C. Gillespie, ed. (New York: Charles Scribner's Sons).

Motte, A. 1995, trans. *The Principia* (Amherst, NY: Prometheus Books). Newton's 1686 masterpiece was translated by Andrew Motte in 1729.

Mozart, W. A., et al. 1966. *The Letters of Mozart and His Family,* vol. 1, 2nd edition, Emily Anderson, trans. (London: Macmillan), p. 130, p. 137.

Mozzochi, C. J. 2004. *The Fermat Proof* (Victoria, Canada: Trafford Publishing).

Mumford, D., Series, C., and Wright, D. 2002. *Indra's Pearls: The Vision of Felix Klein* (Cambridge: Cambridge University Press).

N. E. Thing Enterprises. 1995. *Magic Eye Gallery: A Showing of 88 Images* (Kansas City: Andrews & McMeel).

Nagy, D. 1995. "The 2,500-Year-Old Term Symmetry in Science and Art and Its 'Missing Link' Between the Antiquity and the Modern Age." *Symmetry: Culture and Science,* 6, 18.

Newman, J. R. 1956. "The Rhind Papyrus," in *The World of Mathematics,* J. R. Newman, ed. (New York: Simon & Schuster).

Noether's theorem. http://en.wikipedia.org/wiki/Noether%27s_theorem/.

Nový, L. 1973. *Origins of Modern Algebra* (Prague: Academia).

O'Connor, J. J, and Robertson, E. F. 1996a. www-history.mcs.st-andrews.ac.uk/history/Mathematicians/Bring.html/.

O'Connor, J. J., and Robertson, E. F. 1996b. www-history.mcs.st-andrews.ac.uk/history/Mathematicians/Galois.html/.

O'Connor, J. J., and Robertson, E. F. 1996c. www-history.mcs.st-andrews.ac.uk/history/Mathematicians/Germain.html/.

O'Connor, J. J., and Robertson, E. F. 1996d. www-history.mcs.st-andrews.ac.uk/history/Mathematicians/Cayley.html/.

O'Connor, J. J., and Robertson, E. F. 1996e. www-history.mcs.st-andrews.ac.uk/HistTopics/Non-Euclidean_geometry.html/.

O'Connor, J. J., and Robertson, E. F. 1997. www-history.mcs.st-andrews.ac.uk/history/Mathematicians/Tschirnhaus.html/.

O'Connor, J. J., and Robertson, E. F. 1998. www-history.mcs.st-andrews.ac.uk/history/Mathematicians/Euler.html.

O'Connor, J. J., and Robertson, E. F. 1999a. www-history.mcs.st-andrews.ac.uk/history/Mathematicians/Al-Khwarizmi.html/.

O'Connor, J. J., and Robertson, E. F. 1999b. www-history.mcs.st-andrews.ac.uk/history/Mathematicians/Abraham.html/.

O'Connor, J. J., and Robertson, E. F. 2000a. www-history.mcs.st-andrews.ac.uk/history/HistTopics/Babylonian_mathematics.html/.

O'Connor, J. J., and Robertson, E. F. 2000b. www-history.mcs.st-andrews.ac.uk/history/HistTopics/Egyptian_mathematics.html/.

O'Connor, J. J., and Robertson, E. F. 2000c. www-history.mcs.st-andrews.ac.uk/history/HistTopics/Egyptian_papyri.html/.

O'Connor, J. J., and Robertson, E. F. 2000d. www-history.mcs.st-andrews.ac.uk/history/Mathematicians/Gregory.html/.

O'Connor, J. J., and Robertson, E. F. 2001a. www-history.mcs.st-andrews.ac.uk/history/Mathematicians/Bezout.html/.

O'Connor, J. J., and Robertson, E. F. 2001b. www-history.mcs.st-andrews.ac.uk/history/Mathematicians/Birkhoff.html/.

O'Connor, J. J., and Robertson, E. F. 2002. www-history.mcs.st-andrews.ac.uk/history/Mathematicians/Lie.html/.

Odifreddi, P. 2004. *The Mathematical Century: The 30 Greatest Problems of the Last 100 Years* (Princeton: Princeton University Press).

Olson, S. 2004. *Count Down* (Boston: Houghton Mifflin).

Ore, O. 1953. *Cardano, the Gambling Scholar* (Princeton: Princeton University Press).

Ore, O. 1954. *Niels Henrik Abel: Et geni og hans Samtid* (Oslo: Gyldendal Norsk Forlag). Translated into English in 1957 as *Niels Henrik Abel: Mathematician Extraordinary* (Minneapolis: University of Minnesota Press).

Ore, O. 1972. In *Dictionary of Scientific Biography,* C. C. Gillespie, ed. (New York: Charles Scribner's Sons).

Osborne, H. 1952. *Theory of Beauty: An Introduction to Aesthetics* (London: Routledge & Kegan Paul).

Osborne, H. 1986. "Symmetry as an Aesthetic Factor." *Computers & Mathematics with Applications,* 12B, Nos. 1–2, 77.

Osen, L. M. 1974. *Women in Mathematics* (Cambridge, MA: MIT Press).

Overbye, D. 2000. *Einstein in Love: A Scientific Romance* (New York: Viking).

Oxford English Dictionary. 1978 (Oxford: Oxford University Press).

Pagán Westphal, S. 2003. "Decoding the Ys and Wherefores of Males." *New Scientist,* 21 June, 15.

Pais, A. 1982. *Subtle Is the Lord: The Science and the Life of Albert Einstein* (New York: Oxford University Press).

Palmer, A. R. 1994. "Fluctuating Asymmetry Analysis: A Primer," in *Developmental Instability: Its Origins and Evolutionary Implications,* T. A. Markow, ed. (Dordrecht, The Netherlands: Kluwer), 335.

Palmer, S. E. 1991. "Goodness, Gestalt, Groups, and Garner: Local Symmetry Subgroups as a Theory of Figural Goodness," in *The Perception of Structure: Essays in Honor of Wendell R. Garner,* G. R. Lockhead and J. R. Pomerantz, eds. (Washington, DC: American Psychological Association), 23.

Palmer, S. E. 1999. *Vision Science* (Cambridge, MA: MIT Press).

Palmer, S. E., and Hemenway, K. 1978. "Orientation and Symmetry: Effects of Multiple, Rotational, and Near Symmetries." *Journal of Experimental Psychology,* 4(4), 691–702.

Paraskevopoulos, I. 1968. "Symmetry, Recall, and Preference in Relation to Chronological Age." *Journal of Experimental Child Psychology,* 6, 254–264.

Parry, L., ed. 1996. *William Morris* (New York: Harry N. Abrams).

Pascal, A. 1914. "Girolamo Saccheri Nella Vita e Nelle Opere." *Giornale di Matematica di Battaglini,* 52, 229.

Paterniti, M. 2000. *Driving Mr. Albert: A Trip Across America with Einstein's Brain* (New York: Dial Press).

Patterson, E. M., and Rutherford, D. E. 1965. *Elementary Abstract Algebra* (Edinburgh: Oliver and Boyd).

Pennisi, E. 2004. "Speaking in Tongues." *Science,* 303, 1321.
Penrose, R. 2004. *The Road to Reality* (London: Jonathan Cape).
Perrett, D. I., May, K. A., and Yoshikawa, S. 1994. "Facial Shape and Judgements of Female Attractiveness." *Nature,* 368, 239.
Perrett, D. I., Penton-Voak, I. S., Little, A. C., Tiddeman, B. P., Burt, D. M., Schmidt, N., Oxley, R., and Barrett, L. 2002. "Facial Attractiveness Judgements Reflect Learning of Parental Age Characteristics." *Proceedings of the Royal Society of London B,* 269 (1494), 873.
Pesic, P. 2003. *Abel's Proof: An Essay on the Sources and Meaning of Mathematical Unsolvability* (Cambridge, MA: The MIT Press).
Peterson, I. 2000. "Completing Latin Squares." *Science News Online,* 157, No. 19, http://www.sciencenews.org/20000506/mathtrek.asp.
Peterson, I. 2004. "Heads or Tails?" *Science News,* February 28.
Petsinis, T. 1988. *The French Mathematician* (New York: Walker & Company).
Picard, E., ed. 1897. *Oeuvres mathématiques d'Évariste Galois* (Reprinted in 1951, Paris: Gauthier-Villars).
Pickover, C. A. 2001. *Wonders of Numbers* (Oxford: Oxford University Press).
Pinker, S. 1994. *The Language Instinct* (New York: Morrow).
Pinker, S. 1997. *How the Mind Works* (New York: W. W. Norton & Company).
Portnoy, E. 1982. "Riemann's Contribution to Differential Geometry." *Historia Mathematica,* 9(1), 1.
Putz, J. F. 1995. "The Golden Section and the Piano Sonatas of Mozart." *Mathematics Magazine,* 68, 275–282.
Rahn, J. 1980. *Basic Atonal Theory* (New York: Schirmer Music Books).
Raspail, F.-V. 1839. *Réforme pénitentiaire: Lettres sur les prisons de Paris* (Paris: Tamisey et Champion).
Rees, M. 1997. *Before the Beginning* (Reading, MA: Helix Books).
Revue encyclopédique, t. 55, 568, Septembre 1832.
Ridley, M. 2003. *The Red Queen* (New York: Perennial); originally published in 1993 by Penguin (London).
Rikowski, A., and Grammer, K. 1999. "Human Body Odour, Symmetry and Attractiveness." *Proceedings of the Royal Society of London B,* 266, 869.
Ritter, F. 1895. "François Viète, inventeur de l'algèbre moderne, 1540–1603. Essai sur sa vie et son oeuvre." *Revue Occidentale Philosophique Sociale et Politique,* 10, 234; 354.
Rock, I. 1973. *Orientation and Form* (New York: Academic Press).
Rose, P. L. 1975. *The Italian Renaissance of Mathematics* (Genève: Librairie Droz).
Rosen, J. 1975. *Symmetry Discovered: Concepts and Applications in Nature and Science* (Cambridge: Cambridge University Press).

Rosen, J. 1995. *Symmetry in Science: An Introduction to the General Theory* (New York: Springer-Verlag).

Rosen, M. I. 1995. "Niels Henrik Abel and Equations of the Fifth Degree." *American Mathematical Monthly,* 102, 495.

Rosenbaum, D. E. 1970. "Statisticians Charge Draft Lottery Was Not Random." *New York Times,* Jan. 4, 1970.

Rossi, B., and Hall, D. B. 1941. "Variation of the Rite of Decay of Mesotrons with Momentum." *Physical Review,* 59, 223.

Rothenberg, A., and Hausman, C. R., eds. 1976. *The Creativity Question* (Durham: Duke University Press).

Rothman, T. 1982a. "Genius and Biographers: The Fictionalization of Évariste Galois." *The American Mathematical Monthly,* 89, 2, 84; revised at: http://godel.ph.utexas.edu/~tonyr/galois.html/.

Rothman, T. 1982b. "The Short Life of Évariste Galois." *Scientific American,* 246, 4, 136.

Rotman, J. 1990. *Galois Theory* (New York: Springer-Verlag).

Rotman, J. J. 1995. *An Introduction to the Theory of Groups* (New York: Springer-Verlag).

Rozen, S., Skaletsky, H., Marszalek, J. D., Minx, P. J., Cordam, H. S., Waterston, R. H., Wilson, R. K., and Page, D. C. 2003. "Abundant Gene Conversion between Arms of Palindromes in Human and Ape Y Chromosomes." *Nature,* 423, 873.

Rubik, E., Varga, T., Kéri, G., Marx, G., and Vekerdy, T. 1987. *Rubik's Cubic Compendium* (Oxford: Oxford University Press).

Sabbagh, K. 2003. *The Riemann Hypothesis: The Greatest Unsolved Problem in Mathematics* (New York: Farrar, Straus & Giroux).

Sampson, J. H. 1990–1991. "Sophie Germain and the Theory of Numbers." *Archive for History of Exact Science,* 41, 157.

Sarton, G. 1921. "Évariste Galois." *The Scientific Monthly,* 13, 363; reprinted in 1937, *Osiris,* 3, 241.

Sartre, J.-P. 1964. *Les Mots* (Paris: Gallimard).

Schaal, B., and Porter, R. H. 1991. "Microsmatic Humans Revisited: The Generation and Perception of Chemical Signals." *Advances in the Study of Behavior,* 20, 474.

Schattschneider, D. 2004. *M. C. Escher: Visions of Symmetry* (New York: Harry N. Abrams).

Schoenberg, A. 1969. *Structural Functions of Harmony* (New York: W. W. Norton).

Scholz, E. 1992. "Riemann's Vision of a New Approach to Geometry," in *1830–1930: A Century of Geometry,* L. Boi, D. Flament, and J. M. Salanskis, eds. (Berlin: Springer-Verlag), 22.

Schultz, P. 1984. "Tartaglia, Archimedes and Cubic Equations." *Gazette Australian Mathematical Society,* 11 (4), 81.

Schwarz, C. 2002. *Tales from the Subatomic Zoo* (Staatsburg, NY: Small World Books; www.smallworldbooks.net).

Schwarz, P. M., and Schwarz, J. H. 2004. *Special Relativity: From Einstein to Strings* (Cambridge: Cambridge University Press).

Schweitzer, A. 1967. *J. S. Bach* (vol. 1) (Mineola, NY: Dover Publications).

Shackelford, T. K., and Larsen, R. J. 1997. "Facial Asymmetry as Indicator of Psychological, Emotional, and Physiological Distress." *Journal of Personality and Social Psychology,* 72, 456.

Shubnikov, A. V., and Koptsik, V. A. 1974. *Symmetry in Science and Art* (New York: Plenum Press).

Shurman, J. 1997. *Geometry of the Quintic* (New York: John Wiley & Sons).

Silk, J. 2000. *The Big Bang,* 3rd ed. (New York: Times Books).

Simonton, D. K. 1999. *Origins of Genius* (New York: Oxford University Press).

Singh, D. 1993. "Adaptive Significance of Female Physical Attractiveness: Role of Waist-to-Hip Ratio." *Journal of Personality and Social Psychology,* 65, 293.

Singh, D. 1995. "Female Health, Attractiveness, and Desirability for Relationships: Role of Breast Asymmetry and Waist-to-Hip Ratio." *Ethology and Sociobiology,* 16, 465.

Singh, S. 1997. *Fermat's Enigma* (New York: Anchor Books).

Skaletsky, H., et al. 2003. "The Male-Specific Region of the Human Y Chromosome Is a Mosaic of Discrete Sequence Classics." *Nature,* 423, 825.

Smith, D. K. 1997. "Mathematics, marriage and finding somewhere to eat." http://www.pass.maths.org.uk/issue3/marriage/index.html/.

Smith, T. A. 1996. http://jan.ucc.nau.edu/~tas3/musoffcanons.html.

Smolin, L. 2002. *Three Roads to Quantum Gravity* (New York: Perseus Books).

Solomon, R. 1995. "On Finite Simple Groups and Their Classification." *Notices of the AMS,* 42(2), 231.

Sommerville, D. Y. 1960. *Bibliography of Non-Euclidean Geometry,* 2nd ed. (New York: Chelsea).

Spearman, B. K., and Williams, K. S. 1994. "Characterization of Solvable Quintics $x^5 + ax + b$." *American Mathematical Monthly,* 101, 986.

Stachel, J., ed. 1989. *The Collected Papers of Albert Einstein,* vols. 1 and 2 (Princeton: Princeton University Press).

Starr, N. 1997. "Nonrandom Risk: The 1970 Draft Lottery." *Journal of Statistical Education,* 5, no. 2.

Steiner, G. 2001. *Grammars of Creation* (London: Faber and Faber).

Sternberg, R. J., ed. 1998. *Handbook of Creativity* (Cambridge: Cambridge University Press).

Stevens, P. S. 1996. *Handbook of Regular Patterns* (Cambridge, MA: The MIT Press).

Stewart, I. 1995. *Concepts of Modern Mathematics* (Mineola, NY: Dover).

Stewart, I. 2001. *What Shape Is a Snowflake?* (New York: W. H. Freeman).

Stewart, I. 2004. *Galois Theory* (Boca Raton, FL: Chapman & Hall/CRC).

Stewart, I., and Golubitsky, M. 1992. *Fearful Symmetry: Is God a Geometer?* (Oxford: Blackwell).

Stubhaug, A. 2000. *Niels Henrik Abel and His Times: Called Too Soon by Flames Afar* (Berlin: Springer). This is a translation of the 1996 Norwegian *Et foranskutt lyn, Niels Henrik Abel og hans tid* (Oslo: H. Aschehoug & Co.).

Stubhaug, A. 2002. *The Mathematician Sophus Lie* (Berlin: Springer-Verlag).

Swaddle, J. P., and Cuthill, I. C. 1994. "Preference for Symmetric Males by Female Zebra Finches." *Nature,* 367, 165.

Sylow, L., and Lie, S., eds. 1881. *Oeuvres complètes de Niels Henrik Abel* (Christiania: Grondahl & Son); reprinted in 1965 by Johnson Reprint (New York).

Szénássy, B. 1992. *History of Mathematics in Hungary Until the 20th Century* (Berlin: Springer-Verlag).

Szilagyi, P. G., and Baird, J. C. 1977. "A Quantitative Approach to the Study of Visual Symmetry." *Perception & Psychophysics,* 22(3), 287.

Tannery, J., ed. 1906. "Manuscrits et papiers inédits de Galois." *Bulletin des Sciences mathématiques,* 2nd ser., 30, 246, and 255.

Tannery, J., ed. 1907. "Manuscrits et papiers inédits de Galois." *Bulletin des Sciences mathématiques,* 2nd ser., 31, 275.

Tannery, J., ed. 1908. *Manuscrits de Évariste Galois* (Paris: Gauthier-Villars).

Tannery, J. 1909. "Discours Prononcé à Bourg-La-Reine." *Bulletin des Sciences mathématiques,* année 19, 1.

Tarr, M. J., and Pinker, S. 1989. "Mental Rotation and Orientation-Dependence in Shape Recognition." *Cognitive Psychology,* 21, 233.

Taton, R. 1947. "Les Relations de Galois avec les mathématiciens de son temps." *Revue d'histoire des sciences,* 1, 114.

Taton, R. 1971. "Sur les relations scientifiques d'Augustin Cauchy et d'Évariste Galois." *Revue d'histoire des sciences,* 24, 123.

Taton, R. 1972. "Évariste Galois," in *Dictionary of Scientific Biography,* C. C. Gillespie, ed. (New York: Charles Scribner's Sons).

Taton, R. 1983. "Évariste Galois and His Contemporaries." *Bulletin of London Mathematical Society,* 15, 107.

Taylor, R. E. 1942. *No Royal Road* (Chapel Hill: University of North Carolina Press).

Terquem, O. 1849. "Biographie. Richard, Professeur." *Nouvelles annales de mathématiques,* 8, 448.

Thomas, R. 2001. "And the Winner Is . . ." *Plus,* 16 (news) http://plus.maths.org/.

Thornhill, R., and Gangestad, S. W. 1996. "The Evolution of Human Sexuality." *Trends in Ecology and Evolution,* 11, 98.

Thornhill, R., Gangestad, S. W., and Comer, R. 1995. "Human Female Orgasm and Mate Fluctuating Asymmetry." *Animal Behaviour,* 50, 1601.

Thornhill, R., Gangestad, S. W., Miller, R., Scheyd, G., Knight, J., and Franklin, M. 2003. "MHC, Symmetry, and Body Scent Attractiveness in Men and Women." *Behavioral Ecology,* 14, 668.

Thornhill, R., and Grammer, K. 1999. "The Body and Face of a Woman: One Ornament That Signals Quality?" *Evolution and Human Behavior,* 20, 105.

Tignol, J.-P. 2001. *Galois' Theory of Algebraic Equations* (Singapore: World Scientific).

Tooby, J., and Cosmides, L. 1990. "On the Universality of Human Nature and the Uniqueness of the Individual: The Role of Genetics and Adaptation." *Journal of Personality,* 58, 17.

Tootell, R. B. H., Mendola, J. D., Hadjikhani, N. K., Liu, A. K., and Dale, A. M. 1998. "The representation of the ipsilateral visual field in the human cerebral cortex." *Proceedings of the National Academy of Sciences,* 95, 818.

Toti Rigatelli, L. 1996. *Évariste Galois 1811–1832* (Basel: Birkhäuser Verlag).

Tovey, D. F. 1957. *The Forms of Music* (New York: Meridian Books).

Trudeau, R. J. 1987. *The Non-Euclidean Revolution* (Boston: Birkhäuser).

Turnbull, H. W. 1993. *The Great Mathematicians* (New York: Barnes & Noble Books).

Tyler, C. W. 1983. "Sensory Processing of Binocular Disparity," in *Vergence Eye Movements: Basic and Clinical Aspects,* C. M. Schor and K. J. Ciuffreda, eds. (London: Butterworths).

Tyler, C. W. 1995. "Cyclopean Riches: Cooperativity, Neurontropy, Hysteresis, Stereoattention, Hyperglobality, and Hypercyclopean Processes in Random-Dot Stereopsis," in *Early Vision and Beyond,* T. V. Popathomas, C. Chubb, A. Gorea, and E. Kowler, eds. (Cambridge, MA: MIT Press).

Tyler, C. W., ed. 2002. *Human Symmetry Perception and Its Computational Analysis* (Mahwah, NJ: Lawrence Erlbaum Assoc.).

Tyson, N. D. G., and Goldsmith, D. 2004. *Origins: Fourteen Billion Years of Cosmic Evolution* (New York: W. W. Norton).

van der Helm, P. A., and Leeuwenberg, E. L. 1991. "Accessibility: A Criterion for Regularity and Hierarchy in Visual Pattern Codes." *Journal of Mathematical Psychology,* 35(2), 151.

van der Woerden, B. L. 1983. *Geometry and Algebra in Ancient Civilizations* (Berlin: Springer-Verlag).

van der Woerden, B. L. 1985. *A History of Algebra* (Berlin: Springer-Verlag).

Verdier, N. 2003. "Évariste Galois, Le Mathématicien Maudit." *Pour la Science,* no. 14, 1.

Verriest, G. 1934. *Évariste Galois et la Théorie des Équations Algébriques* (Lonvain: Chez L'Auteur).

Vitruvius. Ca. 27 BC. *De Architectura,* III, I, translated in 1914 by M. H. Morgan; reprinted 1960 by Dover Publications (New York).

Vogel, K. 1972. In *Dictionary of Scientific Biography,* C. C. Gillespie, ed. (New York: Charles Scribner's Sons).

Vucinich, A. 1962. "Nicolai Ivanovich Lobachevskii: The Man behind the First Non-Euclidean Geometry." *Isis,* 53, 465.

Walser, H. 2000. *Symmetry* (Washington, DC: The Mathematical Association of America).

Washburn, D. K., and Crowe, D. W. 1988. *Symmetries of Culture: Theory and Practice of Plane Pattern Analysis* (Seattle: University of Washington Press).

Webb, S. 2004. *Out of This World* (New York: Copernicus Books).

Weinberg, S. 1992. *Dreams of a Final Theory* (New York: Pantheon Books).

Wells, D. 1997. *Curious and Interesting Mathematics* (London: Penguin Books).

Wertheimer, M. 1912. "Experimentelle Studien über das Sehen von Bewegung." *Zeitschrift für Psychologie,* 61, 161.

Weyl, H. 1935. "Emmy Noether." *Scripta Mathematica,* 3, 201.

Weyl, H. 1952. *Symmetry* (Princeton: Princeton University Press).

Wheatstone, C. 1838. "Contributions to the Physiology of Vision. Part the First. On some remarkable, and hitherto unobserved, Phenomena of Binocular Vision." *Philosophical Transactions of the Royal Society, Part 1,* 371 (reprinted in *The Scientific Papers of Sir Charles Wheatstone,* London, 1879, p. 225).

Wheeler, J. A. 1990. *A Journey into Gravity and Spacetime* (New York: Scientific American Library).

Whistler, J. M. 1890. "The Gentle Art of Making Enemies." *Propositions,* 2.

Whiteside, D. T. 1972. In *Dictionary of Scientific Biography,* C. C. Gillespie, ed. (New York: Charles Scribner's Sons).

Wilde, O. 1892. *Lady Windermere's Fan,* act III.

Willard, H. F. 2003. "Tales of the Y Chromosome." *Nature,* 423, 810.

Wilson, D. 1986. "Symmetry and Its 'Love-Hate' Role in Music." *Computers & Mathematics with Applications,* 12B, nos. 1–2, 101.

Wilson, E. B. 1945. "Obituary: George David Birkhoff." *Science (NS),* 102, 578.

Winchel, F. 1967. *Music, Sound and Sensation* (New York: Dover).

Winner, E. 1996. *Gifted Children: Myths and Realities* (New York: Basic Books).

Witelson, S. F., Kigar, D. L., and Harvey, T. 1999. "The Exceptional Brain of Albert Einstein." *The Lancet,* 353, 2149.

Witten, E. 2004a. "Universe on a String." In *Origin and Fate of the Universe* (special cosmology issue of *Astronomy*).

Witten, E. 2004b. "When Symmetry Breaks Down." *Nature,* 429, 507.

Wolff, C. 2001. *Johann Sebastian Bach: The Learned Musician* (New York: W. W. Norton & Company).

Wolfram, S. 2002. *A New Kind of Science* (Champaign, IL: Wolfram Media), 873.

Wood, J. M., Nezworski, M. T., Lilienfeld, S. O., and Garb, H. N. 2003. *What's Wrong with the Rorschach?* (San Francisco: Jossey-Bass).

Wussing, H. 1984. *The Genesis of the Abstract Group Concept* (Cambridge, MA: The MIT Press).

Yaglom, I. M. 1988. *Felix Klein and Sophus Lie: Evolution of the Idea of Symmetry in the Nineteenth Century* (Boston: Birkhauser).

Yardley, P. D. 1990. "Graphical Solution of the Cubic Equation Developed from the Work of Omar Khayyam." *Bull. Inst. Math. Appl.,* 26, 5/6, 122.

Youschkevitch, A. P. 1972a. *Dictionary of Scientific Biography,* C. C. Gillespie, ed. (New York: Charles Scribner's Sons).

Youschkevitch, A. P. 1972b. In *Dictionary of Scientific Biography,* C. C. Gillespie, ed. (New York: Charles Scribner's Sons).

Zahavi, A. 1975. "Mate Selection: A Selection for the Handicap." *Journal of Theoretical Biology,* 53, 205.

Zahavi, A. 1991. "On the Definition of Sexual Selection, Fisher's Model, and the Evolution of Waste and of Signals in General." *Animal Behavior,* 42(3), 501.

Zahavi, A., and Zahavi, A. 1997. *The Handicap Principle: A Missing Piece of Darwin's Puzzle* (Oxford: Oxford University Press).

Zee, A. 1986. *Fearful Symmetry: The Search for Beauty in Modern Physics* (New York: Macmillan Publishing Company).

Zund, J. D. 1983. "Some Comments on Riemann's Contributions to Differential Geometry." *Historia Mathematica,* 10(1), 84.

Zweibach, B. 2004. *First Course in String Theory* (Cambridge: Cambridge University Press).

Credits

The author and publisher gratefully acknowledge permission to reprint the following material:

ART

Figs. 1, 3, 6–9, 12–15, 18–25, 27, 28, 30, 32, 33, 35, 74–77, 79–82, 84, 87, 89, 90, 92, 96, 98, 100–105, and the figures in appendix 1 and appendix 10: by Krista Wildt.

Fig. 2: Alinari/Art Resource, NY; Uffizi, Florence, Italy.

Figs. 4, 5, 25: by Ann Feild.

Fig. 10a: Courtesy Ricardo Villa-Real. From *The Alhambra and the Generalife* by Ricardo Villa-Real.

Fig. 10b: M. C. Escher's "Symmetry Drawing E116 (Fish)," © 2004 The M. C. Escher Company, Baarn, Holland. All Rights Reserved.

Fig. 11a: Morris & Company, London (1861–1940), William Morris, designer (1834–1896): *Apple Wallpaper (blue),* London, designed 1877. Color woodcut on paper, roll 56.0 cm wide. Gift of Haslem & Whiteway Ltd. 2002, Art Gallery of South Australia, Adelaide.

Fig. 11b: Morris & Company, London (1861–1940), William Morris, designer (1834–1896): *St. James wallpaper [fragment]*, 1884, London, designed 1881. Color woodcut on paper, irreg. 38.5 × 17.0 cm. Gift of Scotch College, Torrens Park, Adelaide 1992, Art Gallery of South Australia, Adelaide.

Fig. 16: M. C. Escher's "Symmetry Drawing E97 (Dogs)," © 2004 The M. C. Escher Company, Baarn, Holland. All Rights Reserved.

Fig. 17: Courtesy Thomas M. Brown, NASA and ESA.

Fig. 29: © 2004 Magic Eye Inc./www.magiceye.com/.

Fig. 31: © 2005 Bridget Riley, all rights reserved.

Fig. 34: Photograph © The British Museum.

Figs. 36, 39, 40, 41, 43–46, 78, 83: "Fondo Ritratti" of the Biblioteca Speciale di Matematica "Giuseppe Peano," through the assistance of Laura Garbolino and Livia Giacardi.

Figs. 37, 42, 49, 58, 62, 73: Courtesy of the author.

Fig. 38: B.U.B., ms. 595, N, 7, c. 30 v., Biblioteca Universitaria di Bologna.

Figs. 47, 48, 51: Courtesy of Arild Stubhaug. From *Niels Henrik Abel and His Times: Called Too Soon by Flames Afar*, by Arild Stubhaug.

Figs. 50, 95: Department of Mathematics, University of Oslo, Norway, through the assistance of Yngvar Reichelt.

Figs. 52–56, 59, 68, 70–72: Municipality of Bourg-la-Reine, through the assistance of Philippe Chaplain. Fig. 59: Archives Nationales F17.4176.

Figs. 57, 61, 66, 69: Bibliothèque de l'Institut de France, with the assistance of Norbert Verdier.

Fig. 60: Réunion des Musées Nationaux/Art Resource, NY; Louvre, Paris, France.

Fig. 63: © Photothèque des musées de la ville de Paris/Cliché: Andreani. Musée Carnavalet.

Fig. 64: Bibl. Historique de la ville de Paris, with the assistance of Norbert Verdier.

Figs. 65, 106: Archives de l'Académie des sciences, with the assistance of Norbert Verdier.

Fig. 67: Cent ans d'assistance publique à Paris, with the assistance of Norbert Verdier.

Figs. 85, 88, 93, 94: Private collection of Dr. Elliott Hinkes. "Celestial Harmony: Four Visions of the Universe," an exhibition at the Milton S. Eisenhower Library, Johns Hopkins University, April 26–May 30, 2004.

Fig. 86: Derby Museums and Art Gallery.

Fig. 97: By John Bedke.

Fig. 99: Adapted with permission from Forsman & Merilaita 1999.

TEXT

Appendix 5: Tartaglia's verses, reprinted with permission from Ron G. Keightley.

Appendix 8: The Galois family tree, from the Municipality of Bourg-la-Reine, through the assistance of Philippe Chaplain.

The poem “Chromodynamics”: Reprinted with permission from Cindy Schwarz.

Good faith efforts have been made to contact the copyright holders of the art in this book, but in a few instances the author has been unable to locate them. Such copyright holders should contact Simon & Schuster, 1230 Avenue of the Americas, New York, NY 10020.

Index